BIOTECHNOLOGY
Demystifying the Concepts

Of Related Interest from the Benjamin/Cummings Series in the Life Sciences

GENERAL BIOLOGY
N. A. Campbell, J. B. Reece, and L. G. Mitchell
Biology, Fifth Edition (1999)
N. A. Campbell, L. G. Mitchell, and J. B. Reece
Biology: Concepts and Connections, Third Edition (2000)
R. A. Desharnais and J. R. Bell
Biology Labs On-Line (2000)
J. Dickey
Laboratory Investigations for Biology (1995)
J. B. Hagen, D. Allchin, and F. Singer
Doing Biology (1996)
R. Keck and R. Patterson
Biomath: Problem Solving for Biology Students (2000)
A. E. Lawson and B. D. Smith
Studying for Biology (1995)
J. A. Lee
The Scientific Endeavor: A Primer on Scientific Principles and Practice (2000)
J. G. Morgan and M. E. B. Carter
Investigating Biology, Third Edition (1999)
J. A. Pechenik
A Short Guide to Writing about Biology, Third Edition (1997)
G. I. Sackheim
An Introduction to Chemistry for Biology Students, Sixth Edition (1999)
R. M. Thornton
The Chemistry of Life CD-ROM (1998)

CELL BIOLOGY
W. M. Becker, L. J. Kleinsmith, and J. Hardin
The World of the Cell, Fourth Edition (2000)

BIOCHEMISTRY
R. F. Boyer
Modern Experimental Biochemistry, Third Edition (2000)
C. K. Mathews, K. E. van Holde, and K. G. Ahern
Biochemistry, Third Edition (2000)

GENETICS
R. J. Brooker
Genetics: Analysis and Principles (1999)
J. P. Chinnici and D. J. Matthes
Genetics: Practice Problems and Solutions (1999)
R. P. Nickerson
Genetics: A Guide to Basic Concepts and Problem Solving (1990)
P. J. Russell
Fundamentals of Genetics, Second Edition (2000)

P. J. Russell
Genetics, Fifth Edition (1998)

MOLECULAR BIOLOGY
M. V. Bloom, G. A. Freyer, and D. A. Micklos
Laboratory DNA Science (1996)
J. D. Watson, N. H. Hopkins, J. W. Roberts, J. A. Steitz, and A. M. Weiner
Molecular Biology of the Gene, Fourth Edition (1987)

MICROBIOLOGY
G.J. Tortora, B. R. Funke, and C. L. Case
Microbiology: An Introduction, Sixth Edition (1998)

ANATOMY AND PHYSIOLOGY
E. N. Marieb
Essentials of Human Anatomy and Physiology, Sixth Edition (2000)
E. N. Marieb
Human Anatomy and Physiology, Fourth Edition (1998)
E. N. Marieb and J. Mallatt
Human Anatomy, Second Edition (1997)

ECOLOGY AND EVOLUTION
C. J. Krebs
Ecological Methodology, Second Edition (1999)
C. J. Krebs
Ecology: The Experimental Analysis of Distribution and Abundance, Fourth Edition (1994)
J. W. Nybakken
Marine Biology: An Ecological Approach, Fourth Edition (1997)
E. R. Pianka
Evolutionary Ecology, Sixth Edition (2000)
D.A. Ross
Introduction to Oceanography (1995)
R. L. Smith
Ecology and Field Biology, Fifth Edition (1996)
R. L. Smith and T. M. Smith
Elements of Ecology, Fourth Edition Update (2000)

PLANT ECOLOGY
M. G. Barbour, J. H. Burk, W. D. Pitts, F. S. Gilliam, and M. W. Schwartz
Terrestrial Plant Ecology, Third Edition (1999)

ZOOLOGY
C. L. Harris
Concepts in Zoology, Second Edition (1996)

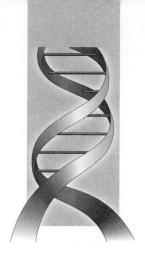

BIOTECHNOLOGY
Demystifying the Concepts

DAVID BOURGAIZE
Whittier College

THOMAS R. JEWELL
University of Wisconsin—Eau Claire

RODOLFO G. BUISER
University of Wisconsin—Eau Claire

An imprint of Addison Wesley Longman
San Francisco • Reading, Massachusetts • New York •
Harlow, England • Don Mills, Ontario • Sydney •
Mexico City • Madrid • Amsterdam

EXECUTIVE EDITOR: Erin Mulligan
SPONSORING EDITOR: Lynn Cox
PROJECT EDITORS: Virginia Simione Jutson, Anne Scanlan-Rohrer
MANAGING EDITOR: Laura Kenney
SENIOR PRODUCTION EDITOR: Angela Mann
SENIOR ART SUPERVISOR: Donna Kalal
ILLUSTRATIONS: Accurate Art, Inc.
PHOTO RESEARCHER: Cindy-Lee Overton
TEXT DESIGNER: Wendy LaChance
COVER DESIGNER: Yvo Riezebos
COPYEDITOR: Mary Roybal
PREPRESS SUPERVISOR: Vivian McDougal
COMPOSITOR: G&S Typesetters
MARKETING MANAGER: Gay Meixel

COVER IMAGES: © copyright 1999 Photodisc, Inc.

Library of Congress Cataloging-in-Publication Data
Bourgaize, David.
   Biotechnology : demystifying the concepts / David Bourgaize,
Thomas R. Jewell, Rodolfo G. Buiser.
      p.   cm.
   ISBN 0-8053-4602-3
   1. Molecular biology.   2. Molecular genetics.   3. Biotechnology.
I. Jewell, Thomas R.   II. Buiser, Rodolfo G.   III. Title.
QH506.B67   2000
572.8—dc21   99-36737

3 4 5 6 7 8 9 10—VOG—03 02 01

Benjamin/Cummings, an imprint of Addison Wesley Longman, Inc.
1301 Sansome Street
San Francisco, CA 94111

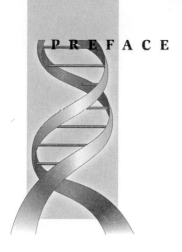

INTRODUCTION

In recent years, there has been a heightened public awareness of the fast-growing field of biotechnology. Each day when we read the newspaper or sit down to watch the nightly news, we are informed of some new scientific breakthrough that is a direct result of biotechnology. In the courtroom—a place with which all of American society is familiar—we ask: What does DNA evidence really prove? We hear the word "biotechnology" and we know that it is important to understand its increasing impact on our daily lives, from the research that leads to breakthroughs in diagnosing and treating medical conditions to the advances in environmental concerns to the rapid growth of career opportunities available in this exciting field. As a result, there has been an increasing interest in biotechnology at the undergraduate level in four-year colleges and universities, as well as two-year community colleges.

The goal of *Biotechnology: Demystifying the Concepts* is two-fold. First, it provides a concise treatment of the underpinnings of biotechnology and explores some of its present and future uses. Our focus is on an explanation of the genetic and immunological foundation of biotechnology known as the "tools" approach. Second, it introduces readers to the basic science behind biotechnology to help them separate the facts from the myths so that they can understand the potential economic, social, ethical, and political impact of this burgeoning field on society. This text has been written for the non-biology major who has little to no science background or course work. In writing this book, we have tried to make the "mysteries" of modern scientific technology understandable and accessible.

ORGANIZATION

This text is organized by introducing the reader to the branches of science (molecular biology, genetics, immunology, and so on) that have come together to result in biotechnology. It then goes on to discuss special topical issues, such as HIV/AIDS, cancer, and the business of biotechnology in the United States. Special features (Issues Boxes and Essays) are woven throughout in order to engage the reader and promote understanding of the content discussed.

The first two chapters (*An Introduction to Living Things* and *Molecular Biology*) explore how cells organize their environments and reproduce themselves through biosynthesis of complex modules. The flow of information from the DNA storehouse to the formation of structural and functional proteins is the primary focus of Chapter 2.

Chapter 3, *Chromosomes, Cell Division, and Sexual Reproduction*, and Chapter 4, *Mendelian Genetics*, provide the chromosomal information and genetic background needed to understand mutations and genetic diseases—issues that are discussed in Chapter 5, *Mutations and Genetic Disease*.

Chapter 6, *Complexities of Genetics*, takes the reader on a journey through some of the complex interactions between genetics and human behavior. Chapter 7, *Biological Control*, discusses life functions from a control point of view. Several known metabolic control systems are described, as is the phenomena of cellular communication. Chapter 8, *Genetic Engineering*, describes the fundamentals of genetic engineering using genetically engineered insulin as an example. Chapter 9, *More Genetic Engineering*, provides information on methods involved in the analysis of DNA and practical applications of genetic engineering in agriculture, medicine, and forensic science.

The basics of *Genetic Disease and Gene Therapy* are discussed in Chapter 10. Issues of privacy, the pros and cons of genetic screening, and some of the ethical, moral, and legal questions surrounding the treatment of genetic diseases are covered in this chapter.

Chapter 11, *Immunology*, provides background information with a focus on prevention: Five major diseases and the vaccines used to prevent them are presented. Chapter 12, *Monoclonal Antibody Technology*, provides the backdrop for rapid serological testing as well as disease treatment using "magic bullets."

Chapter 13, *AIDS and HIV*, discusses the details of HIV/AIDS, and its unique causal virus, as well as methods of diagnosis and treatment. The role of biotechnology as a potential key in the diagnosis and treatment of AIDS is featured. Chapter 14, *Cancer*, focuses on the current and future uses of biotechnology in the diagnosis and treatment of this insidious disease.

Chapter 15, *The Business of Biotechnology in the United States*, discusses the business aspects of biotechnology in pharmaceutical development, agriculture, and environmental issues. Chapter 16, *Biotechnology in the Developing World*, presents the reader with a very different question: How are developing nations affected by the rapid developments in biotechnology?

The last chapter, *Now What?*, discusses the future of biotechnology: its potential uses along with some of the associated risks and ethical issues.

SPECIAL FEATURES

We have included a variety of special features in this text to facilitate comprehension.

Issues Boxes highlight significant controversial scientific topics such as the Human Genome Project, bioethics, and HIV/AIDS that derive from biotechnology. They provide further exploration into the non-science components of the topics presented.

Essays are concepts, concerns, or problems that are explored as sidebars. These essays tend to focus on broader topics such as genetic determinism, vaccine availability, and multiple uses of recombinant DNA.

Icons in the text margin denote principles (scientific principles, techniques, or methods) ✎, social or ethical issues ⚖, and applications ⚗.

End of chapter **Review Questions** stimulate thought and discussion.

A **Glossary** of terms commonly used in biotechnology can be found in the back of the text.

ACKNOWLEDGMENTS

We would like to thank the staff at Addison Wesley Longman/Benjamin Cummings Publishing, especially Rebecca Strehow, Sue Ewing, and Anne Scanlan-Rohrer, who added accuracy and clarity to our efforts. We extend our sincere thanks to all of our fellow educators who reviewed the manuscript and helped to shape this text: D. Gordon Atkins, Andrews University; Stephen Benson, California State University, Hayward; Jim Christamn, Sonoma State University; R. Igor Gamow, University of Colorado, Boulder; Leslie A. Gregg-Jolly, Grinnell College; Greg Johnson, Bethel College; Ann M. Kleinschmidt, Allegheny College/University of Pittsburgh; Don Lightfoot, Eastern Washington University; Jane Magill, Texas A & M University; David Magnus, University of Puget Sound; Sandra G. Porter, Seattle Central Community College; Hildagarde K. Sanders, Villa Julie College; Maureen A. Scharberg, San Jose State University; Brian R. Shmaefsky, Kingwood College; Paul Tavernier, Bethel College; Dan Trubovitz, Miramar College; Steven M. Wietstock, Alma College; Robert J. Wiggers, Stephen F. Austin State University.

A project of this scope requires the skill and support of many people. The authors would personally like to thank the following people:

Dave Bourgaize: I would like to thank my family, in particular my wife, Karen. Her support is directly responsible for my contribution to this book. I would also like to thank my co-authors, Tom Jewell and Rudy Buiser.

Students and colleagues are the reason I am in academia in the first place. For their insights and ideas, I thank them. One in particular, Paul Greenwood, deserves special mention. You have been my closest friend, colleague, and educational role model for over a decade. Thank you, Paul!

Tom Jewell: I would like to thank my wife Kathi for her consistent support and encouragement over the duration of this project. This "thank you" is for all those times when I forgot.

Rudy Buiser: I would like to thank my friends and colleagues who have supported and encouraged me, especially Amy. I never would have made it this far without them.

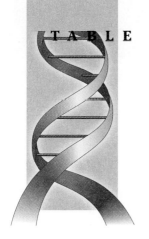

TABLE OF CONTENTS

CHAPTER 1

An Introduction to Living Things

Many say that we are in the midst of a biotechnological revolution. The last twenty years or so have seen the emergence of remarkable new technologies that can produce new drugs for the treatment of disease, alter the genetic makeup of plants and animals, and produce bacteria to help clean up toxic wastes. Never in the history of the world have biological organisms been so valuable to us as commodities that can be bought and sold. What has happened to make this revolution possible? How might it affect our personal lives? What might it do to our society?

The near future holds the prospect of wondrous new medical diagnostic procedures, such that people can be diagnosed as having cancers or Huntington's disease. Medical diagnosis can even be performed on individual human eggs that are fertilized *in vitro,* or outside the body. Fertilized eggs that do not carry such defects can then be implanted in the womb to develop into fetuses. Crops will be engineered that require less fertilizer or pesticide. Previously untreatable diseases will yield to new products and new procedures, sometimes through replacement or repair of defective genetic material. In short, the biological world is full of what are now extremely valuable commodities, from naturally occurring products with medicinal properties found in certain plants to oil-degrading bacteria that can be used to clean up toxic wastes to pieces of genetic material that might be used to cure disease or treat obesity.

But the development of such products and practices is not without potential harm. Should individuals, even fetuses and eggs, be diagnosed and deemed defective? Who should determine this? Should genetic material be altered in individuals? What will be the environmental effects of spreading genetically altered crops or animals? As with any newly developing technology, biotechnology has both positive and negative aspects. Would we choose, in retrospect, to avoid developing the atomic bomb and nuclear technology? Is it possible to choose whether or not new technologies are developed and applied? Would the myriad developments that arose from nuclear technology, particularly medical advances, outweigh the negative aspects? Who makes these decisions, and how? What is an acceptable risk, and how is the risk of an unknown technology estimated?

These questions are of particular importance to biotechnology, in part because some biotechnology developments involve our own health and well-being and in part because some of the more dire consequences of biotechnology, should they occur, would be long-lasting. This is an issue that is often difficult to think about; the inherent risks to populations, both human and others, and the long-term effects on our environment are often far removed from our everyday experience. Maybe it makes sense, then, to consider the various personal risks we already endure and are familiar with and try to build a more expansive view of risk assessment from this vantage point. At the conclusion of this chapter, we will return to the topic of risk and technological development.

Technology is the application of scientific knowledge, skills, materials, labor, and wealth for a practical purpose. The tremendous advances in biological knowledge have allowed the extensive utilization of living things as tools and resources. **Biotechnology**, then, is the application of biological principles, organisms, and products to practical purposes. As such, biotechnology is as old as civilization. Humans have been making bread, wine, and cheese for centuries, as well as breeding desirable plants and animals. These processes all use living things or materials from living things. But biotechnology is also a brand new science, since many of its most powerful tools and techniques have been developed only within the past several decades. Many of these developments have occurred as a result of different sciences—biology, chemistry, physics, and others—coming together in order to solve a particular problem.

One of the goals of this book is to provide a useful understanding of biotechnology and its underlying principles. Another goal is to apply this knowledge to the discussion of issues that you as individuals and we as a society face as a result of the new technology. To help us attain these goals, each chapter interweaves principles with a wide variety of social, political, philosophical, legal, economic, and religious issues that involve the science. The end result should be not just knowledge of scientific facts but an awareness of the inevitable result of using scientific knowledge and an ability to make informed decisions. Many questions, about both the science and its impact, remain unanswered. It is these questions that we hope you find most intriguing.

Since biotechnology involves the manipulation of living things, it is very important that you develop a firm understanding of certain basic principles of life. This first chapter highlights a number of topics related to how living things work. While some material may be review, it should provide you with a framework for understanding how organisms function.

THE CELLULAR BASIS OF LIFE

Such a bewildering variety of life is found on the earth that it is sometimes difficult to imagine that all organisms actually share many common features. Fundamentally, all living organisms must share some characteristics that make them different from nonliving things, such as rocks or water or tennis balls. We would like to identify two characteristics that all living organisms have and nonliving things do not.

- Living things have the ability to reproduce.
- Living things have the ability to extract energy from the environment.

Exactly how organisms accomplish these feats will be the focus of future chapters. But both of these characteristics point toward fundamental similarities in living things. Do these two properties result in any other common features, say, some organizational principle that all living things adhere to that helps both of these things happen?

The second property, the extraction of energy from the environment, actually determines a great deal about how life must be organized. Energy provides the means of accomplishing work, which is, after all, what living things need to do to stay alive. In order to extract energy from materials in the environment, the organism must somehow separate itself from the environment. This allows the accumulation of the energy necessary to perform useful work. It is easy to see how large animals like humans separate themselves; we have skin that covers our body. However, we must look deeper to understand the organizational impact of this necessary separation, for even the most minute organism must be separated from its environment in order to maintain the energy differences that allow useful work to be done.

All living things utilize the same basic organizational principle of being constructed from smaller units. Each "unit of life," the material that serves as the fundamental basis for living things, is an entity unto itself. We call these basic living entities **cells**. Cells are the smallest units of life. In their simplest form, they consist of an interior called the **cytoplasm** surrounded by a barrier called a **cell membrane**. Inside the cell occur all the chemical reactions that serve to extract energy from the environment, as well as all the reactions that make organisms grow, move, reproduce, and do everything else that they must do. Since much of biotechnology deals with what happens inside cells, we need to establish a firm understanding of the structural organization of cells.

ORGANIZATION OF CELLS

There are many different kinds of organisms made of many different types of cells, but in terms of basic cellular structure there are really only two types of cells found in nature: **prokaryotic** (Figure 1.1) and **eukaryotic** (Figure 1.2). The only organisms that consist of prokaryotic cells are the bacteria, single-celled organisms found in virtually every environment on earth. All other organisms, even other single-celled organisms such as yeasts and algae, are eukaryotic. The primary difference between the two cell types is in how the interior of the cell is organized.

Each type of cell is surrounded by a membrane. This serves several purposes, one of which is to act as a selective barrier between the inside and the outside. We will see shortly why the membrane serves so nicely as a barrier. Prokaryotic cells have only one internal compartment, the cytoplasm. Eukaryotic cells, on the other hand, have many internal compartments. Each compartment is surrounded by its own membrane, and each has specific purposes, some of which are indicated in

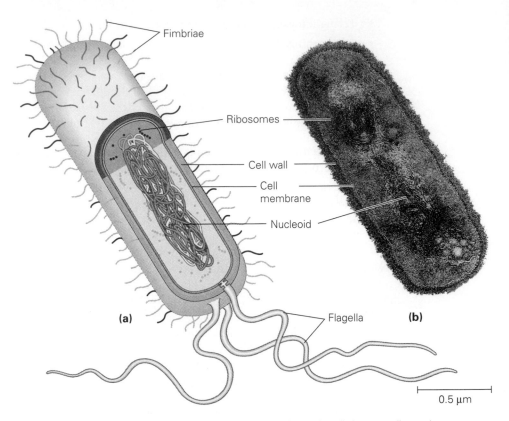

Figure 1.1 A typical prokaryotic cell. (a) Most prokaryotic cells have a cell membrane surrounded by a cell wall. There are no internal compartments, and the DNA occupies a region of the cytoplasm called the nucleoid. Ribosomes are responsible for protein synthesis. Flagella provide movement. Fimbriae help cells attach to each other or to surfaces. **(b)** A photograph of a prokaryotic cell.

Figure 1.2. The compartment we will mention the most in this book is the **nucleus**, the compartment enclosed by the *nuclear membrane* that stores the biological information used by the cell as it goes about its business.

While the existence of internal compartments is the most telling feature of eukaryotic cells, there are other differences as well. Bear in mind that most of these differences are generalizations; there are exceptions. For example, red blood cells in your blood are eukaryotic cells, but they have no internal compartments. During the development of these cells, the internal membranes are dissolved to yield the final form without compartments.

Eukaryotic cells are larger than prokaryotic cells, often ten times the diameter. Some eukaryotic cells, such as many eggs, are so large that you can see them without using a microscope—although such cells are not common. Eukaryotic cells also grow more slowly and reproduce less frequently than prokaryotic cells. This is not surprising since they are usually so large. Some prokaryotic cells can grow and reproduce in fewer than twenty minutes given ideal conditions, whereas the average human cell, even under the best of circumstances, might take twenty-four hours to reproduce.

Eukaryotic cells are often specialized, meaning they have developed specific capabilities. This is certainly true in organisms made of many cells, such as humans. Specialized cells serve thousands of functions within our bodies, such as transmission of electrical impulses by nerves and transport of oxygen in the red blood cells. This implies that a variety of cells must be properly organized in order to provide for all the functions necessary to an organism, and this is indeed the case. Our bodies are organized into functional units called **tissues**. Liver cells, for example, are specialized to perform all the functions of the liver. These functions differ dramatically from those of a skin cell. Specialized cells often have specialized requirements, including the need to have particular neighbors. Nonspecialized cells, on the other

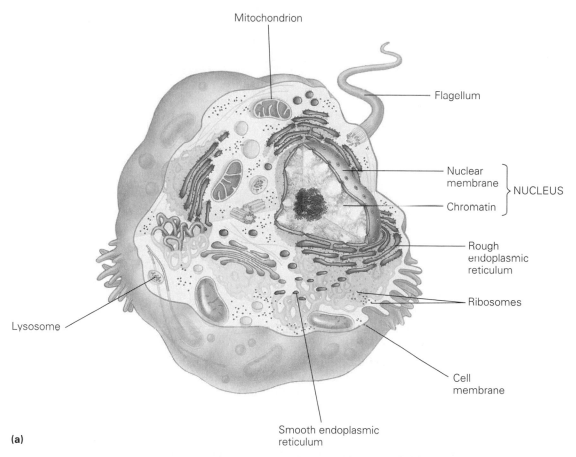

Mitochondrion

Flagellum

Nuclear membrane ⎤
⎬ NUCLEUS
Chromatin ⎦

Rough endoplasmic reticulum

Ribosomes

Lysosome

Cell membrane

Smooth endoplasmic reticulum

(a)

Figure 1.2 Eukaryotic cell. Typical eukaryotic cell, showing some of the major internal compartments of both plant and animal cells. (**a**) Typical animal cell. The DNA is localized within the nucleus. Mitochondria synthesize ATP, the energy currency for the cell. Lysosomes degrade and recycle materials. The endoplasmic reticulum is part of a transport system for molecules within cells.

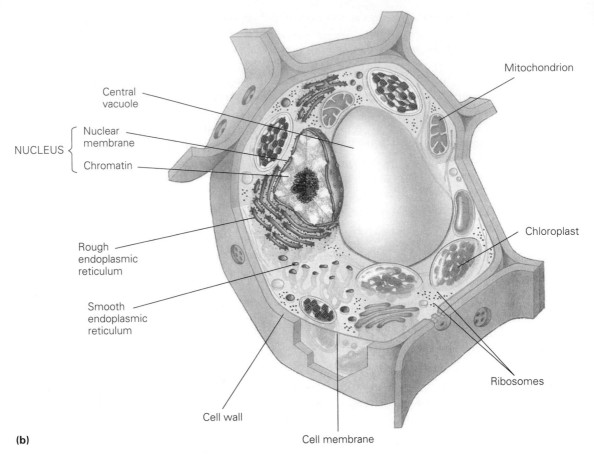

(b)

Central vacuole

NUCLEUS {
Nuclear membrane

Chromatin

Rough endoplasmic reticulum

Smooth endoplasmic reticulum

Cell wall

Cell membrane

Mitochondrion

Chloroplast

Ribosomes

Figure 1.2 (b) A typical plant cell. Plant cells generally have a rigid cell wall and contain at least two compartments: vacuoles, large fluid-filled compartments; and chloroplasts, involved in light-energy capture.

hand, are more self-sufficient and can often exist alone. There are many other differences as well; we will discuss many of them at more appropriate points in the book.

BIOMOLECULES

Molecules are chemical compounds that under normal circumstances cannot be broken into smaller units while maintaining their original characteristics. Many different molecules make up our cells, from lipids that make up cell membranes to DNA, which stores genetic information; from hormones that travel through the bloodstream and elicit special responses to chlorophyll, which absorbs light in plants. Studying these molecules, their structures, and how they function is the domain of biochemistry. Fortunately, it is possible to classify most of the molecules used in biological systems into only a few categories, making an appreciation of their structure and function much easier.

Before we discuss each type of biomolecule in more detail, it is worth noting that many of them share a common feature: They are **polymers**. Most of you have undoubtedly heard of polymers, probably in conjunction with plastic materials. What is a polymer? It is simply a chain of small molecules, generally called *monomers*, chemically linked to form a larger one. A metal chain is a sort of polymer of individual links—the chain is the final product, but it is made of links joined so that they do not normally come apart. The links are the building blocks that make up the final chain. The metal chain analogy is especially good for biological polymers, because the links in a chain almost always appear end to end, in a linear fashion. Each link is connected to only two others, head to tail. Biological polymers are usually made of building blocks attached end to end (Figure 1.3).

You can see how this linked structure makes it easy to manufacture metal chains. With a large supply of the individual building blocks (the links), on hand, any desired chain, of any length or thickness, can be quickly made by using the same machinery. The machinery only has to be designed to accept different building blocks in order to become a "universal" maker of virtually any chain imaginable. The fact that many biomolecules are linear polymers makes sense. If the cell can maintain a steady supply of the appropriate building blocks, then one kind of cellular machine might be able to construct many different molecules depending on the instructions fed to the machine.

Lipids

Consider an oil with which you are familiar, such as cooking oil. What happens if you add some to water? The oil, which may originally disperse into tiny droplets, eventually comes together in larger droplets, or pools, generally on the surface of the water. Why? Oil and water don't mix. The chemical way of describing this is that the oil molecules are **hydrophobic**, or water-hating. Fats and oils, known as **lipids**,

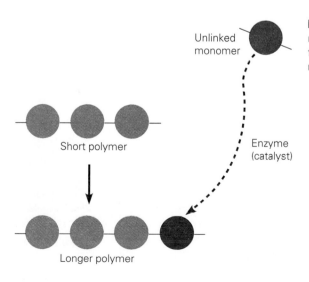

Unlinked monomer

Short polymer

Enzyme (catalyst)

Longer polymer

Figure 1.3 Biological polymers. Polymers are assembled from individual building blocks or monomers.

prefer not to be in contact with water. Their chemistry is more "satisfied" if they are surrounded by other lipid molecules—or even by air—rather than by water. That is why all of the small droplets coalesce into a large one on the surface. This minimizes the contact between the oil and the water. Such hydrophobic forces are a very important part of the chemistry of living things, in large part because they are responsible for the formation of membranes.

The particular lipids involved in cell membrane formation actually have two chemically distinct sides, one that is hydrophobic and another that is not and prefers to be in water. The latter is referred to as **hydrophilic** (Figure 1.4a). Imagine adding lipid molecules of this sort onto a water surface. Initially, a lipid **monolayer** forms on the water's surface, with the hydrophilic ends sticking in the water and the hydrophobic ends poking into the air. Adding more lipid molecules could cause them to congregate around one another, forming a **micelle** (Figure 1.4b). As more and more lipid molecules are added, they eventually will form a spherical structure known as a **bilayered vesicle**. In this way, the hydrophobic parts of the lipid molecules are most satisfied, since they are in contact with other hydrophobic parts of lipids and not water. The hydrophilic parts of the lipid molecules are also satisfied, since they are in contact with water. You can see that a vesicle is essentially a sphere made of two layers of lipids (Figure 1.4b). This bilayering of lipid molecules in sheet-like form is the basic structure of cell membranes.

Imagine such a membrane surrounding a cell. One of the important properties of the membrane is to act as a selective or differential barrier. In order for cells to carry on their business, the internal contents of the cell must be separated from the external environment. Yet the external environment is the source of water, nutrients, and other raw materials that must be able to enter cells somehow, while waste products generated inside cells must be able to leave. The bilayer structure of a lipid

Figure 1.4 Lipids and their interactions with water. (**a**) Representation of lipids involved in the formation of biological membranes. The head groups are hydrophilic; they prefer contact with water. The hydrophobic tails prefer not to be in contact with water. (**b**) The basis of membrane formation in water. As lipids are added to water, they initially form a *monolayer* on the surface of the water. As more lipids are added, *micelles* form, allowing hydrophobic tails to interact with one another and not the water. As more lipids are added, *bilayered vesicles* eventually form, containing two layers of lipids and an internal compartment of water. The hydrophobically connected lipid bilayers form the basic structure of biological membranes.

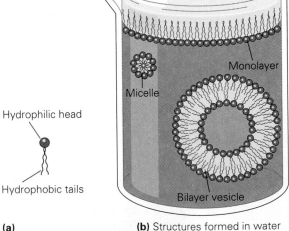

Hydrophilic head

Hydrophobic tails

(a)

Surface of liquid

Monolayer

Micelle

Bilayer vesicle

(b) Structures formed in water

membrane establishes a barrier that prevents the movement of most, but not all, substances. Such a *semipermeable membrane* is a vital part of the cell.

It is easy to see how such a membrane might serve as a semipermeable barrier. Imagine some toxic molecules traveling through the blood. In order for these molecules to exert their toxicity, they must enter a cell and disrupt some aspect of that cell's function. But the toxic molecules are hydrophilic; they much prefer to be in a water-based environment such as blood. How easy is it for these molecules to gain entrance to a cellular cytoplasm by crossing a cell membrane? In order to cross, they must move through what is essentially a water-free zone—the inside of the bilayer, which is hydrophobic. This simply cannot happen. Very few hydrophilic molecules can freely cross membranes, and the few that can (such as water and carbon dioxide) are very small. The cell membrane, simply by having a structure based on lipids that are hydrophilic at one end and hydrophobic at the other, can serve as a protective device, determining which molecules can freely enter cells.

There are many hydrophobic molecules that are not repelled by the lipid bilayer and can freely cross a cell membrane. An example is estrogen, one of many hydrophobic hormones secreted into the bloodstream. Hormones are usually produced in response to a particular need, travel through the circulatory system to particular target cells, and then cross the cell membrane, causing changes in the cell's activity.

If membranes are semipermeable and only a few hydrophilic molecules can enter or exit cells freely, how do all the necessary nutrients get into cells? Most of the nutrients the cell needs are very hydrophilic. There must be a way of getting these molecules into the cell. This leads us to the next group of biomolecules, the proteins.

Proteins

Proteins are the most diverse class of biological polymers. They can serve as transporters in a variety of ways, moving molecules such as oxygen and cholesterol through the blood, or sugar and salt across a cell membrane. They can also function as structural elements (hair is made of protein, as is the lens of the eye). But one of the primary functions of proteins is to perform and control the innumerable chemical reactions that occur within cells as materials are processed and made into various products. It is in this role that we will become most familiar with proteins.

How does one group of molecules perform such a diverse set of functions? The answer lies in the variety of available building blocks. The metal chains we used as examples of "polymers" are always constructed from a single type of building block—a link of a particular size, made of a particular metal, and having a certain color and strength—to suit the intended application. In contrast, proteins can be made from twenty different building blocks that can be attached in any order. These building blocks (monomers), called **amino acids**, provide the raw material for all protein production in all organisms.

Each of the twenty amino acids has several chemical groups that we will learn more about later. Some of these groups are involved in linking amino acids to each other in the protein chain; each amino acid has two ends that can join to other amino acids. Unlike the links of a metallic chain, each amino acid also has a particular chemical group that lends the amino acid its identity. No matter what order the amino

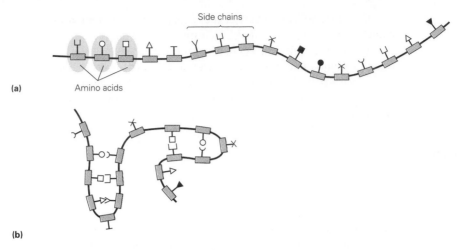

(a)   Amino acids

(b)

Figure 1.5 Folding a chain of amino acids. (**a**) A geometric representation of an amino acid chain. Each symbol represents a side chain, a chemical group that extends from the individual amino acids that make up the main chain. (**b**) Favorable interactions between individual amino acid side chains dictate how the chain will fold into a specific shape. Interactions of this type determine the structure of individual proteins.

acids are connected in, the individual chemical groups of the amino acids, called *side chains*, are found extending from the sides of the protein chain (Figure 1.5a). All of these chemical groups can interact with nearby neighbors or with those farther away—the protein chain can be quite flexible (Figure 1.5b). For example, a chemical group that possesses a negative charge might interact with a nearby group having a positive charge. Since negative and positive charges attract each other, the two groups will be pulled closer together. Or hydrophobic side chains can be forced together. Thus, unlike our metal links, the individual building blocks of the protein chain are not inert. Each retains a chemical identity and can perform a chemical function even within the protein chain. It is the collection of amino acid chemical groups and the sequence in which they are arranged that ultimately gives a protein its final three-dimensional shape and its particular function.

How many different proteins can potentially be formed from the twenty different amino acids? Consider an average-size protein, which might consist of 300 amino acids linked into one chain. Since each of the 300 positions of this chain could be filled with any one of the twenty amino acids, there are 20^{300} possible combinations, an extremely large number! Clearly, the number of possible proteins is very large. Nature, however, uses only a limited subset of all possible protein chains.

The sequence of amino acids determines the relative positioning of the individual chemical groups of each building block, which in large part determines which groups interact with each other and thus the final structure and function of the molecule. For instance, consider a hypothetical transport protein that is found embedded in the membrane but in contact with both the external and the internal environments. Molecules outside the cell that are being brought in will react to the chemical groups present on the outside of the protein. The part of the protein within the membrane must contain hydrophobic regions that can exist within the lipid bilayer. The region of the protein exposed on the inside of the cell must be hydrophilic and contain the proper chemical groups for internal interactions.

Thus, of primary importance to understanding the nature of a particular protein is knowing its amino acid sequence, since this sequence determines what function the protein will perform in the cell. Some of the fundamental biological questions we will deal with in future chapters are how the information for the amino acid sequences of proteins is stored in a usable fashion in cells and how this information is used by the cell to construct actual proteins.

Nucleic Acids

Nucleic acids are primarily responsible for the storage and transmission of biological information. Nucleic acids, like proteins, are linear polymers, built from building blocks called **nucleotides**. Four different primary nucleotides are found in each of the two kinds of nucleic acid molecules, DNA and RNA. Nucleotides are successively linked to each other in a particular manner, yielding a final "chain" of nucleotides. Like the amino acids of proteins, each nucleotide possesses a unique chemical group—a nitrogenous base—that extends from the nucleotide chain. The particular structures assumed by nucleic acid molecules, particularly DNA, are the subject of more elaborate description in the next chapter.

Carbohydrates

Carbohydrates, also known as sugars and starches, comprise a large and diverse group of molecules that serve several functions. Some carbohydrates are purely structural elements, such as the cellulose that provides much of the strength and rigidity of plant tissues. Others serve a role in communication or recognition, such as the molecules that determine your ABO blood type. Some carbohydrates are used as building blocks; the nucleotides described above all contain a carbohydrate component. But the most prevalent role of carbohydrates is as a source of energy in cells. Carbohydrates are the preferred source of energy in most cells. We will have more to say about the role of carbohydrates in providing energy later in this chapter.

CHEMICAL FORCES IMPORTANT
TO BIOMOLECULES

What holds the various building blocks together in a biomolecule? Exactly what kinds of chemical forces operate to give a protein or a nucleic acid its final structure? Such forces are key to understanding how cells work—not only do they determine the structure and function of individual molecules, but they also govern molecular interactions. We will start by describing the forces responsible for holding individual atoms together and work our way up through larger and larger molecules until we can understand some of what is necessary to maintain the structure of large biomolecules and ensure their proper functioning.

Molecules are made of *atoms*, the smallest chemical entities. Only a handful of different atoms are important to our understanding of most biological molecules. Carbon plays a central role in almost all biological molecules because of certain chemical properties it possesses: It can bond to many different kinds of atoms, and these bonds are generally quite stable under conditions found inside the cell. Oxygen, hydrogen, nitrogen, sulfur, and phosphorous are the other prominent atoms in biomolecules. Each atom brings with it particular chemical properties.

Atoms can react with each other in many ways, sometimes forming stable interactions called **bonds**. These bonds take many forms, but three of the most important in biological systems are the ionic bond, the covalent bond, and the hydrogen bond.

Ionic Bonds

Ionic bonds form between an atom that easily gives up one of its electrons and another atom that readily accepts that electron. One of the atoms becomes positively charged, and the other becomes negatively charged. Charged atoms are called **ions**. A classic example of a solid crystal compound formed by ionic bonds is sodium chloride (table salt). The sodium and chlorine atoms are present in salt in a one-to-one ratio: For every sodium atom there is a corresponding chlorine atom. A sodium atom readily donates an electron to a chlorine atom. However, the structure of sodium chloride is such that there is no direct connection between each pair of atoms even though the sodium ion has a positive charge and the chloride ion has a negative one. Rather, they are arranged in a three-dimensional pattern of alternating sodium and chloride ions. Every sodium ion is surrounded by chloride ions, and every chloride ion is surrounded by sodium ions (Figure 1.6). Thus, in the crystal of sodium chloride, each positive charge (sodium ion) is surrounded by a "shield" of negative charges (chloride ions), and each of these negative charges is surrounded by a corresponding "shield" of positive charges. Overall, each positive charge is negated, or balanced by a negative charge, such that overall the sodium chloride crystals have no net charge.

An interesting thing happens to most ionic compounds when they are put into water, for example, when table salt is dissolved in water. One of the unique properties of water is that it can form its own "shields" around charges. As a salt crystal is placed in water, individual ions begin to come out of the crystal structure. As the ions move out of the crystal, they become surrounded and shielded by water molecules.

Figure 1.6 The crystal structure of sodium chloride. Within a crystal of sodium chloride—table salt—each atom of sodium (Na^+) or chlorine (Cl^-) is surrounded by six of the opposite atoms.

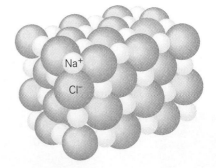

The individual ions no longer need to be associated with the others. Eventually, all of the ions move into the water, and there is no crystal left. This is what happens when compounds dissolve: Water breaks apart the normal structure and forms *hydration shells* around the individual components.

We can begin to get an idea of the relative strength of ionic bonds by considering an example. Take a few crystals of table salt and put them onto a regular tablespoon. Then put another spoon on top. Using your thumb, try to crush the crystals in the first spoon by grinding down on them. What you will find is that you can take the relatively uniform salt crystals found in salt shakers and break them into smaller crystals without much effort. This should not be too surprising—nothing in the structure of the sodium chloride crystal suggests a preference for a particular size. Table salt is packaged and sold as very uniform crystals of useful size; it would be rather unwieldy at the table to add salt by chiseling off a bit from a 2-foot-by-2-foot crystal in the middle of the table! The point is this: When you break a crystal into smaller ones, you are breaking ionic bonds. It doesn't take much input of energy (a bit of thumb power will easily do it) to break the ionic bond between sodium and chloride. Some ionic bonds are stronger than others. The basic point, however, remains true: Ionic bonds are relatively weak bonds, especially in water. The fact that salt crystals break apart into individual ions when they are put into water supports this notion. Thus, ionic bonds can be characterized (in biological systems, where everything is in water) as relatively weak bonds occurring between atoms that can easily become charged.

Covalent Bonds

Covalent bonds, on the other hand, can be relatively strong. **Covalent bonds** occur when atoms come together and, instead of completely donating electrons, share them. In contrast to what happens when atoms form ionic bonds, the electrons in covalent bonds are not garnered by one of the participating atoms and lost by the other. Instead, they are more equally shared, acting as a part of both atoms (Figure 1.7a, b). Thus, there are no ions, and the result can be a much stronger bond. Depending on the individual atoms involved, covalent bonds can be difficult to break, requiring a significant input of energy. Most compounds formed from covalent bonds do not ionize in water unless the covalent bonds are very weak. Still, they can dissolve in

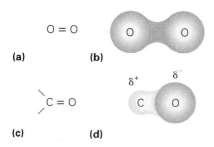

(a) $O = O$

(b)

(c) $\diagdown C = O$ $\diagup$

(d) δ^+ δ^-

Figure 1.7 Covalent bonds. Atoms engaged in covalent bonding share electrons. (**a**) A representation of a covalent bond. The lines represent pairs of electrons being shared. (**b**) A more graphical representation of a covalent bond, showing the electron "cloud" surrounding the two atoms. Note that, since the atoms are identical, the cloud is shared equally between the two. (**c**) A representation of a polar covalent bond. (**d**) In this polar bond, oxygen has a greater affinity for electrons than does carbon; therefore, it attracts them to a greater extent. This is illustrated by the electron cloud being more concentrated around the oxygen atom than around the carbon atom. This unequal distribution of electrons results in the partial charges indicated by δ^- and δ^+.

water if the water molecules create a hydration shell around the entire molecule. Hydration shells are complete envelopments of a molecule by a host of water molecules that effectively shield the enclosed molecule from any other interaction.

What kinds of atoms form covalent bonds? An atom's ability to attract an electron provides a good measure of whether it will form covalent or ionic bonds, or both. Chemists express this likelihood as a measure of the ability of an atom to attract the shared electrons in a covalent bond, or the **electronegativity**. Ionic bonds are formed by atoms with very different electronegativities, that is, between one atom that has a great attraction for an electron and another atom that readily gives one up. Covalent bonds are formed when at least one of the substituent atoms prefers to share the electron and thus has an intermediate electronegativity. For example, the electronegativities of sodium and chlorine are quite different, reflecting the tendency of sodium to easily lose an electron and of chlorine to readily gain one. Carbon has an electronegativity in the middle, meaning that it forms covalent bonds almost exclusively.

There are some fine points involving covalent bonds. For instance, not all covalent bonds are the same. For example, if a carbon shares an electron with another carbon, all else being equal the two atoms participating in the bond have equal electronegativities, and the electron should therefore be equally shared. On the other hand, if a carbon atom shares an electron with an oxygen atom, the two atoms have different electronegativities. Oxygen has a much greater tendency to accept an electron, so the electron is not equally shared but spends more of its time associated with the oxygen atom. Thus, a partial negative charge is associated with the oxygen atom. There must also be an equal and opposite partial positive charge associated with the carbon atom. Since there is not a complete loss of the electron by the carbon, the result is not an ionic bond, but the bond has a partial ionic character. The bond is said to be **polar** (Figure 1.7c, d). It might be compared to a bar magnet, with one end being north and the other south. The bond acts like a bar magnet of charge, with some negative charge at one end, around the oxygen, and some positive charge at the other end, around the carbon.

Even though this bond is not ionic, it will act according to its electronic properties. In other words, should a strong negative charge come near the bond, the carbon atom will be attracted to it, and the oxygen atom will be repelled. In this way, polar bonds can play a significant role in establishing the shape and function of biological molecules. Many of the amino acids, for example, have side chains with polar bonds, while others have side chains that can participate in ionic bonds.

Hydrogen Bonds

Now that we understand polar bonds, we can see how water molecules shield charges so well. Water is a compound made up of covalent bonds—one oxygen atom bonded to two hydrogen atoms. Each of these two bonds is covalent, but the oxygen atom has a much stronger tendency to attract electrons than do the hydrogen atoms. Thus, the oxygen atom becomes partially negatively charged, and the hydrogens become partially positively charged. The shape of the water molecule is something like a shallow V, because of the additional electrons around the oxygen atom that essentially

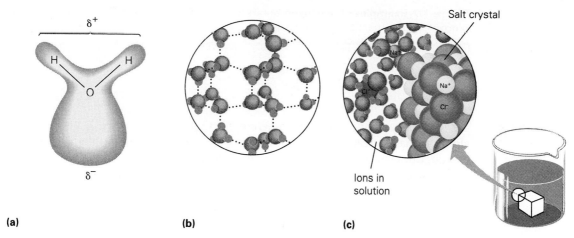

Figure 1.8 Hydrogen bonding in water. (**a**) The shape and polarity of a water molecule. Each of the hydrogen–oxygen bonds is a polar covalent bond, giving rise to the indicated distribution of charge. (**b**) Water molecules hydrogen-bond to each other readily. (**c**) When salt crystals are added to water, the polar water molecules surround each ion, shielding its charge and allowing it to leave the crystal.

force the hydrogen atoms toward each other (Figure 1.8a). The net result of the distribution of the partial charges over this shape is a molecule with one end partially negatively charged and one end partially positively charged.

Interactions between two different polar covalent bonds are an incredibly important part of cellular chemistry. The charges associated with water molecules cause them to interact with one another. The negatively charged oxygen on one water molecule will be attracted to positively charged hydrogens on other molecules (Figure 1.8b). **Hydrogen bonds** are very weak compared to covalent bonds; they actually consist of an attraction between partially negative oxygen atoms and partially positive hydrogen atoms.

Hydrogen bonding in water results in properties such as surface tension. You can, if you are careful, fill up a glass of water so that the water bulges over the lip of the glass but does not spill. What prevents the water from running over the lip? Why can water spiders scoot across the surface of ponds and rivers and not sink? The answer is *surface tension*—a consequence of the hydrogen bonding between individual water molecules.

The polarity of water molecules helps explain how water acts to disrupt ionic bonds and form hydration shells: The polar nature of the water molecule results in an organized arrangement being formed around any charged or polar atom or molecule that is in water (Figure 1.8c). Just think: All biomolecules inside cells are essentially "dissolved" in water. There must be water interacting with the outside surfaces of virtually all biomolecules. It is important to remember that a layer of water surrounds every biomolecule and that this water needs to be moved out of the way if two molecules are to interact.

Figure 1.9 Examples of hydrogen bonds in biomolecules. The different chemical groups illustrated occur in all biomolecules: proteins, nucleic acids, carbohydrates, and lipids. All biomolecules contain numerous chemical groups that can hydrogen-bond to water or to other chemical groups.

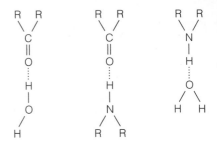

Hydrogen bonding is not limited to water. Whenever a hydrogen atom with a partial positive charge can be oriented in a line with another atom possessing a partial negative charge, hydrogen bonds can form. Some other examples of hydrogen bonds are shown in Figure 1.9. As we will see, even these very weak hydrogen bonds are used in nature to maintain structures and perform functions.

Hydrophobic Forces

It is possible now for us to more precisely understand the nature of what happens when lipids are placed into water. Since the water molecules are so polar, they like to have other polar molecules around them. Adding a nonpolar molecule, such as a lipid, to water makes the water molecules disrupt their normal interactions in order to accommodate the hydrophobic molecule. If more and more hydrophobic molecules are added, they eventually find one another and clump together, not because of much innate attraction between them but because there is less disruption of the water if they are together than if they are separate. This is called the **hydrophobic force**, and as we have already seen it drives the formation of membranes in cells.

Hydrophobic forces are important in biomolecules besides lipids. Recall that proteins are made of many individual amino acid building blocks. Extending from the amino acid chain are numerous hydrophobic groups. When in water, these hydrophobic groups are driven around by water until they find one another and clump together, with the water trying to make them as unintrusive as possible. A protein chain often gets folded and crunched as these hydrophobic groups assemble in the center of the protein. For a great many proteins, collectively referred to as globular proteins, a basic structure is arrived at in which a protein chain folds back on itself to create a core region of hydrophobic amino acid side chains surrounded by a layer of the protein chain that contains the more polar and ionic side chains. This external layer is in contact with the water, a situation more acceptable to the molecules involved, since they can then hydrogen-bond.

THE KEY TO LIFE

Recall that one of the primary activities that distinguishes living things from nonliving things is the ability to extract energy from the surrounding environment. In one sense, different organisms accomplish this in a great variety of ways: We eat large chunks of food, which are subsequently broken down (digested) into usable pieces. Many bacteria inhabit the intestines of mammals, wherein they are constantly supplied with nutrients—as long as the host continues to eat. Bacteria in the soil similarly obtain appropriate nutrients from right around them. Plants send out roots into the soil that can then take up nutrients and send them along to the rest of the plant.

So the process of gaining nutrients and raw materials varies greatly in nature. But ultimately, no matter how the nutrients are gathered, they must end up inside individual cells. We call the set of chemical reactions that occur in living cellular organisms **metabolism**. Despite the great variety of individual metabolic reactions, there are some common chemical themes.

Oxidation-Reduction Reactions

How does the cell obtain energy from raw materials (nutrients or foods)? Cells need energy to do virtually everything, from growing larger to signaling their neighbors, and very elaborate methods of obtaining this energy have evolved.

Let's begin with an example of energy extraction: how we heat our homes in the winter. Oil, gas, coal, and wood are all organic materials on which people in different parts of the world depend for fuel. In certain parts of the United States, people burn wood to warm their homes. We take wood from our environment and use it to provide heat energy. But what is the exact source of the energy?

We know that burning wood or gas releases heat; fire is hot, and heat is a form of energy. When we burn a log, a great deal of heat is given off—but at what expense? Afterward, there is no log, only some ashes, soot, and smoke that has gone up the chimney. What precisely does *burning* mean in chemical terms? Burning is a kind of chemical reaction called **oxidation**. Chemists tell us that oxidation is the loss of electrons from a material. A burning log needs air to burn. Why? Because air contains oxygen, which is the real requirement in the burning process. Large polymers such as cellulose, which make up the bulk of the wood, react chemically with oxygen and are transformed into carbon dioxide and water vapor during burning.

In the fire, the carbons from the cellulose undergo oxidation. Remember what the chemists say. In the fire, as an example of oxidation, a carbon atom is chemically "attacked" by oxygen, which replaces all the other atoms formerly bound to the carbon. The most stable end product consists of two oxygens bound to one carbon, or carbon dioxide. Carbon dioxide cannot react anymore with oxygen; it cannot be further oxidized, that is, it cannot share more of its electrons. Therefore, the carbon atom is fully oxidized; it has two of its electrons pulled toward oxygen atoms—a "loss" of electrons.

The oxygen atoms, on the other hand, have been **reduced**, the opposite of being oxidized. We can visualize this as each oxygen atom "gaining" an electron (from the carbon atom). Whenever there is an oxidation reaction, there must be an accompanying reduction. Before burning, the carbon in the wood was reduced but then became fully oxidized, and concurrently the atmospheric oxygen (O_2) started out as fully oxidized but was reduced. Why concern ourselves with oxidation and reduction? Very often, oxidation gives off energy. Oxidizing the carbons in burning wood gives off heat energy.

Let's continue our wood-fire analogy. In a stove, we purposefully control the rate at which the wood burns. We can make it burn very slowly by restricting the oxygen (closing the damper); if we open the damper, too much oxygen will be allowed to react, the wood will burn more quickly, and stifling heat will be released. So we control the rate at which carbon atoms are oxidized and release heat energy.

Back to biology: Cells use, in essence, the same process to derive energy from the environment—oxidation of carbon-containing fuels. What fuels do cells use? Most **heterotrophic** organisms, those that need preformed organic food, can use a variety of fuels, including the six-carbon sugar, glucose. As previously indicated, cellulose is made entirely of glucose molecules, as is starch, which is found in many plants, including potatoes.

So cells use glucose (from cellulose or starch) and other reduced carbon compounds as energy sources. Cells oxidize glucose to release energy as heat; for living cells, the remaining energy released during oxidation is captured by a compound called adenosine triphosphate, or ATP. ATP acts as an energy intermediate, or energy "currency," and can transfer energy released from the oxidation reactions.

Just as with our wood-fire example, cells must regulate the rate at which their fuel is burned to capture some of the released energy. How do living cells regulate oxidation? Cellular oxidations occur in little steps. Carbons are not completely oxidized immediately; instead they are taken through a number of intermediate steps, in which a reduced carbon is gradually oxidized until it becomes fully oxidized, resulting in the formation of carbon dioxide. At several steps during the oxidation process, as partial oxidation reactions occur, the energy being released is partially recovered as ATP.

Consider a particular carbon that is joined to two other carbons and two hydrogens, a common structure found in biomolecules. This atom can become oxidized through several steps: First one hydrogen gets replaced by an oxygen, next the other hydrogen is removed, and then the bonds to the other two carbons are replaced by bonds to a second oxygen. This carbon is now completely oxidized. Each step oxidized the carbon a bit more until it became fully oxidized as CO_2. During a wood fire, these steps occur almost instantaneously. In cells, they are more ordered and controlled.

Activation Energy

Consider how a fire in a wood stove gets started. Logs do not burn spontaneously. You can put a log in the stove, and it will sit there indefinitely, not burning. To get the log burning, you must set fire to it—giving it an energetic kick in the pants, as it were. Usually you do this by using a smaller flame obtained in some other fashion (match, lighter, and so on). To get the log to burn, you initially have to add a bit of energy.

This energy is called the **activation energy** (Figure 1.10a). It represents the energy necessary to get the chemical reaction under way. Oxidation reactions don't always happen spontaneously; if they did, fuel materials would disappear instantly. Like burning logs, burning glucose or other fuels inside cells requires an input of energy. We can view this as a barrier that requires an input of energy to be crossed.

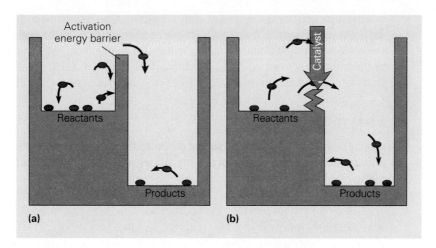

Figure 1.10 Activation energy. (**a**) Imagine putting "jumping beans" into a well. They will rarely have the energy necessary to jump out of the well. This situation is analogous to molecules with insufficient energy being unable to react and form products. The barrier to reaction is the activation energy. (**b**) A catalyst lowers the activation energy, essentially lowering the barrier for the reactants. This makes it much easier for the molecules to react; therefore, many more of them will react.

However, cells use a different approach to overcoming the activation energy than do people burning wood. This is where our wood-burning analogy falls apart. Lighting a smaller fire to the wood in the stove adds just enough energy to the system to overcome the activation energy barrier. Cells, instead of adding the necessary energy, lower the activation energy—a completely different approach to solving the problem. The question now becomes: How can an activation energy barrier be lowered?

Catalysis

For two atoms or compounds to react, they must first be brought into close proximity and be put into the appropriate positions for the reaction. The activation energy represents the amount of energy necessary to achieve these particular goals for a given reaction. Cells lower the activation energy by making it much easier for these steps to occur. Cells take the reacting compounds and, using a sort of "chemical scaffold," line them up properly and make the start of the reaction more favorable. This process is called **catalysis**. A **catalyst** is something that causes a reaction to occur more readily by lowering the activation energy (Figure 1.10b). Although the catalyst may participate in the reaction, it is never permanently changed by the reaction; at the end, the two reactants are altered, but not the catalyst. Thus, a catalyst can be used over and over again to catalyze the same reaction.

Catalysts are common, and they are in use in everyday chemical events. The hardening of epoxy glue depends on a catalyst. You may know that epoxy comes in two tubes; one is the glue itself, and the other is the "hardener," actually a catalyst. The glue does not function as a glue when it comes out of its tube; it will not harden or cure properly by itself. The chemical reactions necessary to make it a strong glue do not happen spontaneously. The hardener, when it comes in contact with the glue, initiates the chemical reactions that will result in the formation of the hard, durable, permanent epoxy cement. The hardener does this by catalyzing reactions—by lowering the activation energy.

Enzymes

The biological catalysts cells use are called **enzymes**, and they are nearly always protein molecules. Enzymes catalyze virtually every chemical reaction that occurs inside cells: They join reactants in such a way that a reaction between them becomes more favorable.

Each individual reaction that occurs in a cell employs a specific enzyme made for the express purpose of catalyzing that reaction or a group of related reactions. Remember our discussion of burning carbohydrates to yield water and carbon dioxide. While this may be a simple reaction in a wood stove, in the cell it is more complex; thirty to thirty-five steps or chemical reactions occur between the start of this process and the final carbon dioxide product. Each step involves a unique enzyme that functions to catalyze that specific reaction.

Enzyme uniqueness is critical for many reasons, of which we will list two. First, it assures that only the proper reactants will come together, preventing the occurrence of mistakes that might result in improper, perhaps toxic, chemical products. Second, enzymes can be controlled by the cell, meaning that the rate of a particular reaction can be controlled by the activity of the catalyst. If a cell needs more of a product, it can speed up the reactions leading to that product. If the cell has an overabundance of a product, it can slow down or stop the reactions leading to that product. Cells control enzymes sometimes by adjusting the amount of enzyme present or more often by changing the ability of enzymes to bind or transform reactants. The cell can fine-tune the various reactions in order to maintain the proper levels of chemical intermediates. This is vital to the health of the cell; remember that if the "burning" or oxidation of fuel goes too quickly the cell will not be able to recover all of the energy it might need, and if it cannot occur quickly enough the cell will not have as much energy as it needs.

Catabolic and Anabolic Reactions

We have couched our coverage of enzymes and metabolism in terms of burning fuels to obtain energy. Many other chemical reactions occur inside cells as well. We can divide these reactions into those responsible for breaking things down, such as the oxidation reactions of glucose, and those that build things up or put things together, including the reactions that form proteins from amino acids and DNA from nucleotides. Reactions that break things down are called **catabolic**. Reactions that build things up are called **anabolic**. In a cell at any given time, both anabolic and catabolic reactions are occurring; the cell is constantly dealing with its energy demands by breaking down fuel and making ATP (catabolic reactions) as well as breaking down other cell components that are no longer needed. The cell is constantly replacing components that become depleted and making new materials for growth (anabolic reactions), which requires energy. Catabolism provides the raw materials and useful energy for anabolism. Thus, the cell maintains an exquisite balance of breakdown and growth. All of these reactions are controlled by the activities of the enzymes that catalyze the reactions. As we will see later, throwing off the normal balance of metabolism, by whatever means, can have disastrous effects on the cell.

The manifestations of most diseases arise because of some alteration in the metabolism of cells. One of the more important areas of biotechnology research aims to understand how disruptions in metabolism can cause disease and what can be done to restore the processes to normal. We should begin thinking about cells in terms of controlled sets of chemical reactions now, so that later we can see how alterations in this normal balance become threatening.

ISSUES: RISK AND NEW TECHNOLOGY

At the beginning of this chapter, we briefly discussed the inherent risks of developing new technologies. It is unfortunate, but risk is a necessary component of such development. It takes various forms, from the venture capital risks taken by investors to the personal risks assumed by individuals who volunteer as part of a testing program for a new drug. Corporations take huge risks by investing heavily in specific products and procedures, hoping that they will eventually find their way to market and become useful and profitable. Municipalities assume risks by agreeing to allow research to be conducted that might prove harmful to the region. In short, the development of biotechnology has seen and will continue to see its share of risky ventures.

Does this mean the technology should not be pursued? Such a question is difficult to answer, and the answer probably depends on your perspective. A theme that becomes apparent quite frequently during discussions of risk is that individuals perceive risk differently from society as a whole. What might seem to some people to be beneficial to society as a whole is often perceived as risky by some individuals, and vice versa. Since living in a society must be a compromise among all the individuals involved, this apparent conflict seems quite reasonable. However, it is not always easy to decide, either personally or as a part of society, what risks to assume.

Every day, we subject ourselves to numerous risks that could be avoided. Most of us in the United States drive far more frequently than is absolutely necessary, incurring some level of risk. Many choose to live at high altitudes, exposing themselves to slightly higher levels of ultraviolet radiation. Many of us live in or near large cities, where the air often contains potentially harmful substances. Yet we are used to all of these risks—they are a normal part of our lives. It doesn't always matter that virtually all of us are aware of the risks of smoking or practicing unsafe sex—numerous individuals choose to engage in these high-risk behaviors. Others will do everything in their power to avoid unnecessary risk. How is society to come to grips with risk when there are so many different individual responses? Unfortunately, that is exactly the situation being thrust upon us by the new and powerful developments of biotechnology. As we proceed in our study of biotechnology, it will be interesting to step back now and again to revisit this issue.

• •

SUMMARY

Living organisms exist as either free-living or groups of discrete units, or cells. Although there are many different types of cells, all are organized as either prokaryotic or eukaryotic cells. The primary distinction between these two cell types is the absence or presence of membrane-bound compartments.

Biomolecules make up the materials found within cells. Most of these biomolecules are polymers, chains of relatively simple molecules linked together to create larger, more complex molecules. The identities of the individual building blocks determine the ultimate structure and function of the molecules.

A number of different forces act to hold molecules together in various shapes and to govern the interactions particular molecules may have with other molecules. Such interactions are important to any biological process, since it is through interactions that molecules function. Often, these interactions allow chemical reactions to occur, leading to structural changes in the molecules. Barriers that prevent reactions from occurring can be overcome through the action of catalysts. Enzymes are biological catalysts that provide the means for cells to create and modify many biomolecules.

REVIEW QUESTIONS

1. Name two characteristics shared by all living organisms that separate them from nonliving entities.

2. What are the most important differences between eukaryotes and prokaryotes?

3. How does the structure of the lipid bilayer (cell membrane) explain its semipermeable nature?

4. Explain how an almost infinite number of different proteins can be made using only twenty different amino acids.

5. Describe each of these types of chemical bonds by drawing a labeled diagram: ionic, covalent, and hydrogen.

6. Use a molecular diagram to explain how table salt dissolves in water.

7. Why do water molecules have a "shallow V" shape?

8. Explain why the statement "enzymes are organic catalysts" is accurate.

9. Give two reasons why enzyme uniqueness is important.

10. Why are catabolism and anabolism linked within living cells?

CHAPTER 2

Molecular Biology: The Flow of Information Within Cells

In the last chapter we examined cells and the major types of molecules that make them up. One of the central ideas that emerged is that biological information is required to direct the production of biomolecules. This is particularly significant when we are trying to understand how the various proteins work, which in large part determines the ultimate structure and functions of the cell. This chapter is concerned with biological information—how it is stored and utilized. **Molecular biology** is the name given to the study of the storage and flow of information within a cell.

There is such a variety of living things in the world that it is surprising at first to realize how much all living things have in common. In fact, they have so much in common that it is possible to study a great deal of biology by concentrating just on the similarities among organisms, rather than on the differences between them. Most of these similarities occur at the molecular level, a level smaller than the cell itself.

What sorts of features do all organisms share? We've already seen that all cells have membranes made of lipids and proteins and that all cells use enzymes to catalyze chemical reactions. In addition, all cells use the same material, **DNA** (*Deoxy-riboNucleic Acid*), for information storage. All cells also use a system in which the information contained within the DNA is copied into RNA and then turned into proteins by molecular machines called ribosomes. At the molecular level, cells have many more similarities than differences.

We will see both in this chapter and later in the book that maintaining the integrity of information is vital to a cell. Without proper information storage and retrieval systems, the normal activities of cells change dramatically—so much so that they often die. The information contained within a cell is probably its most valuable commodity.

The explosion of knowledge in the past twenty-five to thirty years in cell and molecular biology, molecular genetics, and immunology has provided the necessary background for the rapid development of biotechnology. The manipulation of living organisms in ways unimaginable thirty years ago is now possible because of our knowledge of how cells store and use information. In other words, biological information is also now a substantial commodity for us, as manipulators of the biological world. This reality can be measured by the current emphasis placed on patenting biological information. The ownership of such information and the production of the genetically altered organisms that result from it have become an economic issue. Who, if anyone, owns the information contained within genes?

First, let's look at the scientific principles. Since this chapter is devoted to a study of the flow of biological information within cells, we'll begin by examining how biological information exists within cells. Then we will consider how the information is used.

THE STRUCTURE OF DNA

Imagine that you are a cellular engineer trying to devise an information storage system. What properties would be most useful in such a system? You would want the system to be permanent, such that it would not necessarily be easy to introduce changes into the information. You would want some sort of backup system, so that if one copy were damaged the information could be restored or replaced. You would want the information to be accessible. And, since you know that many cells will ultimately be derived from one original cell, you would want a storage system that could be easily and accurately replicated. Nature settled on the use of DNA as an information storage system in all living cellular beings. We will examine DNA with an eye toward understanding how its structure allows it to function as an information storage molecule.

Building Blocks

The chemical structure of DNA is uniquely suited to perform its function: storing genetic information in a virtually permanent, unchanging form. As we examine various levels of DNA structure, we'll remind ourselves along the way how each structural feature contributes to the stability of the final molecule.

Recall that DNA is one of the polymer molecules discussed earlier and thus is made up of long chains of building blocks, called nucleotides, linked together end to end. Each nucleotide is made up of three simpler chemical constituents: a sugar, a phosphate, and a nitrogen base (Figure 2.1a). In DNA, the sugar is **deoxyribose**—hence the name DNA, for deoxyribonucleic acid. The four different nucleotides of DNA differ only in their bases. We typically designate each building block—each nucleotide—as A (adenine), C (cytosine), G (guanine), or T (thymine). A nitrogen base or an entire nucleotide containing that nitrogen base is often referred to in the same way, so the nomenclature can be confusing. (*A* stands for the nitrogen base adenine and also for the nucleotide composed of adenine, deoxyribose, and a phosphate group.)

The nucleotides are joined into a polymer by covalent chemical bonds between the phosphate and the sugar unit (Figure 2.1b, p. 26). The polymer chain, then, actually consists of alternating sugar-phosphate-sugar-phosphate- and so on. This is

GENERAL STRUCTURE

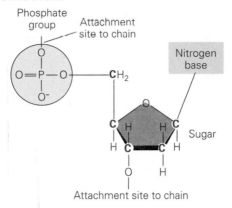

AVAILABLE NITROGEN BASES

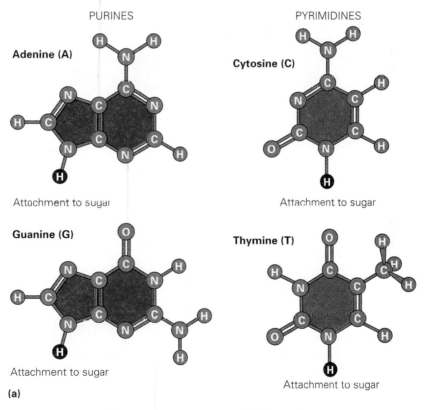

(a)

Figure 2.1 Structure of nucleotides and a DNA chain. (**a**) Nucleotides are composed of a phosphate group, a sugar (deoxyribose), and one of four nitrogen bases.

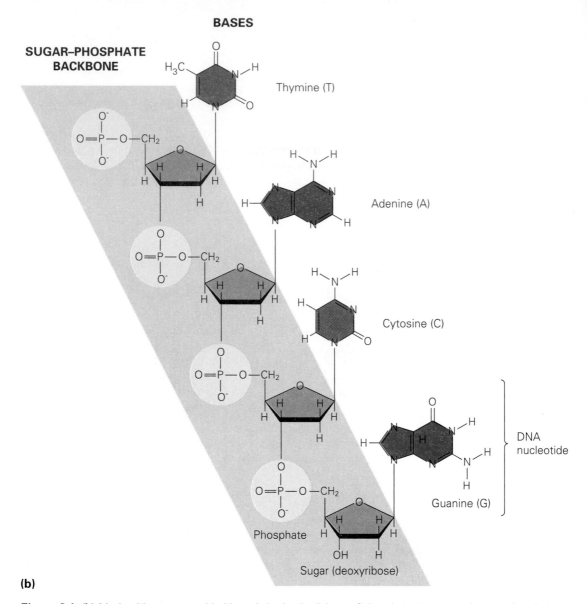

(b)

Figure 2.1 (**b**) Nucleotides are assembled into chains by the linkage of phosphates to sugars. A sugar-phosphate backbone results, with the bases extending out from the sides. Any base can be incorporated at any position.

often called the *backbone* of the DNA. Notice that the bases extend to the side of the chain. They are chemically joined only to the sugar unit in their nucleotide, not to each other.

Since each nucleotide has the same sugar-phosphate structure, any of the four nucleotides could be built into the chain at any location with no change in the overall structure of the polymer. Thus, the specific sequence of nucleotides, or rather the sequence of nitrogenous bases sticking out from the DNA backbone, is how information is stored. Just as in our alphabet we use sequences of letters (along with spaces and various punctuation marks) to store information in written form, cells use the four nucleotides of DNA to store biological information in chemical form.

Most DNA molecules are extremely long, containing literally millions of nucleotides. Note that there are never any branches to the DNA backbones. They are always long strings of nucleotides joined by covalent bonds. However, not all DNA molecules are linear; some are circular.

Since the bonds between the individual building blocks in a strand are covalent, they are very strong. With few exceptions, they are permanent and thus do not come apart under normal cellular conditions. In fact, it is even difficult to break these bonds in isolated DNA in a test tube. However, during certain situations cells require these bonds to be broken, notably during repair of existing DNA.

The Double-stranded Structure

The simple polymer chain of DNA we have pictured and described is not the complete DNA molecule. Two strands form a molecule of DNA. The interaction of these two strands creates a structure that ensures the permanence of the information and allows it to be readily copied and transmitted. Thus, it is essential to understand the chemical interaction between the two strands.

Recall hydrogen bonding—the tendency of hydrogen atoms, when attached to oxygen or nitrogen atoms, to be shared with other electronegative atoms. Hydrogen bonding is the key to double-stranded DNA. A number of potential hydrogen-bonding structures are present in each of the bases. You may have noted some chemical similarities between different bases. For example, A and G have somewhat similar structures, as do C and T (Figure 2.1a). Looking carefully at the structures of the bases, you see that G and C can be arranged so that three hydrogen bonds can form between the two bases (Figure 2.2, p. 28). Similarly, A and T can be arranged to form two hydrogen bonds. This represents a base-pairing rule: A always base-pairs with T, and C always base-pairs with G. C and G are said to be complementary, as are A and T. Interestingly, when these base-pairs are superimposed, they occupy the same space.

What does this base-pairing rule really mean? DNA normally consists of two strands, side by side, with the two strands lying in such a way that the bases from each strand extend outward toward the other strand. What if the sequence of each strand were such that directly across from a C on one strand would be found a G on the other, and directly across from each A would be found a T? Then every base could hydrogen-bond with its associate, or its complement, from the opposite strand. In other words, the strands would be base-paired to each other (Figure 2.3). DNA can be visualized as a long zipper; this analogy is not too farfetched, as we will see later.

How strongly are the two strands held together? Each base-pair consists of two or three hydrogen bonds. Individual hydrogen bonds are very weak and easily broken—recall that they are the bonds responsible for giving water its surface tension. But, given the sheer number of hydrogen bonds in any given DNA molecule, the overall contribution to structural stability is great. Think of Velcro®: One side of the tape has little fibers sticking out, with tiny hooks on them. The other side of the tape has

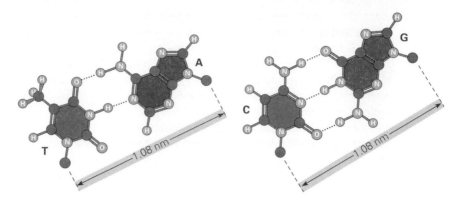

Figure 2.2 Base-pairs in DNA. A–T and G–C base-pairs. The attachment points to the sugar-phosphate backbones are shown as darkened circles. The hydrogen bonds between bases are shown as dotted lines. Note that the A–T base-pair fits into the same space as the G–C base-pair.

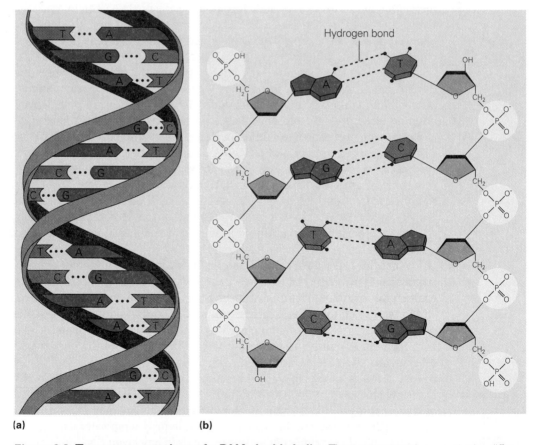

(a)   (b)

Figure 2.3 Two representations of a DNA double helix. These representations emphasize different aspects of the double helix. (**a**) This model illustrates the backbones (of the double helix) and the bases extending into the middle. (**b**) This model draws attention to the *antiparallel* relationship (note the orientation of the sugar molecules) between the two strands.

loops onto which the hooks can attach. Individually, the hooks provide very little strength and are easily pulled apart. But collectively they can create a substantial force. The same is true of a zipper, which has small pieces of plastic or metal on one side stuck between two others on the other side. Individually, these pieces are not very strong, but put many of them together and it becomes much more difficult to separate them.

DNA exists in water inside cells, and water can also hydrogen-bond. Why does one strand of DNA have a preference for hydrogen-bonding to another complementary strand instead of to lots of water molecules? Actually, the principal forces that hold two DNA strands together are not the hydrogen bonds but rather hydrophobic forces. The nitrogenous bases of the DNA are somewhat hydrophobic, particularly when they are base-paired. Base-pairing neutralizes much of the polarity of the bases, resulting in fairly hydrophobic structures. By pairing with another DNA strand, these hydrophobic bases are to some degree "hidden" within the structure so that water has less access. Because of the hydrophobic nature of the bases, a double-stranded structure is far more stable in water than a single-stranded structure. It is the combination of these forces—hydrophobic forces and hydrogen bonding—that yields the double-stranded structure of DNA. Hydrophobic forces drive strands together, but they cannot attain a stable structure unless base-pairs can hydrogen-bond. Hydrogen bonding thus provides the specificity of interaction between two strands. As we indicated previously, the key to specificity of interaction is that the base-pairs A–T and C–G are specific because *three* hydrogen bonds form between C and G whereas only *two* hydrogen bonds form between A and T (Figure 2.2).

The fact that DNA preferentially exists as a double-stranded molecule does not mean that the two strands of DNA cannot be separated. Mechanisms exist inside the cell for separating these two strands when necessary—for example, during the copying of DNA.

We now need to consider another noteworthy structural feature. Recall our simple model of a single DNA strand: sugar-phosphate-sugar- and so on. It turns out that there is a chemical direction to this chain; that is, one end does not look chemically like the other (Figure 2.1b). One end is terminated by a phosphate group (PO_4) and the other by a hydroxyl group (OH). The two strands that come together to form DNA are oriented in opposite directions—we refer to them as **antiparallel strands** (Figure 2.3b). The fact that the two strands run in opposite directions will take on greater significance when we examine the process of copying information from DNA.

Let's review the different levels of DNA structure. First, we considered the sequence of bases in one strand of the molecule. Second, we noted the hydrogen bonding between complementary strands of DNA. Remember that because of complementary base-pairing the sequence of one strand determines what the sequence of the other will be. Finally, we noted that the two strands of nucleotides are antiparallel.

The Helix

We will briefly consider one additional level of structure at this point. We've compared the DNA molecule, with its two strands held together by many hydrogen bonds, to a zipper, and in some ways this is a good analogy. However, in terms of shape, it is not—DNA molecules are not thin and flat. They form a helix, a twisting telephone-cord type of arrangement (Figure 2.3a). This occurs simply because of the nature of the chemical bonds involved in DNA. Torsion in each bond results in a twisted, helical structure that is assumed no matter which base-pairs are in the sequence. Since each base-pair occupies the same space, any combination of them can be included within the helical structure. This double helix is important historically; before the structure of DNA was known, the first experiments to reveal anything about its structure showed that it must be a helix of some sort.

The helix structure has functional consequences, the most critical of which involves *sequence recognition*. Cells can extract information from the sequence of bases within the DNA helix, by physically separating the two strands and "reading" the sequence of exposed nitrogenous bases. Strand separation and "reading" happen when information is copied for use elsewhere in the cell. However, this is not the only way sequence information can be obtained. Cells need to be able to recognize the appropriate information to use before the strands are separated. When we examine the double helical structure of DNA (Figure 2.3a), we can see that the "sides" of the base-pairs are accessible in the grooves of the helix. Because the "sides" of the different base-pairs are chemically different, cells have access to the information contained within a DNA molecule without having to unzip the DNA structure—an important feature that enables the cell to regulate the utilization of information.

Review of Structure and Function

What biological properties can result from the unique structure of DNA? First, the structure can accommodate any sequence of bases. This allows great potential for information storage. Second, it provides essentially two copies of the information—actually, two strands of DNA that both contain the same information, one directly and the other as the complement. If something should happen to one strand, the information (or its complement) exists to direct replacement or repair. The existence of two complementary strands also allows for straightforward replication of the DNA. Third, the structure itself is hardy. It is strong, not easily disrupted, and allows for the stable maintenance of information within a very chemically reactive environment, the inside of a cell. DNA would seem to be ideally suited to serve as an information storage molecule.

THE FLOW OF BIOLOGICAL INFORMATION

Now that we understand the structure of DNA, let's concentrate on what the information is used for in cells. Since proteins do virtually all of the "work" of the cell—providing structure, transporting nutrients, catalyzing reactions—the particular proteins present within a cell will ultimately determine the properties and activities of that cell. Thus, a cell will use the biological information stored as a sequence of bases in DNA to produce the proteins necessary to the functioning of that cell. An overview of this process is shown in Figure 2.4. DNA stores a tremendous amount of information—more than the cell needs to use at any particular time. So small bits of

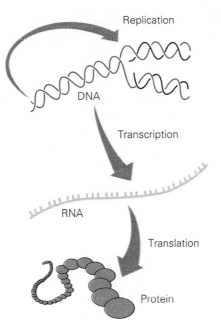

Replication

DNA

Transcription

RNA

Translation

Protein

Figure 2.4 The flow of biological information within cells. Information stored in DNA is copied into RNA form during the process of *transcription*. The RNA then directs the synthesis of a protein during *translation*. Copies of DNA are made during the process of *replication*.

the sequence information are first copied into temporary molecules (called **messenger RNA**, or **mRNA**) during **transcription** (the process of assembling mRNA using a DNA template). RNA is a nucleic acid polymer similar to DNA except for three primary differences: The base uracil (U) is used instead of T, ribose is used as the sugar unit instead of deoxyribose, and RNA molecules are generally single-stranded. Following transcription, the sequence of bases in the mRNAs is *translated* into a sequence of amino acids, resulting in the formation of a specific protein.

The Language of DNA

The flow of information through transcription and translation is a very complex process, involving much of a cell's time and energy. In order to understand it, we will consider each step separately, beginning with how the DNA molecule stores biological information in chemical form. To do this, we will make references to one of our own systems of information storage—letters and words. Our alphabet consists of a number of characters that, when placed in combination with others, signify words. Words can then be combined into sentences, with the proper organization and punctuation. Related sentences become paragraphs, and so on. There are an infinite number of possible combinations of the letters of our alphabet, and although we don't use all of them, we do use many.

Our language example has some similarity to the biological language. The DNA alphabet has only four letters—A, C, G, and T—and they can be arranged in any order. Ultimately, arrangements of these letters will be used to determine the individual amino acids in proteins. Unlike our language, the words of the DNA language are all the same size. Since there are twenty different amino acids that need identification, there must be at least twenty different DNA words. A word consisting of a single base could code for only four different amino acids, since there are only four letters in the DNA alphabet. A set of two bases could code for only sixteen amino acids (4×4). And a set of three bases would be able to code for sixty-four amino acids ($4 \times 4 \times 4$). A sequence of three nitrogen bases on either DNA or mRNA that is used to code for an amino acid is a **codon**. When this information is used, it is first converted into RNA words (equivalent except that U replaces T). Each codon specifies a particular amino acid, with the exception that three codons, called "stop" or termination codons, are the equivalent of periods in our language (Figure 2.5a). Most of the twenty amino acids are encoded by more than one codon, some by as many as six. We refer to this as the **redundancy** of the **genetic code**. What we mean by the *genetic code* is actually a printed table of triplets of nitrogenous bases from mRNA that enables us to determine what specific amino acid is required for emplacement in a protein.

The genetic code is nearly universal. A GCC codon on mRNA orders the placement of the amino acid alanine in a bacterial cell as well as in a cat, and AAC orders the placement of the amino acid asparagine in fungi as well as in squirrels. The impact of this point on biotechnology cannot be overstated, for, at least in theory, it allows genetic information from any organism to be used in the cells of any other organism.

How are codons put together to make information for proteins? There is a one-to-one correspondence between a codon and an amino acid (Figure 2.5b). During the processes of transcription and translation, which we will study in detail shortly, the series of bases in DNA first serves as a *template*, or pattern, for a series of equivalent RNA codons. In turn, the mRNA codons determine the specific order of amino acids in a protein. In this manner, the information contained in the DNA is decoded and translated into the language of proteins. The unit of information really consists of the entire string of codons necessary to produce a protein. In many ways, this is a partial molecular definition of a **gene**: the linear sequence of DNA that contains all the information necessary for the production of one separate string of amino acids, or one protein. Often this means an entire protein, but some proteins are made of multiple pieces, for example, hemoglobin, which is made of four separate protein chains.

The concept of the gene is very important; however, the strict definition of a gene can be difficult to pin down. Since proteins are the smallest functional units in cells and each protein comes from a different gene, we can think of genes as the functional units of information within cells. Now that we have an understanding of what a gene is, we can examine the details of the processes that allow genes to be used.

Transcription

Remember that mRNAs are temporary messengers that carry copies of the information contained in a gene or set of genes from the DNA to the apparatus that synthesizes proteins. Why should such copies be necessary? Why can't the information simply be used directly from the DNA? The reasons may not be obvious, but let's explore some.

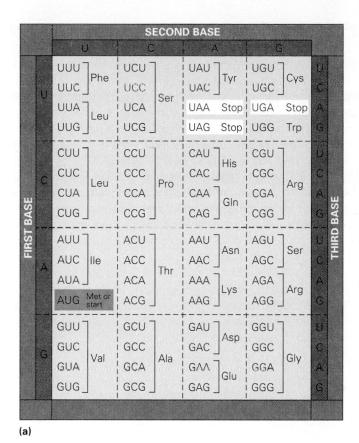

(a)

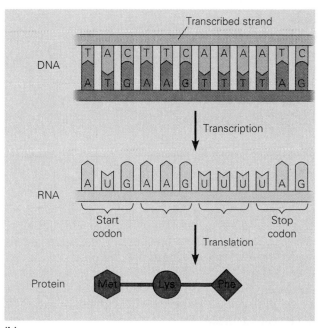

(b)

Figure 2.5 The genetic code. (a) Each codon, or group of three bases, "codes" for an amino acid—the ribosome and associated apparatus will interpret a particular codon to mean that the associated amino acid should be put into the growing amino acid chain. The code helps us understand the relationship between the codons and the amino acids. By locating a codon within the printed code, we can determine the amino acid for which it codes. The first base of the codon is listed on the left side of the code, the second base across the top, and the third base on the right side. (b) How the genetic code is used to direct the synthesis of chains of amino acids.

33

Why Use mRNA Intermediates? Cells go to lots of trouble to protect their DNA, and rightly so. The more access the cells give to various molecules and enzymes, the more likely it is that the DNA itself will be altered, damaged, or destroyed. The permanence of DNA is crucial, because it is the source of biological information in the cell. It must be preserved intact. Other than DNA replication, a very accurate process, the only other exposure of DNA is during transcription, or mRNA production. Such limited access is one way to protect cellular DNA.

Also, there is a serious regulatory phenomenon at work. Virtually all cells of an individual organism contain the same DNA, but not all genes are used by every cell. The extra step of transcription allows a variety of regulatory features to be added to the process of information flow. Individual cells can more easily control which genes are being used at any one time and can govern the ultimate production of proteins more carefully by regulating transcription.

Finally, at least in eukaryotic cells, there is the issue of cellular location. DNA is present in the nucleus, and protein synthesis takes place in the cytoplasm. Thus, mRNAs serve as mobile carriers of the information present within DNA, moving from the nucleus to the cytoplasm.

Construction of mRNA   How is mRNA made? An enzyme called **RNA polymerase** is responsible for the synthesis of RNA molecules. RNA polymerase recognizes a section of DNA that contains a gene and begins the process of information transfer. In essence, it produces a strand of RNA that is complementary to one of the strands of the DNA and thus contains the same sequence information as the opposite strand. We will now examine a number of features of transcription.

How does the RNA polymerase enzyme know where to start transcribing mRNA? As we've outlined, mRNAs are produced from only small portions of the DNA, the actual genes. The rest of the DNA does not directly code for proteins. A signal, called a **promoter**, is built into the DNA. It is a special DNA sequence found just before the gene. RNA polymerase binds tightly to DNA at precisely this promoter sequence, setting in motion a repetitive series of events (Figure 2.6). Once RNA polymerase attaches to the promoter, it unwinds the DNA helix in that region and begins to "read" one strand of the bases it exposes. Each time it reads a base, it takes a complementary nucleotide from the surrounding environment and adds it to the growing polymer of RNA nucleotides. The RNA polymerase moves along the DNA as it does this, and each time it completes a base-pair it moves to the next DNA base and repeats the process. This continues until the RNA polymerase finds a transcription **terminator**, a DNA sequence located after the gene that tells the RNA polymerase to stop what it is doing. The whole complex—DNA, RNA polymerase, and mRNA molecule—then falls apart. The DNA zips back together again. It is unchanged as a result of transcription; it has simply supplied information by being copied and is now restored to its original state. The RNA polymerase is also the same—recall that, as an enzyme, it does not become altered during the chemical reactions it catalyzes. The only thing new is the mRNA molecule, a single-stranded piece of RNA containing the same genetic information as one of the strands of DNA. The information contained within a gene has been copied to mRNA.

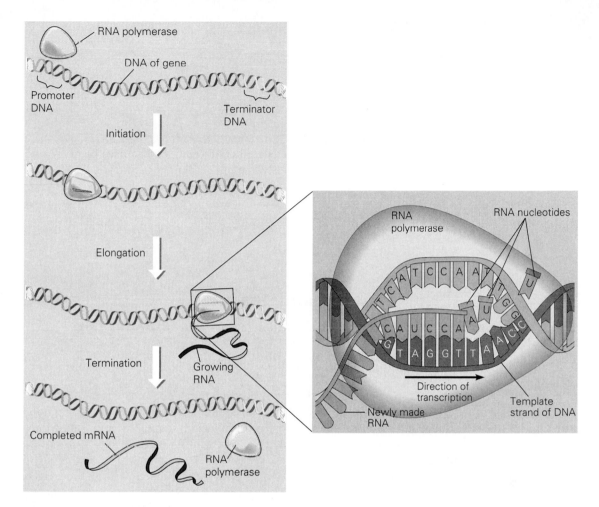

Figure 2.6 Transcription. RNA polymerase initially binds to DNA at a promoter. The DNA helix is temporarily opened (see expanded view), and the RNA polymerase synthesizes a chain of RNA with a sequence complementary to the DNA strand being read. As the RNA polymerase moves forward, the DNA strands rejoin. This process continues until a terminator is reached, at which time the complex disassembles, releasing the mRNA.

In prokaryotic cells, there are no compartments, so mRNAs are immediately used to direct protein synthesis. In most cases, the synthesis of a protein begins before the mRNA is even finished being transcribed! However, in eukaryotes, this is not the case, for at least two reasons. First, mRNAs are synthesized within the nucleus of a eukaryotic cell and must be moved outside the nucleus in order to be used in protein synthesis. The transcription of the mRNA must be completed before it can be moved. Second, eukaryotic genes usually contain **introns**, regions of noncoding DNA (sequences that do not specify amino acids in a protein), interspersed with the **exons** (sequences that do specify amino acids in a protein). Introns need to be removed before a functional mRNA can be formed (Figure 2.7).

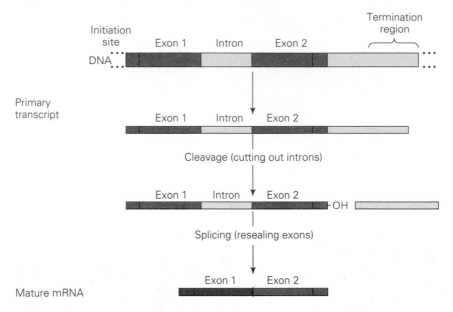

Figure 2.7 General steps in the formation of eukaryotic mRNA.

Regulating Transcription Continuous, total gene transcription is not desirable, because not all proteins are needed at all times. Some genes are continuously transcribed, but some are transcribed only at certain times, some are transcribed only once, and others are not transcribed at all. Which genes are transcribed at any given moment depends on the cell type, its continuous adjustments to changing environmental conditions, and signals it receives from the environment, from other cells, and from its own internal control system. For example, consider the bacterial population (*Escherichia coli*) that inhabits your intestines. The availability of nutrients can change rapidly and often, depending on what and when you eat. *E. coli* can rapidly transcribe certain genes to produce specific nutrient-digesting enzymes when that particular nutrient becomes available. Conversely, these bacteria can shut down synthesis of the same enzyme when the nutrient becomes scarce by inhibiting transcription. As we will see in Chapter 7, regulation of transcription is a common way of controlling which genes are "expressed" (transcribed and translated).

One way to regulate transcription is to use other proteins. Proteins in addition to RNA polymerase are required at the promoter to initiate transcription. Such accessory proteins are called **transcription factors**, a term that identifies them as participants in transcription. They are modulators of the basic transcription process. Some genes become active only when certain transcription factors bind to the promoters. These transcription factors can themselves be subject to all sorts of controls, for example, by hormones. A hormone acting on a particular cell might cause the activation of a particular transcription factor, which then would, in conjunction with RNA polymerase, cause the transcription of a particular gene or genes. In this way, the cell can alter its activity as dictated by the hormone.

Other regulatory proteins can affect the transcription process, for example, **repressor proteins**. Repressor proteins bind to the promoter regions of particular genes and block the access of RNA polymerase. If RNA polymerase cannot bind to a promoter, it cannot begin transcription, and that particular gene will not be transcribed. Repressor proteins are a very effective means of regulating the expression of genes. We will encounter them later when we discuss regulation of gene activity more completely in Chapter 7.

Translation

We've seen how information is stored and how a copy is made that can be used by the cell. How is the copy used to make a protein? This is accomplished through the process of protein synthesis, or **translation**. Why the linguistic name? In essence, what is accomplished is a translation from the nucleic acid (RNA and DNA) language of four different bases arranged in codons into the protein language of twenty different amino acids.

What Is Required for Translation? Protein synthesis occurs on **ribosomes**, precise functional structures of some fifty-plus proteins and several RNA molecules. Ribosomes are large assemblies, and cells have many of them (a single bacterial cell may have 20,000 ribosomes—making up about half the dry weight of the cell). A ribosome is actually made of two parts, or subunits: a smaller one and a larger one. When the ribosomes are inactive, the two subunits are separated.

Another critical component of protein synthesis is **transfer RNA (tRNA)**. tRNAs are the molecules that accomplish the actual "translation" between the two languages. They are small molecules that have two vital features. At one end is a three-base sequence of RNA, called the **anticodon** (Figure 2.8). The anticodon of any particular tRNA is complementary to one of the codons on an mRNA. At the other end of the tRNA is a chemical group to which a specific amino acid can be attached. The particular amino acid that gets attached to an individual tRNA depends, among other things, on the sequence of the anticodon. tRNAs thus serve as the "decoding" molecules; they read the codons present on the mRNA, bind to them, and bring to the ribosome the amino acid specified by that particular codon.

Steps of Translation Now let's trace the steps of protein synthesis. After an mRNA is synthesized by RNA polymerase, it travels to the cytoplasm. There it joins with a small ribosomal subunit, accessory proteins called **initiation factors**, and a specific tRNA that carries the first amino acid of the future protein (Figure 2.9a). In most proteins, this amino acid is methionine, since AUG is almost always the start codon on mRNA. Once this complex forms, it signals the beginning of protein synthesis, and the large ribosomal subunit binds to the mRNA plus the smaller subunit (Figure 2.9b).

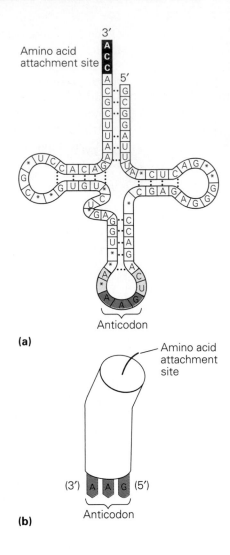

Figure 2.8 Transfer RNA. (**a**) A structural diagram of tRNA. (**b**) The tRNA symbol used in this book.

Amino acid attachment site

Anticodon

(a)

Amino acid attachment site

(3') A A G (5')

Anticodon

(b)

Without further movement of the ribosome, a tRNA whose anticodon is complementary to the second codon binds (step 1 in Figure 2.10, p. 40), bringing with it the amino acid specified by the second codon. Now the first two amino acids of the future protein are very close together. The ribosome enzymatically joins these two amino acids together by peptide bond formation (step 2) by removing the first one from its tRNA and attaching it directly to the second with a covalent bond. The "empty" tRNA in the first position is ejected by the ribosome, after which the ribosome pulls the mRNA down one codon and sets up for the next codon to be read (translocation, step 3). The correct tRNA, with its attached amino acid, is recruited to this third codon, and the process continues. The amino acid chain attached to the tRNA now in the first position gets transferred onto the amino acid on the tRNA in the second position.

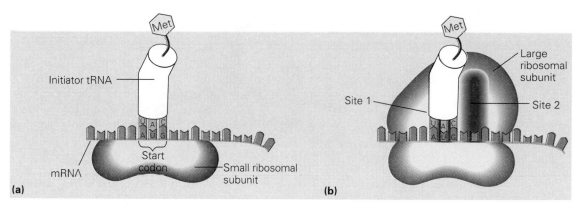

Figure 2.9 Initiation of protein synthesis. (**a**) In this process, the small ribosomal subunit, the mRNA, and the initiator tRNA all come together (with the aid of a number of accessory proteins called initiation factors) to place the initiator tRNA on the start codon. (**b**) Once this complex is formed, the large ribosomal subunit joins, establishing the sites where tRNAs will decode the message.

In this fashion, the ribosome moves along the mRNA, recruiting tRNAs as it goes and forming an ever-lengthening chain of amino acids attached to the tRNA left on the ribosome. At the end of the coding sequence, the ribosome encounters one of the three codons that signal "stop." Specific proteins, not a tRNA, recognize the stop codon and cause the ribosome to release the mRNA.

Several signals are necessary for protein synthesis. We have already described the stop codons; they indicate the end of a coding sequence, that is, where the ribosome should stop. The start codon really serves two purposes. First, it determines the beginning of the mRNA sequence that will be translated by the ribosome. Second, it determines the **reading frame** (or the way the message is divided into codons). Since the information is encoded by sets of three bases, three reading frames are possible depending on which base is the start of the message. Consider the following message:

. . . AUGCCUGUCAAACAUGAU . . .

This message could be divided into codons and read as follows:

. . . AUG CCU GUC AAA CAU GAU . . .

Alternatively, it might be separated into codons in other ways:

. . . A UGC CUG UCA AAC AUG AU . . .

. . . AU GCC UGU CAA ACA UGA U . . .

Each of these messages is an entirely different sequence of codons and would result in the synthesis of a completely different protein during translation. By using a specific start codon, AUG, the ribosome knows both where to start and which reading frame to use.

Structure of a Gene What must the actual gene in the DNA include for this whole flow of information to work properly? The gene must contain many things. First, it

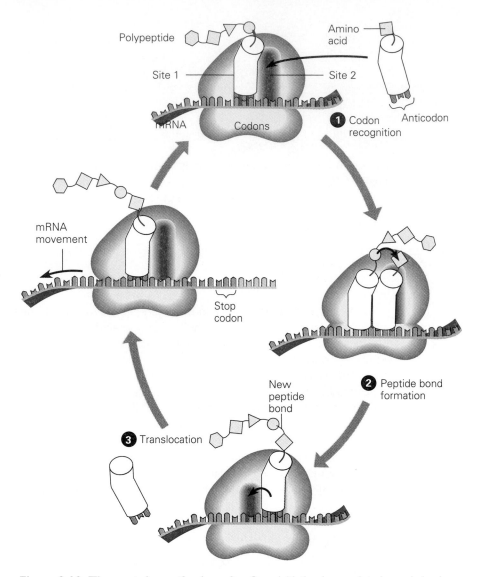

Figure 2.10 The protein synthesis cycle. Once initiation is completed, a cycle begins that brings appropriate tRNAs to site 2 (step 1), transfers the growing amino acid chain to the amino acid on tRNA-2 (step 2), and ejects tRNA-1 (step 3). The ribosome then moves to the next codon (step 3). This cycle continues until a stop codon is reached, at which time the complex disassembles and the newly synthesized protein is released.

must contain the actual coding sequence that is translated into mRNA. The mRNA must include both a start codon and a stop codon. In front of the start codon, the mRNA usually has an untranslated sequence (used in attaching the mRNA to the ribosome, not to code for amino acids). The information for the start codon, the stop codon, and the untranslated sequence must be in the actual gene for it to become a part of the mRNA. Then, in front of this sequence that will be transcribed into mRNA, there must be control sequences such as the promoter region, where the RNA polymerase binds and begins transcription. Thus, the gene must actually contain quite a bit of information besides the coding sequence (Figure 2.11). The process needs to be as efficient and smooth-running as possible, because protein synthesis is an energetically intensive process for the cell. For example, it is estimated that in bacterial cells, during normal growth periods, at least 50 percent of all of the cells' energy goes directly into the synthesis of new proteins.

The Complexity of Protein Synthesis Why is protein synthesis such an energy-intensive task, and why are so many different transcriptional and translational components necessary? In large part, the answer lies in the necessity for specificity of the chemical reactions, that is (at least in part) the complementary base-pairings that occur. Let's return briefly to our wood-burning example from Chapter 1. A cellulose molecule in the wood reacts with oxygen if the two come together in the proper orientation and if the activation energy can be surmounted. Whether the two come together is simply a function of random events—oxygen in the air randomly bumps into the molecules of the log.

Protein synthesis cannot be random. The cell cannot tolerate just any amino acid

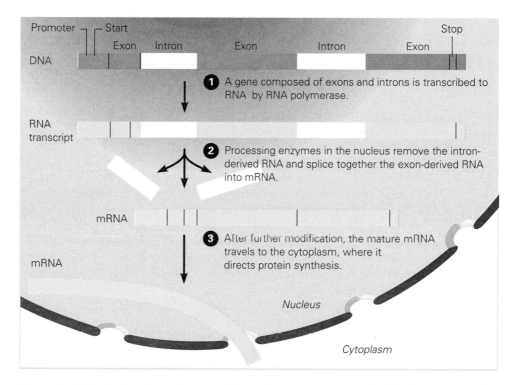

Figure 2.11 Typical eukaryotic gene. The coding sequence is carried within the gene. An initial RNA transcript is made, after which splicing reactions remove the introns and attach the exons to each other. The final mRNA contains exons only.

within a growing protein chain—it requires a particular one. tRNAs cannot have just any amino acid attached—only one specific amino acid may be attached to each tRNA. Thus, the chemical reactions are not random; they are specific, as initially dictated by the DNA sequence. Achieving such specificity requires additional components and energy in order to prevent random reactions.

Replication of DNA

We now need to examine what happens to the information in DNA when cells reproduce. Every time a cell divides, a new copy of the DNA must be produced. **Replication** is the process that produces this new set of information for the new cell.

Replication must produce exact replicas of the existing DNA so that the information content is preserved and can be passed to the next cell or the next organism. But how can such a huge molecule as DNA be accurately replicated? The key is the base-pairing properties that establish the structure of DNA and are used in transcription and translation. When DNA is to be replicated, a set of proteins binds to the DNA and begins to separate the strands, unzipping them. **DNA polymerases**, enzymes responsible for replicating DNA, then bind to the now single strands of DNA and begin to "read" the exposed individual bases. DNA polymerase takes a nucleotide from the surrounding solution that will properly base-pair with the base being read and attaches it to the growing chain of nucleotides (Figure 2.12). This is how a totally new strand of DNA is produced. The entire process results in the formation of a new double-stranded molecule where there was but a single strand to begin with. This new double-stranded molecule will have the same sequence as the original DNA molecule. As replication is occurring on one strand, it also occurs on the other. The net result is the formation of two double-stranded molecules from the original double-stranded molecule. Each of the new molecules has one strand of "old" DNA and one strand of "new" DNA. The original double-stranded DNA molecule has been duplicated exactly.

However, DNA polymerases sometimes make mistakes. Most mistakes are immediately corrected by the enzyme itself, but occasionally a mistake is not rectified. Such mistakes result in **mutations**, or changes in the sequence of bases in the DNA that can be inherited. As you will see in Chapter 5, mutations can result in disease. Fortunately, the accuracy of DNA polymerases is very high, with about one mistake in every 10^9 nucleotides incorporated—an exceedingly low rate of error. We can think of this as evidence of the importance of the replication process: An exceedingly accurate system replicates the most important biological information molecule that exists—the DNA molecule.

DIFFERENCES BETWEEN PROKARYOTIC AND EUKARYOTIC CELLS

We've hinted at the differences between prokaryotic and eukaryotic cells. Now we will delineate these differences, as well as some predominant similarities.

The primary mechanisms of information storage and utilization are the same for both prokaryotes and eukaryotes. DNA stores genetic information, and RNA inter-

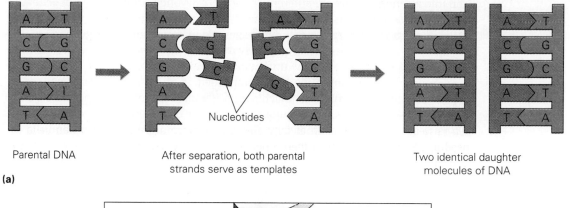

Parental DNA

After separation, both parental strands serve as templates

Two identical daughter molecules of DNA

(a)

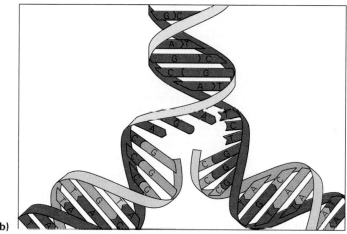

(b)

Figure 2.12 Replication of DNA. (**a**) The parental strands of DNA must first be separated. Nucleotides are assembled by DNA polymerase in a complementary fashion to create new strands. (**b**) The final result is the production of two identical DNA molecules from one.

mediates carry copies of the information to ribosomes, which use a genetic code to construct proteins. In many ways, the processes are similar for prokaryotes and eu karyotes. But there are significant differences.

First, eukaryotic cells have internal compartments, including a nucleus that houses the DNA. This structure introduces complications into the flow of information, since the mRNA must move to a completely different compartment to be used by ribosomes. Eukaryotic DNA is also packaged somewhat differently, since the much larger amount of DNA present in these cells requires more extensive packaging schemes than prokaryotic cells. Chapter 3 has more details about how DNA is packaged.

Transcription works the same way in both kinds of cell, although the machinery and the signals used are different. The enzymes involved—the RNA polymerases—are different. They perform the same function—synthesizing mRNA or other RNA molecules—but prokaryotic enzymes will not work in eukaryotes, and vice versa.

The transcription signals are different as well; eukaryotic promoters do not generally work in prokaryotic cells, and vice versa. This is a salient point in our future vision of genetic engineering. As we manipulate genetic material and move genes from one organism into an organism in a completely different group, we have to remember the specific signals necessary for the proper use of the transferred genetic material.

Similarly, while translation is accomplished in the same general manner in prokaryotes and eukaryotes, the machinery differs. Prokaryotic ribosomes are different from those in eukaryotes. Fortunately for genetic engineering, the genetic code itself is nearly universal, theoretically allowing a piece of genetic information to be used in either cell type given the proper signals and machinery.

ISSUES: WHO OWNS OUR GENES?

Thinking of biological materials as commodities has fueled a number of legal battles in recent years. Most notable is the effort to patent genes derived from humans. Patenting is a process designed to encourage the development of new products and processes: The receiver of a patent has exclusive production and distribution rights for that product or process for seventeen years. In this way, the developer can reap some reward by recovering the costs of development and testing during this time of protection.

In 1991, Craig Venter, a scientist at the National Institutes of Health, filed two patent applications covering several thousands of genes found in the human brain. Although this was not the first patent application pertaining to biological material, it was the first involving specific genes. Awarding such a patent would essentially confer ownership of these genes.

Can a part of a human body or something derived from the human body be owned by others? Insulin, a hormone involved in the control of blood sugar levels, is required by many diabetics. At one time, insulin used by these patients was derived from animals, but now it is made from human genes that have been transferred to bacteria. It is human insulin, although it is not produced in humans. Patents cover the ways insulin is produced outside the body. Thus, even though the insulin is identical to human insulin, it is not made in humans, and its patenting generated little controversy.

On the other hand, numerous samples taken from many different types of medical patients each year are studied further in the laboratory and are sometimes used to produce specific products. Should these materials be patented? Who should be the beneficiary of the patent, the patient or the scientists who isolate the samples and use them in the laboratory?

An interesting line of ongoing research is designed to genetically alter animals so they produce specific valuable proteins in their milk. Further, applications have already been received by the European Patent Office concerning human females who might someday be genetically altered to produce products in their milk. In other words, there have already been attempts to claim ownership of human beings!

In some sense, these applications are in the intended spirit of patent law. A great amount of technical expertise and scientific knowledge is involved in all of these ventures. Should these approaches ultimately yield products appropriate for the market, developers should be able to recover the tremendous costs involved in their development. On the other hand, can ownership of a person, even in this limited sense, be granted by a government to an individual or a corporation?

The line becomes blurrier when we consider individual genes. Any genes that turn out to be important commercially are not unique; we all have much of the same genetic information within our cells. Do we allow patenting of the actual gene, of the information it contains, or simply of the processes leading to its use for commercial purposes?

Government and legal experts readily agree that the law has not kept pace with scientific developments in this field. The U.S. Patent Office is floundering, awaiting legal guidance and agreement concerning biological patents. This is certainly one of the best examples, though not the only one, of the fact that developments in biotechnology have occurred too rapidly for some segments of our society to keep up.

SUMMARY

Our understanding of biological information has increased dramatically in recent years. This understanding will allow us to manipulate cells and organisms in ways we never dreamed possible. At the heart of biological information is DNA, the molecule that encodes the information. The structure of DNA is ideally suited to its role of information storage; it can accommodate immense amounts of information, and its double-stranded structure allows for ease of information use and replication.

The informational component of DNA consists of sequences of bases, organized into groups of three. Ultimately, each group of three determines the identity of a particular amino acid within a protein. Through the processes of transcription and translation, this information is used to produce the cellular proteins.

The principle that allows information to be stored and used in this way is the concept of base-pairing. Simple base-pairing rules govern the structure of DNA, the process of transcription, and the production of proteins. The nearly universal nature of the genetic code and its reliance on the base-pairing rules enable us to understand how information flows in all organisms.

REVIEW QUESTIONS

1. List the shared characteristics of all living cellular organisms.

2. How does the structure of DNA facilitate replication, transcription, and repair?

3. Explain how hydrogen bonding and hydrophobic forces together cause the formation of a DNA double helix.

4. List three major differences between DNA and mRNA.

5. Explain how specific tRNAs allow a message in mRNA codons to become a chain of amino acids.

6. State three reasons why mRNA is used as an intermediate in translation.

7. List the series of steps involved in transcription.

8. Explain how tRNA acts as a decoder in translation.

9. List the differences between prokaryotic and eukaryotic cells.

10. It might seem strange that the same genetic code is used in cell types as different as bacterial and human. What might explain this? How can this be useful in biotechnology?

11. Why is it significant that some of the bonds holding the individual building blocks of DNA together are very strong, while others are rather weak?

12. DNA replication is a very accurate process, generally producing errors in only one of 10^9 bases replicated. Transcription is not nearly as accurate, producing errors once in every 10^4 bases. Why does DNA replication need to be so accurate? Why can transcription be less accurate?

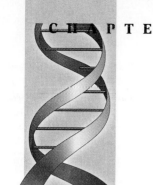

CHAPTER 3

Chromosomes, Cell Division, and Sexual Reproduction

One of the more heated debates in which our society has recently engaged concerns the issues of sexual identity and sexual preference. While the nonphysiological distinctions between males and females have become blurred, it remains a biological fact that for human reproduction to occur one member of each sex must perform very precise functions: Males must deliver sperm cells, and females must harbor eggs. The use of sperm banks and artificial insemination does nothing to change the fact that both sperm and eggs are required to create a new human being.

What determines the sex of an individual? Physiologically, sex is controlled by genetic information, specifically, the presence or absence of a particular chromosome. But what of sexual attitudes? To what degree are they determined by genes? Is homosexuality a part of the genetic information of certain individuals? Is it a genetic alteration, or simply one of several potential sexual identities that has been buried by social dictum?

One way to enter the debate about sexual preference is to ask why two different sexes are necessary. What advantage does sexual reproduction hold over other kinds of reproduction? The answer lies in the ability of sexual reproduction to generate variability in genetic information, a necessary feature for evolutionary change. This variability arises from the union of distinct sets of genes that occurs during the fertilization of an egg by a sperm.

One of the most amazing events in the biological world occurs when millions of sperm cells swim toward an egg. Against all odds, after swimming a tortuous course, a single sperm fertilizes the egg. This tiny sperm cell encounters the much larger egg and penetrates the protective layers around it. Eventually, the sperm contacts the cell membrane of the egg and initiates fertilization. The sperm and egg fuse, and the biological information contained within the sperm and egg cells joins and is copied and distributed as the fertilized egg divides. Eventually, the division of this fertilized egg forms a new body, each cell of which contains the same genetic information as the original fertilized egg—information that is different from that of either parent.

Chapter 2 introduced us to the flow of biological information within cells, from the DNA that serves as the information warehouse for cellular proteins to all the machinery that helps produce the proteins and other molecules that constitute life. We also encountered the DNA replication machinery and found it to be highly effective at producing exact replicas of DNA. In this chapter, we will see how these replicas are moved precisely into the progeny that result from cell division. We will also examine the more specialized cell division that produces reproductive cells, the unique eggs and sperm necessary for sexual reproduction.

The total genetic constitution of a cell is referred to as its **genome**. A cell's genome is contained in one or more very long molecules of DNA. Each DNA molecule, which contains from hundreds to thousands of genes, is called a **chromosome**. Cell division involves equal distribution of the genetic information contained in the chromosomes to two daughter cells. Thus, during cell division, all of the DNA molecules that exist inside a cell must be replicated and sorted in order to accurately distribute all the genetic information. Because we know that DNA molecules are extremely long and thin, we might wonder how the chromosomes manage to avoid getting tangled. Let's examine how cells package DNA.

THE PACKAGING AND STRUCTURE OF CHROMOSOMES

Chromosomes are the DNA molecules in cells. The structure of chromosomes changes during different stages of a cell's life. DNA is associated with various proteins that maintain the structural integrity of the chromosome and help control gene activity. As we will see during our examination of cell division, when cells are actively growing and the information contained within genes is in heavy use, the chromosomes are less compact and more diffuse. In this condition, the DNA (with its associated proteins) is called **chromatin**. It is only during cell division, when the chromosomes must be separated and moved, that a very compact or condensed chromosomal structure is assumed. Just like everything else in a cell, the structure of chromosomes is not static; it is always changing.

In Chapter 2 we painted DNA as a long, double-stranded, information-carrying sequence of bases that ultimately directs the production of cellular proteins. We deliberately kept the structural representation of DNA simple by assuming that it was all stretched out and naked, that is, not twisted or covered with proteins or other molecules. This simplified view helped us understand how information flows without having to consider too many complications. However, it is now time to more accurately portray the structure that DNA assumes within cells.

If the nucleus of a human cell were the size of a basketball and you were to take all the DNA molecules, extend them, and lay them end to end, the result would be a strand of DNA 20 miles long! How can cells contain these extremely long DNA molecules?

The simple answer is that the DNA is coiled up into what looks like huge knots (Figure 3.1). Imagine a very long, coiled, and kinked telephone cord. This cord represents a DNA double helix. Imagine that two friends hold the ends. Now, imagine

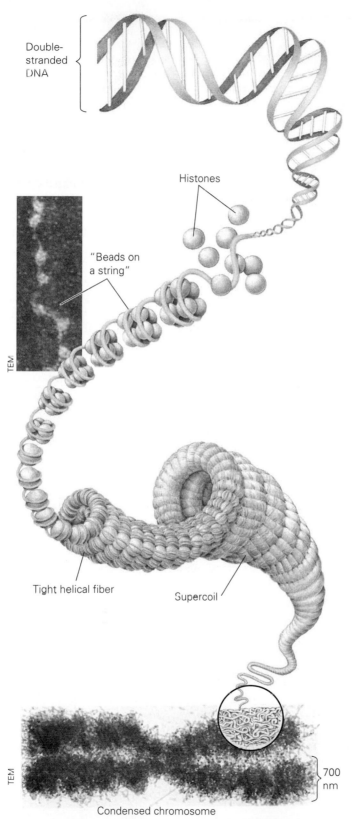

Double-stranded DNA

Histones

"Beads on a string"

TEM

Tight helical fiber

Supercoil

TEM

700 nm

Condensed chromosome

Figure 3.1 The complex organization of DNA in chromosomes. DNA is made more compact through a variety of methods. Histones serve as core particles around which DNA becomes tightly wound, forming what is known as "beads on a string." As chromosomes are further twisted, they form tight helical fibers and supercoils. Loops of DNA are then arranged along a protein scaffolding. The most compact form of a chromosome is remarkably smaller than the DNA molecule it contains.

that somewhere in the middle of the cord you stick your finger into one of the coils and start twisting it in such a manner that the cord wraps a few times around your finger. You take all your fingers and do the same thing at different locations—every few inches. Every time the coils are wrapped around a finger, they become more closely packed, shortening the total length of the cord.

In what other ways could you compact the original cord? Choose a spot in the middle of the cord. Pull the cord up and stick it to a piece of duct tape. About a foot farther down the cord, grab it again and pull this spot up to the duct tape, sticking it right next to the first spot. Continue this for the length of the cord. The cord is now a series of loops held together by the duct tape.

Are there other ways to condense the cord? What happens if the ends of the cord are twisted? If you keep twisting, eventually a knot will form as the cord "jumps" into a new shape to relieve the torsion you are applying. Keep twisting, and more kinks form. Eventually, the cord will get so twisted and knotted that it can't be easily pulled apart—but it will be much more compact!

All three of these methods of compacting a telephone cord are somewhat analogous to what actually happens to DNA in cells (Figure 3.1). A number of structural proteins called **histones** serve a role similar to that of the fingers in the coils; the DNA helix wraps twice around a group of histones. This is the primary packaging activity. There are many groups of histones, probably an average of one group for every 200 base pairs of DNA, making the DNA look like a string with beads on it (Figure 3.1). Other histones serve to connect groups of histones together and further compact the DNA. Still other proteins serve as points of attachment, analogous to the tape, such that the DNA becomes looped. Additionally, enzymes called **topoisomerases** (so named because they alter the topology, or shape, of the DNA) introduce or remove twists in the DNA molecule itself, generating further torsion that helps coil the DNA. In all these ways, a cell manages to twist and condense DNA into packages we call chromosomes.

Most cells do not have only a single DNA molecule. Just as different organisms contain different amounts of DNA, organisms also have characteristically different numbers of chromosomes. For example, yeast cells have eleven different pairs of chromosomes, fruit flies have four pairs, and humans have twenty-three. There appears to be no obvious reason why organisms have the particular number of chromosomes they do. Why do they exist in pairs? All eukaryotic organisms that reproduce sexually have at least two sets of chromosomes, each of which was derived from one parent. Organisms containing two sets of chromosomes as we do are called **diploid**. The chromosomes of each pair, although not exactly the same, are very similar and contain information for the same characteristic; they are called **homologous chromosomes**.

We can visualize condensed chromosomes through a process known as **karyotyping** (Figure 3.2). In this process, chromosomes, usually obtained from white blood cells, are first spread onto a microscope slide and then stained with a dye. Characteristic features of individual chromosomes can be discerned, such as differing lengths and particular patterns of alternating light and dark bands. A photograph is taken, and the images of individual chromosomes are cut out and displayed in or-

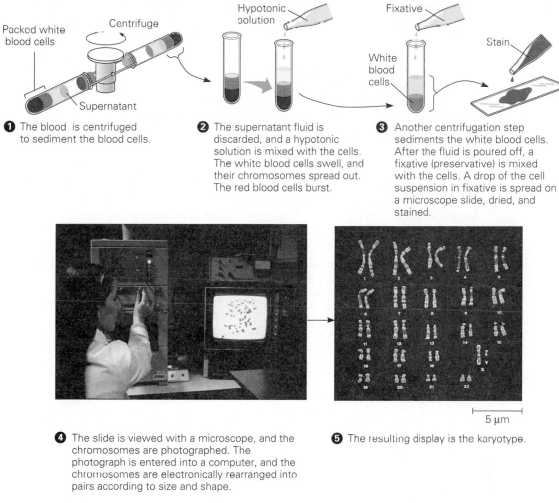

① The blood is centrifuged to sediment the blood cells.

② The supernatant fluid is discarded, and a hypotonic solution is mixed with the cells. The white blood cells swell, and their chromosomes spread out. The red blood cells burst.

③ Another centrifugation step sediments the white blood cells. After the fluid is poured off, a fixative (preservative) is mixed with the cells. A drop of the cell suspension in fixative is spread on a microscope slide, dried, and stained.

④ The slide is viewed with a microscope, and the chromosomes are photographed. The photograph is entered into a computer, and the chromosomes are electronically rearranged into pairs according to size and shape.

⑤ The resulting display is the karyotype.

Figure 3.2 Preparation of a karyotype. A human karyotype is prepared from DNA isolated from blood. Each pair of autosomes is assigned a number. In this karyotype, one each of the X and Y chromosomes is displayed.

der to create the karyotype. With some practice, it is possible to detect alterations in chromosomes, some of which may signal disease.

In many organisms, including humans, one pair of chromosomes functions somewhat differently from the rest. These two chromosomes are called the **sex chromosomes**, and the other twenty-two pairs are called autosomal chromosomes, or **autosomes**. Humans can have two different sex chromosomes (Figure 3.2).

Females have two X chromosomes, while males have one X and one Y. As we will see later, the Y chromosome is ultimately responsible for causing the differences in the sexes.

We have seen that DNA within a chromosome is not naked; the chromosome is a mixture of DNA and protein molecules, intertwined in a very complex yet highly organized fashion. Although this packaging allows DNA to become very compact within cells, isn't it likely to interfere with the use of the DNA as a template? For example, consider transcription, during which RNA polymerase makes an mRNA copy of the base sequence of a gene. RNA polymerase must interact specifically with the DNA, first at the promoter region and then all along the length of the gene, in order to read the base sequence of the DNA. Now we find that the DNA within cells is actually covered with lots of proteins that twist and contort the DNA into a variety of shapes. These kinks and the proteins themselves interfere with RNA polymerase as it tries to gain access to the DNA. We don't yet understand all the changes in DNA structure and packaging that must occur to allow all these proteins—structural and enzymatic—to interact with DNA at different places and at different times during cell growth. But we can appreciate the fact that a very complex organization and ordering of events must take place during the normal functioning of a cell.

CELL DIVISION

Every human being begins as a single cell. Then each of us, through repeated cell divisions, becomes an adult composed of about 60 trillion cells. Many of your cells are still dividing, even though you may not be still growing. Skin, hair, and blood cells are constantly being made to replace those that are lost or damaged. Some cell types, notably muscle and nerve cells, cease dividing. Cell division is composed of two processes; *mitosis*, or nuclear division, and *cytokinesis*, or cytoplasmic division. The role of mitosis is to guarantee that the daughter cells contain exactly the same genetic information as the parent cell.

Eukaryotic Cell Cycle

Studies of the many different cell types in multicellular organisms have led to the understanding that all eukaryotic cells go through a very well-defined cell cycle, as illustrated in Figure 3.3. The cell cycle is composed of two major stages: interphase and M phase (mitotic phase). **Interphase** is the "normal" state of a cell, in which primary metabolic events such as protein synthesis, energy conversion, and DNA replication occur. These events occupy the typical cell throughout most of its cycle. During interphase, the DNA is in the form of chromatin; this is the form that resembles beads on a string. Only in preparation for cell division does the DNA condense into chromosomes, a process called condensation.

The remaining portion of the cell cycle is made up of M phase, during which cell division occurs. M phase is made up of two processes: mitosis and cytokinesis. Many individual steps must occur following DNA replication before a cell can reproduce. First, the replicated DNA must be distributed in such a way that each daugh-

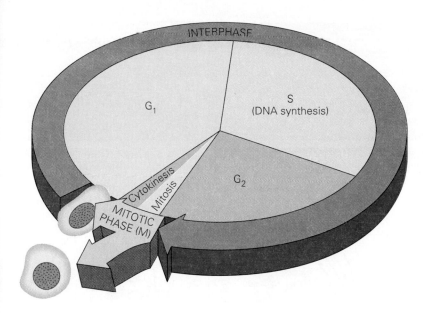

Figure 3.3 The eukaryotic cell cycle. All actively growing eukaryotic cells exhibit this general cell cycle. Interphase, the period encompassing the normal growth and activity stages of the cell, is divided into three stages. During G_1, the first growth stage, cells synthesize materials and grow. S stage represents the replication of DNA, or DNA synthesis. G_2, the second growth stage, allows further growth and function following replication. During mitosis and cytokinesis, or M phase, the cell actually carries out the act of division. The divisions of mitosis are described in the text and shown in detail in Figure 3.4. Note that, for most cells, M phase represents a very small period of time, while interphase might occupy a cell for more than 90 percent of its time.

ter cell will obtain the proper set of chromosomes. This is **mitosis**. Then new cell membranes must be produced to physically separate the two new cells. This is **cytokinesis**.

Mitosis

As we examine mitosis in some detail, we will focus our attention on how the primary goal of cell division, that of properly distributing the replicated chromosomes into both daughter cells, is accomplished. Mitosis itself can be divided into five phases based on recognizable features or events that help us understand how the cell accomplishes its goal: prophase, prometaphase, metaphase, anaphase, and telophase (Figure 3.4).

Prophase Prior to prophase, the cells have replicated their DNA, but it is still in the chromatin form characteristic of interphase. In a human cell, the result is twenty-three pairs of duplicated chromosomes. Each replicated chromosome, instead of being a distinct molecule, is actually attached to its sister at a constricted region known as the **centromere**. The two joined replicas are called **sister chromatids**. During prophase, replicated chromosomes undergo condensation (Figure 3.4). Shortly, the cell will separate the sister chromatids from each other.

Also during prophase, a system of **spindle fibers** is set up that will ultimately move the chromosomes to opposite sides of the cell. The spindle fibers are formed

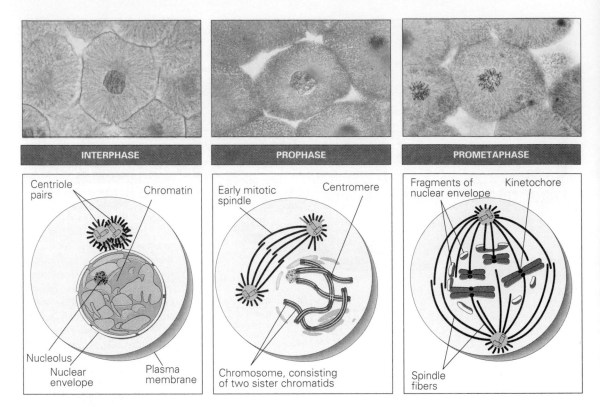

Figure 3.4 Mitosis and cytokinesis. The events associated with the various stages of the cell cycle are described in the text. The accompanying micrographs of stained cells show the progression of chromosome compaction, nuclear membrane dissolution, and chromosome movement.

and organized by **centrioles** outside the nucleus. Early in prophase, the two pairs of centrioles begin moving to opposite ends of the cell. As they migrate, spindle fibers form between them. The centrioles also serve as attachment or anchor points for the spindle fibers. The entire network of fibers and anchors is referred to as the **spindle apparatus** (Figures 3.4 and 3.5, p. 56). Finally, the nuclear membrane (or nuclear envelope) is broken down, allowing the spindle fibers to gain access to the condensed chromosomes.

Prometaphase During prometaphase, the centriole pairs reach the poles of the cell, and the spindle apparatus is completely formed between them. Protein structures called **kinetochores** form on both sides of the centromere on every duplicated chromosome. The kinetochores serve as attachment points for the spindle fibers. Spindle fibers from each pole reach out and attach to each individual chromosome at the kinetochore. Note in Figure 3.5 that one spindle fiber from each pole attaches to each duplicated chromosome. This ensures that one copy of each chromosome is distributed to each daughter cell. After the spindle fibers attach to the chromosomes, they

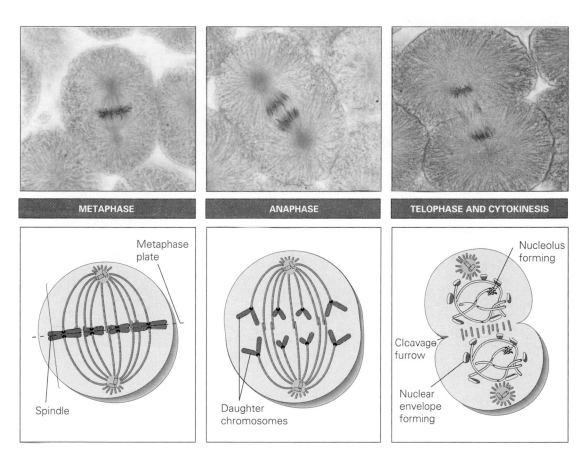

Figure 3.4 (*cont.*)

begin pulling on the chromosomes. What you would see under a microscope at this point are the chromosomes moving back and forth between the poles of the cell as they are pulled by the spindle fibers.

Metaphase Eventually, the pulling action of the spindle fibers causes the duplicated chromosomes to align along a plane midway between the poles of the cell. When the chromosomes are aligned in this manner, the cell is said to be in metaphase. The plane along which the chromosomes are aligned is referred to as the **metaphase plate**. How does the cell know to attach each of the replicated chromosomes to different centrioles so they will later be separated into the two daughter cells? The answer turns out to be a question of balance. If sister chromatids are not attached to opposite centrioles, they will not be pulled into the midline. Should this sort of mistake be made, the cell will break down the attachment to the chromosomes and form a new one, repeating the process.

At the end of metaphase, the centromere of each chromosome splits in two, liberating the two sister chromatids from each other. This separation occurs simultaneously for all the chromosomes. Now that the sister chromatids are separated from

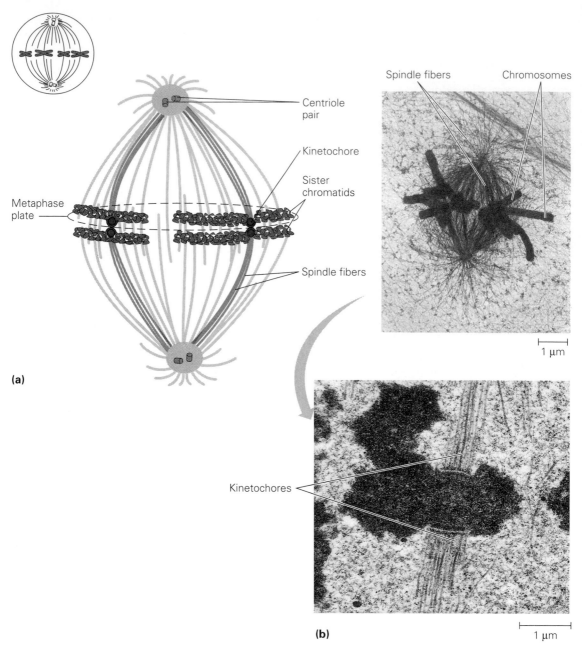

(a)

(b)

Figure 3.5 The network of fibers responsible for chromosome movement. (**a**) Fibers extend from centrioles at either end of the cell to attachment points (kinetochores) on each chromosome. Fibers from each centriole are attached to each pair of chromosomes, and tension pulls the chromosome pairs into the middle of the cell. Only after all chromosome pairs are aligned do they separate and become pulled toward each end of the cell. (**b**) Electron micrographs illustrate the spindle fiber network (top) and the attachment of spindle fibers to kinetochores (bottom).

each other, each is considered an individual chromosome. And, since they are no longer joined at the centromere, they can be physically moved apart in the next phase of mitosis.

Anaphase Anaphase is perhaps the most spectacular stage of mitosis, because it involves the movement of individual chromosomes toward the two poles of the cell (Figure 3.4). The individual chromosomes become V-shaped as their centromeres lead the way to opposite ends of the cell, with the arms of the chromosomes following. This movement is very rapid.

Telophase Following the movement of chromosomes to opposite ends of the cell, new nuclear membranes are synthesized around each set of chromosomes, forming daughter nuclei. This stage is called telophase (Figure 3.4). The spindle fibers are disassembled into their component parts to be used for other functions or to be broken down and recycled through metabolism.

Cytokinesis

As telophase nears completion, the cell begins the second process involved in cell division: cytokinesis. In animal cells and other cells that lack a cell wall, cytokinesis occurs by a process known as **cleavage**. During cleavage, a structure called a **contractile ring**, made up of **microfilaments**, forms along the midline of the cell. Cleavage is carried out by constriction of the contractile ring. The microfilaments slide past each other as the ring contracts, pinching the cell and forming a **cleavage furrow** around the cell's circumference (Figure 3.4). As cleavage continues, the cleavage furrow deepens until the cell eventually pinches in two. The end result is the division of one parent cell into two daughter cells, each of which contains a copy of all the chromosomes in the parent cell.

Mitosis in Context

The primary purpose of mitosis is to ensure the accurate replication of genetic material and its directed distribution into two new daughter cells. But what of the rest of the cellular material? Is it sorted or separated in any specific manner? No. Where a particular entity lies with respect to the midline of the cell determines which daughter cell it will reside in.

The process of mitosis is a costly one for cells in terms of energy expenditure. Many new proteins must be synthesized to cause condensation of chromosomes and construction of the spindle fibers. Energy is necessary for the rapid movement of the condensed chromosomes. Also, membranes are being alternately dissolved and synthesized.

Since each cell must contain the proper biological information, it is not surprising that mitosis is an elaborate process. Were it not imperative for each cell to inherit precise copies of each chromosome, less elaborate methods of cell division surely would suffice. In fact, before it was known that chromosomes were the storehouses of cellular information, scientists had difficulty understanding why such an elaborate process existed for their distribution during cell division.

Occasionally, mistakes are made during the division process, at a low but significant rate (perhaps one mistake per 1,000 divisions). For example, sister chromatids sometimes do not separate during anaphase, resulting in one cell receiving three copies of one chromosome and the other cell receiving only one. This is referred to as **mitotic nondisjunction** and may be responsible for the formation of certain kinds of tumors. We will describe other types of nondisjunction in Chapter 5.

Single-celled organisms such as bacteria (prokaryotes) undergo a process called *transverse binary fission*, producing two new identical organisms from a single parent. This is in stark contrast to most multicellular organisms, which use different reproductive strategies, including sexual reproduction. During the sexual reproduction of complex organisms, two parents each donate genetic material to each offspring with the result that every new individual has some genetic variability. This reproductive strategy requires a different method of cell division.

SEXUAL REPRODUCTION

Our knowledge of mitosis helps us understand how individual cells produce identical copies of themselves such that a cell initially designated to become a liver can grow and become a full-size liver, for instance. However, we are still left without an understanding of how multicellular organisms distribute genetic information to offspring. How did your parents create reproductive cells that resulted in your having the correct number of chromosomes? Why are two biological parents necessary to produce a single child? Extending our study of inheritance to consider whole organisms presents us with many new questions that cannot be answered by what we have covered thus far.

Sexual reproduction, while not the sole method of reproduction found in nature, is the only way humans and many other organisms can reproduce. And, while a variety of mechanisms are used by different organisms in nature to accomplish sexual reproduction, the necessary cellular events all involve a particular form of cell division that produces unique reproductive cells.

Sexual reproduction in humans involves two kinds of reproductive cells, or **gametes**: eggs produced by the female and sperm produced by the male. Remember that nonreproductive cells are diploid, with two sets of chromosomes. What genetically distinguishes egg and sperm cells from nonreproductive cells is that they are **haploid**, which means that they each have only a single copy of each chromosome. Gametes are the only haploid cells; all other cells of the body are diploid. During sexual reproduction, a sperm cell from the father fuses with an egg cell from the mother. The resulting fertilized egg contains the genetic material from each cell, and thus is now diploid; it has two sets of chromosomes. The DNA present in the fertilized egg is not exactly the same as that in either biological parent. From this single, fertilized egg cell, all other cells of the offspring will be derived through mitosis.

The most pertinent cellular questions concerning sexual reproduction are how haploid reproductive cells are produced and why. First we will consider why. Sexual reproduction involves the fusion of two cells, egg and sperm, to form a **zygote**. If the egg and sperm cells were diploid, the resulting zygote would have twice as many chromosomes as required. Next we will consider how. Nature has developed a kind of cell division called **meiosis**, two successive nuclear divisions with corresponding cell divisions that produce gametes. Instead of producing two genetically identical diploid cells from a single cell, meiosis produces four haploid cells. What's more, each of these four cells contains unique genetic information. Sequentially, meiosis consists of one round of DNA replication followed by two rounds of cell division (Figure 3.6). The resulting cells, therefore, are haploid.

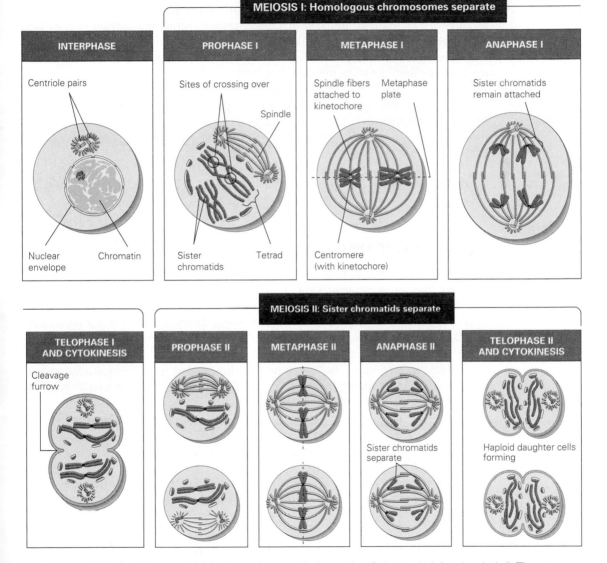

Figure 3.6 Meiosis. The two divisions that make up meiosis are identified as meiosis I and meiosis II. The stages are quite similar to those of mitosis, with some exceptions. Recombination can occur between homologous chromosomes during prophase I. Chromosome pairs are moved together during anaphase I. Finally, there is no additional DNA replication prior to prophase II.

Each of the two cell divisions that occur during meiosis differs from mitosis. Most notably, during the first meiotic division, there is an extended period of time when the replicated homologous chromosomes line up very precisely alongside each other (Figure 3.7). This alignment is called **synapsis**. A specific protein complex is formed that helps join these two DNA molecules. During synapsis, pieces of the homologous chromosomes can be exchanged with each other. In such cases, specific enzymes cut the DNA of each homologue, and a reciprocal exchange of segments occurs (Figure 3.7). The process can be thought of as a sort of genetic cut-and-paste mechanism in which some region of DNA on one chromosome is exchanged for the same region in its associate. This process is known as **recombination**, or **crossing over**; the DNA molecules are said to recombine with each other.

No counterpart to recombination occurs in mitosis. If it were to occur, it would generally be detrimental to the organism, because the alteration of cellular information during mitosis would be counterproductive to the goal of producing offspring cells that are exactly the same. However, during meiosis, recombination does occur. It is a salient feature of sexual reproduction, forming the basis of the genetic uniqueness of each individual. A general comparison of the two methods of cell division is provided in Figure 3.8, p. 62.

Why has a method of reproduction evolved that ensures that every offspring will be genetically distinct? What is advantageous about sexual reproduction? The answers to these questions require a review of the basics of evolutionary theory.

THE BENEFITS OF GENETIC VARIETY

In 1859, Charles Darwin proposed a theory of evolution, couched in a description of how new species might arise. One of the most important features of the theory was his description of a mechanism, called **natural selection**, through which evolution could occur. The principles he put forth may seem rather simple now, but remember that Darwin had no idea what genes were, let alone any detailed knowledge of how characteristics were inherited. His theory is based on a remarkable set of straightforward ideas that have stood the test of time and have found application in a far greater context than that in which they were originally proposed. Three principles make a good first statement of evolutionary theory:

1. Principle of variation: Within a population, characteristics of individuals are variable.
2. Principle of heredity: Characteristics tend to be transmitted to offspring.
3. Principle of selection: Some of the variable inherited traits contribute more to the reproductive capabilities of the organism than others.

What would be the result of the action of these three principles within a population of organisms? Advantageous traits, since they enhance reproductive capability in some way, would be preferentially transmitted to offspring. Therefore, they would be present in a higher proportion of the population; that is, those traits would get selected from one generation to the next.

In this way, the genetic makeup of a population changes, or evolves, over the

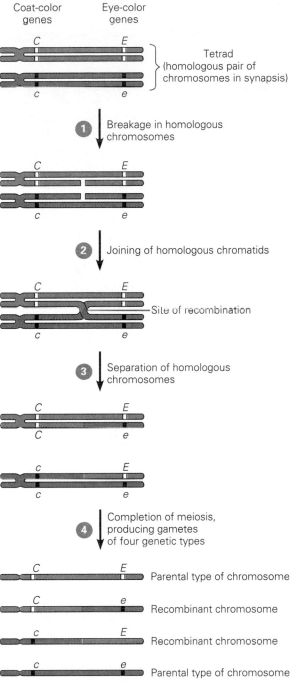

Coat-color genes   Eye-color genes

Tetrad (homologous pair of chromosomes in synapsis)

1 Breakage in homologous chromosomes

2 Joining of homologous chromatids

Site of recombination

3 Separation of homologous chromosomes

4 Completion of meiosis, producing gametes of four genetic types

Parental type of chromosome

Recombinant chromosome

Recombinant chromosome

Parental type of chromosome

Gametes of four genetic types

Figure 3.7 Chromosomes during synapsis. During synapsis, homologous chromosomes associate closely. During this close association recombination can occur.

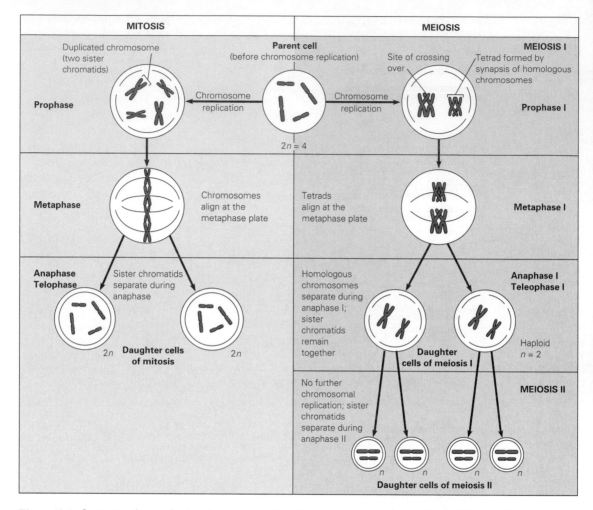

Figure 3.8 A comparison of mitosis and meiosis. There are two primary functional differences between mitosis and meiosis. First, synapsis and recombination can occur during meiosis, but not during mitosis. Second, during meiosis there are two actual cell divisions but only one replication of DNA. The results of the two processes are different: Mitosis results in the production of two progeny that are identical to the parent, while meiosis results in the production of four progeny, each of which is haploid and may contain recombinant chromosomes.

course of time. Note that an infinite number of traits can enhance reproductive capability. For example, simply living longer might allow more reproduction to take place. Having better vision, hearing, or sense of smell might allow an individual to evade predators and thus live longer and reproduce more often. Growing more fur might help an animal survive colder temperatures. Being taller might help an animal reach higher into trees, allowing it to obtain more food. We can imagine many other examples of advantageous traits.

Natural selection is a mechanism of evolution that explains the adaptation of organisms to their environment. Changes in an environment exert a selective pressure on the organisms living within it. Perhaps temperatures become cooler or hotter, or there is less rainfall, or a source of food disappears, or a new kind of predator crosses a mountain range and enters a different territory. The environment can affect the reproduction of individuals in many ways (Figure 3.9). Natural selection predicts that certain individuals will be better equipped to cope with or adjust to environmental stress because they exhibit specific traits that help them survive and reproduce.

This theory suggests that, for a particular population of organisms, there is an inherent advantage to having a wide variety of traits. The more variation that exists within a population, the greater the possibility that some individuals of the population will survive environmental stresses. The less variation there is, the less likely a population will be to survive such stresses. Note that variation does not necessarily mean improvement. Some variations will decrease, rather than increase, reproductive capability. Variation itself is simply a random difference, neither good nor bad until it is put into the context of a particular environment. If a variation is helpful, so much the better for that organism.

Organisms that reproduce sexually have the capability of shuffling existing DNA during meiosis. By rearranging homologous chromosomes and taking material from two different sources, each offspring is guaranteed a unique set of information (with the exception of identical twins, who inherit identical information). Thus, sexual

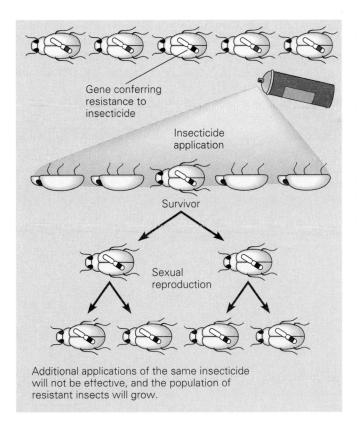

Gene conferring resistance to insecticide

Insecticide application

Survivor

Sexual reproduction

Additional applications of the same insecticide will not be effective, and the population of resistant insects will grow.

Figure 3.9 Natural selection. Natural selection is illustrated for the case where an insect population is sprayed with insecticide. Most of the insects will be killed by the insecticide (which is a change in environment for the insects), but a few individuals may survive because of some genetic resistance to the insecticide. As these individuals reproduce and regenerate the population, the new population will be resistant to the insecticide. The environmental change results in the selection of survivors with particular resistance genes. Since only these individuals survive, all subsequent members of the population will also contain these genes.

reproduction and meiosis allow the generation of much greater diversity than would simple mitosis.

Variation is reduced when individuals mate within their own families. Why? Two related individuals, even though they will have many genes that are different, will also have many that are the same. Therefore, when they mate, their offspring will obtain some genes from the two biological parents that are the same, effectively reducing the variation in the offspring. Continued inbreeding further reduces variation in a population. Can this ever become a serious problem for a species? The answer is yes. A current example of a population so affected is that of cheetahs.

Cheetahs are highly inbred because their population is dwindling. Individuals exhibit significantly less genetic variability than in any other population that has been studied. The lack of variation that now exists in this population is sufficient to cause health problems: Infant mortality rates are very high, and the population is very susceptible to infectious disease. It is hoped that special breeding programs designed for cheetahs will increase their genetic variation and begin to restore health to the population.

Let's consider another example of the effects of lack of variation in a population. Corn productivity has increased greatly over the years, in large part because of the development of hybrid strains. These strains are bred to contain traits that will result in the greatest yield in the shortest time. Once a successful hybrid strain is found, it is used for the development of new strains. Thus, over the years, corn seed has come from an increasingly inbred population that is much less genetically variable than a typical wild population. While higher productivity is achieved utilizing these hybrids, unforeseen problems have resulted. In 1970, an epidemic of Southern leaf blight fungus struck the United States. All told, it was responsible for the destruction of more than 12 percent of the corn crop—more than all other diseases combined. Needless to say, the financial effects were tremendous.

Why did this epidemic occur? The fungus that causes Southern leaf blight was not new; it had always been around, contributing in a small measure to crop loss. However, by successfully breeding a particular trait (specifically, a factor that prevents self-fertilization) into hybrid strains, geneticists had unknowingly also bred a genetic susceptibility to the fungus. As more and more of the genetically similar seeds were planted, the fungus found a huge population of susceptible hosts for infection. In a more mixed and varied population, infection would not have occurred as often, since more individuals would have been able to resist the fungus. Clearly, maintaining genetic diversity has some inherent biological advantages.

ISSUES: GENETIC DIFFERENCES BETWEEN SEXES

Let's consider one more topic related to the genetics of sex: What is it that genetically distinguishes males and females? We know the only chromosomal difference is that females have two X chromosomes and males have one X and one Y chromosome. Do we know what it is about these chromosomes that causes individuals to become male or female?

A specific protein called testis determining factor (TDF) is produced from information present on the Y chromosome. Through the use of genetic engineering techniques that we will cover later, this particular gene can be moved into an organism that normally does not contain the gene, say, a normal female mouse that has no Y chromosome. The addition of this single gene to the female's genetic information results in the full development of male genitals in these otherwise female mice. We do not yet understand how TDF accomplishes this, nor do we know whether the product of this single gene is ultimately responsible for other "male" traits, such as hormonal or behavioral differences. But, because a major component of the sex-determining machinery has now been identified, studies to help explain more completely the genetic differences between males and females should move forward rapidly.

You may have noticed that only the genetic differences between the sexes have been discussed to this point. What fundamentally constitutes "maleness" and "femaleness" is not clear. Certainly the appearances and the roles of the two sexes have changed rapidly within Western society, and traditional "traits" that may have been associated with females and males in the past, including sexual orientation, may no longer be appropriate. Is it even possible to define a set of biological traits (including behaviors) that are male or female? Can a genetic basis for all of these traits then be found?

Interestingly, some highly publicized scientific research has claimed to have found evidence for a biological basis, and in some cases a genetic basis, for homosexuality. Further, some research on twins has reported an increased incidence of homosexuality among both twin siblings compared to nontwin siblings or adopted children. Those performing the studies claim that this higher incidence implies a genetic basis for sexual preference. The studies are not universally accepted; many critics have pointed out flaws inherent in such difficult studies. An in-depth discussion of this particular work is beyond the scope of this chapter, but it is important to note that even if a convincing genetic component is found for homosexuality, genetic factors are not the sole determinant of sexual preference. Many sets of identical twins that have been studied consist of one homosexual individual and one heterosexual individual. If genetic factors were the sole determinant, then both twins should always be homosexual if either is.

Another study reported that an area of the brain within the hypothalamus (located at the base of the brain) is smaller in homosexual men and heterosexual women than in heterosexual men. The authors of the study suggest that this size difference is related to a preference for male partners, and although they offer no evidence they suggest that there is a genetic reason for the size difference. Again, a critical discussion of this work is beyond our scope at this point; suffice it to say that there is great disagreement as to the validity of this work and even about its interpretation should the observation be correct.

The purpose of this discussion is to emphasize early on the powerful social role that any science or pseudoscience involving genetics can play. Throughout many of the chapters of this book, you will encounter situations in which advances in science may have an impact on individuals and society. No new technology is free of social effects. Certainly the ability to smash an atom has resulted in great harm to a large number of people, as well as drastic changes in world politics. However, good has come from this capability as well, including vast improvements in medical care. Technology can be applied in a variety of ways, and the perception of good or bad effects can change over time. This is only the first chapter on genetics; we will be leading ultimately to some of the more important ways in which the science of genetics has affected social thought and action. But before we can appreciate the role genetics can play in society, we need to learn more about inheritance, or how genetic information flows from one organism to another.

ESSAY: GENES, BEHAVIOR, AND PREJUDICE

There is no doubt that possible differences in brains and genes that might govern sexual orientation are of interest and lead to healthy debate. However, we would like to ask a broader question at this point: Why is there such a great interest in establishing a strictly biological basis, and in particular a genetic basis, for so many human characteristics? Ruth Hubbard, in her book *Exploding the Gene Myth*, asks why it would be desirable to explain homosexuality on the basis of genetics. A large portion of the gay population believes that finding a genetic cause of homosexuality would, as the gay journalist Randy Shilts has said, "reduce being gay to something like being left-handed, which is in fact all it is" (D. Gelman, D. Foole, T. Barrett, M. Talbot, "Born or Bred?" *Newsweek*, Feb. 24, 1992). Proponents of this viewpoint believe that a simple biological explanation accounts for the phenomenon and that it should be perceived as normal, although perhaps somewhat rare. However, the analogy to left-handedness is an interesting one: The perception of left-handedness as being "normal" is a fairly recent development. Not long ago, left-handed people were punished if they did not conform to the right-handed world.

A genetic explanation of a behavior does not necessarily result in the elimination of bias or prejudice, as many would hope. Our history of racism is ample evidence to the contrary. If anything, a biological explanation can result in more stigmatization and contribute even more to the perception of a behavior as a flaw. There appears to be no pressing need to explain genetically other aspects of sexual preference, for example, why individuals might find certain physical features of the opposite sex more attractive. To our knowledge, no one is particularly interested in investigating the genetics of foot fetishes, which are accepted as nothing more than a somewhat unusual individual preference. But homosexuality has a social stigma attached. Were a universally accepted genetic contribution to be found for homosexuality, it might be interpreted by many as a flaw or a genetic disease. The stigma would be legitimized by scientific "fact." And once a stigma is legitimized, prejudicial behavior becomes acceptable, perhaps even the norm. Our society has a bad track record of tolerance for disease or perceived defectiveness.

SUMMARY

Cells face an enormous task as they divide to produce progeny. The complex structure of DNA—its association with numerous proteins and its compactness—presents numerous challenges to the accurate production of identical cells. Therefore, it is not surprising that mitosis is a complex process.

Although mitosis serves well as a method of generating identical progeny cells, sexual reproduction requires the production of unique reproductive cells. A combination of meiosis and sexual reproduction introduces variation into a population of organisms. Such variation is important as a source of genetic raw material with which to adapt to changing environments.

R E V I E W   Q U E S T I O N S

1. Outline the major steps involved in mitosis and cytokinesis.

2. Explain why the condensation of DNA is important during cell division.

3. What do haploid and diploid states have to do with sexual reproduction?

4. Describe the eukaryotic cell cycle.

5. Describe the series of events known as meiosis.

6. Why is the packaging of DNA important?

7. What factors ensure that the daughter cells produced by cell division contain all of the required genetic material?

8. What are the advantages and disadvantages of sexual reproduction?

Mendelian Genetics

The significance of predictions cannot be overestimated. Science rests on its ability to predict specific phenomena. The accuracy of our understanding is tested by our being able to predict: Accurate predictions give us confidence that we understand; inaccurate predictions convince us that we do not.

Later chapters will build on our understanding and use of genetics as a predictive tool. Here we consider how the first predictive genetic theories, developed by Gregor Mendel in 1865, came about.

Mendel was an Augustinian monk who was trained as a scientist. Monasteries at the time were havens of learning in all fields, and Mendel was encouraged by his superior to pursue his biological interests. He began a program of research designed to study plant *hybrids*, plants resulting from the matings of plants with different traits. Hybridization was used not only for plants but also for animals.

So began a nine-year study of inheritance in peas. Mendel's results, published in 1865, consisted of the development of two basic laws regarding the transmission of characteristics to offspring. He only briefly speculated about the cause of these characteristics, which he referred to simply as hereditary "factors." Not until almost half a century later did the importance of his two laws begin to be fully appreciated.

Our primary concern in this chapter is with Mendel's laws and how they help explain the flow of information from organism to organism, from parents to offspring. We will explore how Mendel conducted his experiments and see how he arrived at his conclusions.

Mendel's laws can be understood in the context of what we already know from previous chapters, and there are several reasons for devoting so much attention to them. First, it is fitting to try to re-create the discovery process. Everyone should have the opportunity to discover something that no one else knows. Trying to understand how others have made discoveries makes the process more worthwhile for ourselves. Second, it is difficult to overestimate the contribution of Mendel's laws. They gave birth to the science of genetics, a cornerstone of biotechnology. Third, many genetic diseases precisely follow Mendel's laws. Understanding these laws will allow us to more fully appreciate the method of transmission of these diseases—as well as help us understand the limitations of genetics in explaining others. And finally, Mendel's laws, through the early development of genetics, formed the basis for one of the first widespread political and social programs based on inheritance—eugenics. Since much of biotechnology has great political and social impact today, it is well worth our time to explore such an episode of interaction between genetics and society from our not-too-distant past.

The last two concerns will be addressed in the next two chapters. Here we will concentrate on how Mendel actually came to an understanding of inheritance.

THE LAW OF EQUAL SEGREGATION

Mendel's research involved the controlled matings of pea plants with different characteristics. He chose peas for a variety of reasons, including their availability, their ease of growth, and, particularly, the ease with which the mating of individual plants could be controlled. Plants could be self-pollinated, with a single plant serving as both male and female parent; alternatively, a given plant could easily be pollinated by another. Over the nine years during which he performed his experiments, Mendel studied a large number of characteristics and eventually decided to concentrate on seven (Figure 4.1). These seven characteristics were easy to see and measure and were essentially all-or-nothing in nature. For example, a seed was either round or wrinkled; seldom was a seed found that was in between.

Inheritance of Traits

Mendel initially observed that some plants were **pure-breeding** for particular traits. For a given trait, a pure-breeding plant, if mated with itself or any of its relatives, always gave plants with the same trait. For example, a pure-breeding purple-flowered plant when mated with itself always yielded purple-flowered plants. Similarly, pure-breeding white-flowered plants, when mated to their own lineage, produced only white-flowered plants. Why was it so important for Mendel to begin his experiments with pure-breeding plants? He used them to set up a situation in which a known, constant pattern of inheritance would occur so that deviations from this pattern would be easily observed. This represents a **control experiment**, one that will reproducibly generate expected results. For example, seeds of purple-flowered plants produce only purple-flowered plants. Control experiments provide a baseline against which other experimental results can be measured.

Mendel then proceeded to mate pure-breeding pairs that differed from each other in a single trait and observed the offspring. Such a mating is referred to as a **monohybrid cross**. In one experiment, he mated a pure-breeding white-flowered plant with a pure-breeding purple-flowered plant. Interestingly, all the plants of the next generation (called the **first filial generation**, or the F_1 generation) were purple-flowered (Figure 4.2).

How could this be explained? Mendel decided he needed more information about the F_1-generation plants. So he took F_1 plants (all of which had purple flowers) and

Figure 4.1 The seven characteristics of peas studied by Mendel. Each characteristic can appear in either of two forms, with no intermediates. For example, with regard to stem length, plants are either tall *or* dwarf—there are no plants of intermediate stem length. Each trait was easy for Mendel to observe and measure, allowing him to perform detailed comparisons of different pea plants.

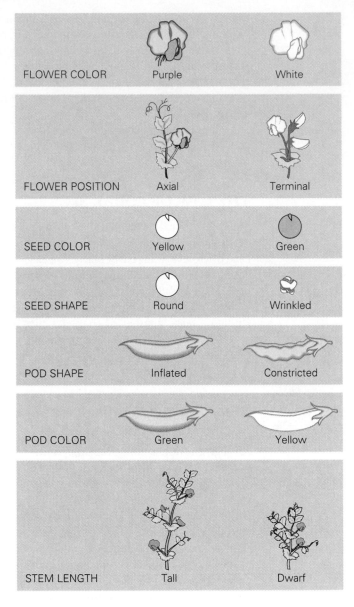

self-pollinated them. Since only a single parent was involved, no new hereditary information could have been introduced during this mating. What flower color would the next generation exhibit?

Surprisingly, the offspring of this self-pollinated F_1 generation, called the F_2 generation, contained some white-flowered plants, even though there were no white-flowered plants in the F_1 generation. When Mendel counted the number of white- and purple-flowered plants, he found that purple-flowered plants were predominant; in one study, he found 705 purple-flowered plants and 224 white-flowered plants, a ra-

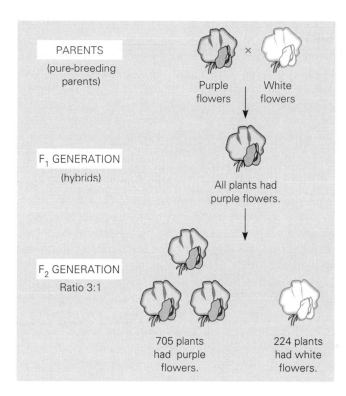

Figure 4.2 Results of a monohybrid cross. When a pure-breeding purple-flowered plant is crossed with a pure-breeding white-flowered plant, all the resulting plants have purple flowers. The F_2 generation is made up of the offspring of self-pollinating plants of the F_1 generation. In this way, no additional genetic information is introduced into the F_1 or subsequent plants. If the initial cross was between pure-breeding purple- and white-flowered plants, then the F_2 generation consists of three times as many purple-flowered plants as white-flowered plants.

PARENTS
(pure-breeding parents)

Purple flowers   ×   White flowers

F_1 GENERATION
(hybrids)

All plants had purple flowers.

F_2 GENERATION
Ratio 3:1

705 plants had purple flowers.

224 plants had white flowers.

Table 4.1 Mendel's Analysis of F_2 Generations from Pure-Breeding Plants

Appearance of Parents	F_1 Appearance	F_2 Appearance	F_2 Ratio
Purple × white flowers	Purple	705 purple, 224 white	3.15:1
Axial × terminal flowers	Axial	651 axial, 207 terminal	3.14:1
Green × yellow seeds	Yellow	6022 yellow, 2001 green	3.01:1
Round × wrinkled seeds	Round	5474 round, 1850 wrinkled	2.96:1
Inflated × constricted pods	Inflated	882 inflated, 299 constricted	2.95:1
Green × yellow pods	Green	428 green, 152 yellow	2.82:1
Tall × dwarf stems	Tall	787 tall, 277 dwarf	2.84:1

tio close to 3:1 (Figure 4.2). Mendel repeated this kind of analysis for each of the other six pairs of traits he had chosen to study, mating pure lines to give F_1 offspring. In each case, only one of the contrasting pair of traits was exhibited. For example, in matings involving seed shape, all of the F_1 offspring exhibited round seeds; none had wrinkled seeds. He allowed these F_1 plants to self-pollinate and then analyzed the F_2 generation. For all seven of the characteristics he studied, he found that the plants of the F_2 generation exhibited a 3:1 ratio of the two possible characteristics (Table 4.1).

What conclusions did Mendel reach? First, in the case of flower color, the purple-flowered plants of the F_1 generation must have had the potential to form white-flowered plants, since they did so in the next generation without the input of any new genetic material. None of the F_1 plants was pure-breeding. Where did the information for white color in these plants come from? It could only have come from the original white-flowered parent two generations earlier. This meant that plants of the F_1 generation must be carrying at least two sets of information for flower color. Since each trait Mendel studied exhibited two possible appearances, it seemed likely that only two different versions of information could be present. Thus, he described these organisms as diploid, that is, having two copies of genetic information for a given trait.

Second, a relationship between these two sets of information, for example the "factor" for purple color and the "factor" for white color, was evident. The white trait must have been hidden or masked by the purple trait in the F_1 plants. Mendel used the word **dominant** to describe a trait that is able to mask another. Thus, he found that purple flower color in peas is dominant over white, as are round seeds over wrinkled, green pods over yellow, and so on. The hidden traits he termed **recessive**; if purple flower color is dominant, white flower color is recessive. The different forms of a trait, for example, purple and white color, are determined by alternate forms, or **alleles** of the gene for flower color. Different alleles result in different phenotypes. Incidentally, many traits, such as human eye color, are determined by more than just two alleles.

Significance of Statistical Ratios

What is the significance of the $3:1$ ratio that appeared in the F_2 generation for each of the seven traits Mendel studied? Mendel continued his analysis by creating an F_3 generation through self-pollination of purple-flowered F_2 plants. He again found a $3:1$ ratio of purple-colored to white-colored plants. Based on these results, he determined that one-quarter of the F_2 plants were purple and pure-breeding, meaning that they produced only purple-flowered plants in future generations. Another one-quarter of the plants turned out to be pure-breeding white plants. The remaining one-half of the F_2 plants were purple-flowered but were not pure-breeding purple plants like the F_1 generation, meaning that they produced both purple- and white-flowered plants in subsequent generations. Thus, where Mendel found a $3:1$ ratio of flower color, he found a $1:2:1$ ratio of pure-breeding lines (Figure 4.3).

We can now see that there is a distinction between what an organism looks like and what information it may be carrying. **Phenotype**, a word derived from the Greek meaning "form that is shown," is used to describe the outward appearance of an organism. Phenotype essentially means observable trait. A purple-flowered plant is said to have a purple phenotype. A white-flowered plant has a white phenotype. Note, however, that the phenotype does not necessarily reflect all the information a plant carries for a particular trait: A plant having a purple phenotype may produce either purple or white plants. Only breeding experiments can reveal whether additional information is present in the purple-flowered plant, as Mendel demonstrated.

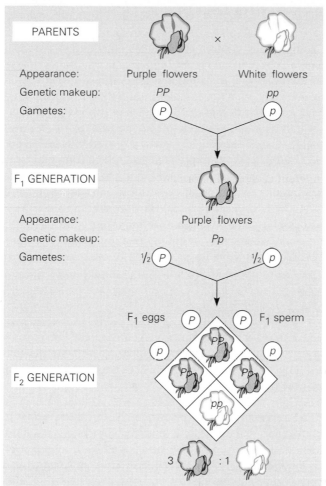

PARENTS

Appearance: Purple flowers   White flowers
Genetic makeup: *PP*   *pp*
Gametes: (P)   (p)

F₁ GENERATION

Appearance: Purple flowers
Genetic makeup: *Pp*
Gametes: ½(P)   ½(p)

F₁ eggs (P)   (P) F₁ sperm
(p)   (p)

F₂ GENERATION

PP
Pp   Pp
pp

3 : 1

Figure 4.3 Phenotypic and genotypic ratios. The F₂ offspring of a monohybrid cross display the typical 3:1 ratio of phenotypes. In terms of genotype, however, there are two classes of purple-flowered plants— homozygous (*PP*) and heterozygous (*Pp*).

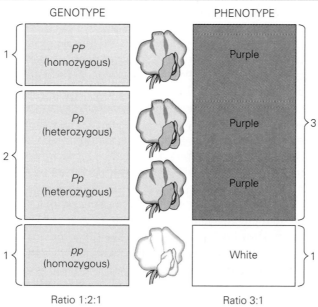

GENOTYPE   PHENOTYPE

1 { *PP* (homozygous)   Purple

2 { *Pp* (heterozygous)   Purple } 3
    Pp (heterozygous)   Purple

1 { *pp* (homozygous)   White } 1

Ratio 1:2:1   Ratio 3:1

The actual complement of information contained within an organism is its **genotype**. The genotype of an organism results in its phenotype, but a phenotype does not always reveal the genotype. Knowing that a plant has purple flowers does not mean that we know exactly what flower-color information it carries.

Mendel still needed to explain the origins of the $3:1$ phenotypic ratio, as well as the $1:2:1$ genotypic ratio. He drew on his knowledge of reproduction to devise a simple model that would explain all he had observed thus far. He knew that two cells must fuse in order to generate a fertilized egg, which through many cell divisions ultimately turns into a mature organism. He had already postulated that organisms are diploid and therefore the fertilized egg from which the organism develops must be diploid as well. Mendel postulated that the only way the amount of genetic information can be held constant at two copies through reproduction is if each reproductive cell contains only a single copy of the information. In other words, the reproductive cells must be *haploid*. In this way, the union of male and female reproductive cells results in two copies of genetic information in the fertilized egg.

Diagraming Inheritance

How did the haploid-diploid distinction help explain the ratios of characteristics Mendel observed? Our genetic analysis will be easier if we develop a diagrammatic way of examining patterns of inheritance. Let's begin by reexamining one of the original matings that Mendel performed, that of a pure-breeding purple plant with a pure-breeding white plant. We know that these plants contain pairs of genes. We will use uppercase *P* to represent a gene for purple color and lowercase *p* to represent the gene for white color. The upper- and lowercase letters are standard genetic symbols for traits; the dominant trait is always designated by an uppercase letter and the recessive trait by a lowercase letter. Since the purple-flowered plant is pure-breeding, it must contain information only for purple flowers. Therefore, both its genes for color must code for purple. We represent this plant genetically as *PP* and call it **homozygous**, since both genes are the same. Similarly, the pure-breeding white plant must be homozygous (*pp*). Note that we are now describing the actual genes of the plants, not their appearances. In other words, we are describing the genotypes of the plants, not their phenotypes. When we introduced the idea of phenotype, we saw that the phenotype cannot reliably be used to determine the genotype. However, if we know the dominance relationship of the genes, the genotype allows us to unambiguously determine the phenotype.

Our diagrammatic representation of this mating is shown in Figure 4.4. The genotypes of the parents are *PP* and *pp*. Since the reproductive cells, gametes, of the parents are haploid, we know that every gamete produced by the purple plant must contain a single *P* gene. Similarly, every reproductive cell produced by the white plant must contain a single *p* gene. Can we now predict the genotypes and the phenotypes of the offspring in the F_1 generation? They all must receive *P* from one parent and *p* from the other, making them *Pp* (or *pP*, which is the same). When the two genes of a pair are different, the plant (or the trait itself) is called **heterozygous**. Since *P* is dominant, the genotype *Pp* will result in a purple phenotype. Note that every indi-

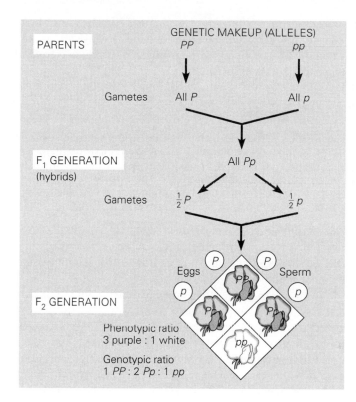

Figure 4.4 Diagram of a cross or mating. Letters signify the individual traits carried as genetic information. Uppercase letters represent dominant traits, while lowercase letters represent recessive traits. The chart shows the possible phenotypes and genotypes in the F_2 generation.

vidual in the F_1 generation of this mating will possess this genotype. Mendel observed this outcome; every member of the F_1 generation was purple.

Let's continue with the development of our genetic analysis chart by analyzing subsequent matings that Mendel performed. He next mated the F_1 plants with themselves. Since each F_1 plant is *Pp*, and the reproductive cells receive only one copy of genetic material, it follows that half the gametes will be *P* and the other half will be *p* (Figure 4.4). This will be true for both the male and female gametes since they are derived from the same plant.

Mendel correctly assumed that the fusion of reproductive cells is random; that is, any male gamete can potentially fuse with any female gamete. There are then four possible genetic combinations that can result, as illustrated by the matrix in Figure 4.4. Half of the time the male *P* reproductive cells will combine with female *P* reproductive cells to result in *PP* offspring. The other half of the time, male *P* cells will fuse with female *p* cells, resulting in *Pp* offspring. Similarly, male *p* cells will combine with female *P* cells half the time and with female *p* cells the other half. Since *pP* is identical to *Pp*, we can determine the final ratios of possible genotypes: 1/4 *PP*, 1/2 (or 1/4 + 1/4) *Pp*, and 1/4 *pp*. Since *P* is dominant, all plants containing one or more *P* genes will be purple-flowered, resulting in 3/4 of all the plants being purple-flowered and only 1/4 white-flowered. This is the 3:1 ratio that Mendel observed in his F_2 generation. In addition, this genotype corresponds to

1/4 pure-breeding purple-flowered, 1/2 impure-breeding purple-flowered, and 1/4 pure-breeding white-flowered plants, the 1:2:1 breeding ratio that Mendel observed. Thus, Mendel's model of diploid organisms that reproduce through haploid gametes successfully explains his observations.

Developing a Model

The statistical rule derived from the mating experiments is so important that it has been named Mendel's first law, or the **law of equal segregation**:

> The two sets of information for a given characteristic segregate from each other into the reproductive cells, such that half of the reproductive cells carry one set of information and the other half carry the other set of information.

As with all simple descriptions of the real world, we must remember that what Mendel developed was only a model. Did it represent the way things really were? A good model generates predictions that can be tested. Could his model be tested by examining other mating combinations?

Mendel went a step further in his mating experiments in order to do exactly this kind of test. He mated two plants, one of which was Pp (not pure-breeding purple) and the other of which was pp (pure-breeding white). Figure 4.5 shows the genetic analysis of this mating. What is the prediction for the phenotypic ratio? Half of the offspring should be purple, and half white. If Mendel's proposed ideas about the genetic content of reproductive cells and the presence of two sets of information in normal cells are true, then the predicted results should be observed when such plants are actually mated. Not surprisingly, he obtained these results. Thus, the test supported his model.

Figure 4.5 Predicting the results of a cross. The results of a cross between two different plants, one Pp and the other pp, are shown. Mendel obtained the results he predicted with his models.

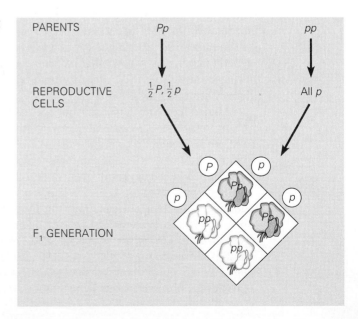

THE LAW OF INDEPENDENT ASSORTMENT

Mendel's first law predicts the inheritance patterns of genes for single traits. Can similar predictions be made for more than one trait at the same time? For example, consider the traits seed color and seed shape. The seeds can be either of two colors, yellow or green, and either of two shapes, round or wrinkled. Is there any general relationship that governs the simultaneous inheritance of these two traits?

Observing Multiple Traits

Mendel performed experiments to address that very question. Each trait Mendel chose to study exhibited only two different phenotypes, such as yellow or green, round or wrinkled.

For pea seeds, yellow color is dominant, so the genetic symbolism for the alleles that determines the trait is Y for yellow color and y for green. Round shape is dominant over wrinkled, so we designate the alleles as R for round and r for wrinkled. Mendel began with two plant lines, one of which was pure-breeding for both yellow seed color and round shape, the other of which was pure-breeding for both green seed color and wrinkled shape. A mating between individuals that differ from each other with respect to two traits is referred to as a **dihybrid cross**. Because the lines Mendel began with were pure-breeding, we know that they must have been homozygous for their respective traits. Thus, one line must be $RRYY$ and the other $rryy$.

Mendel crossed these two plant lines and analyzed the offspring (Figure 4.6). The F_1 generation produced seeds that were all round and yellow. This should not be too surprising, since we already know that each of these traits is dominant. In one experiment, when he self-pollinated the F_1 generation he observed four combinations of these two traits: Of the 556 seeds examined, 315 seeds (about 9/16 of the total) were round and yellow, 108 (about 3/16) were round and green, 101 (about 3/16) were wrinkled and yellow, and 32 (about 1/16) were wrinkled and green. The ratio of these numbers is close to 9:3:3:1 (Figure 4.6).

Let's try to make sense of these numbers. First, consider only the trait of seed shape, which can be manifested as round or wrinkled. Of the 556 seeds, 423 were round and 133 wrinkled, a ratio of approximately 3:1. What about the color trait? The ratio of yellow to green was 416:140, also approximately 3:1. Again, this should not surprise us, since we know from Mendel's studies of single traits that the F_2 generation always exhibits this ratio.

Developing a Model

Mendel concluded that since both traits are still being individually expressed in a 3:1 ratio in the F_2 generation, the same results he saw when he examined one trait at a time, the traits must be acting independently of each other. In other words, whether or not a seed is wrinkled or round has absolutely no effect on the color one way or another. Whatever alleles for seed shape are present, they do not affect the normal inheritance patterns of the seed color. How can this independence be explained? Our symbolic table method of analysis works just as well for two traits as for one

Figure 4.6 Results of a dihybrid cross. Results of a cross involving two different characteristics showing the parents, the F_1 generation, and the F_2 generation.

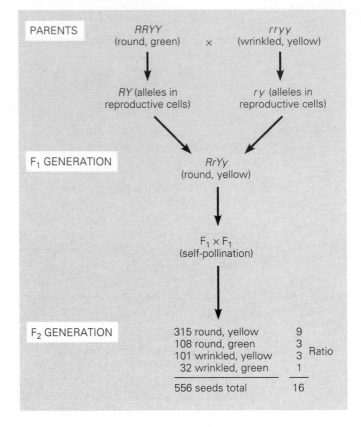

(Figure 4.7). Based on this table, a 9:3:3:1 phenotypic ratio is predictable for the F_2 generation, as is a 3:1 ratio for each individual trait. Mendel concluded that the alleles for each trait must segregate independently during the formation of gametes. For example, in a *RrYy* plant, the reproductive cells have an equal probability of being *RY*, *Ry*, *rY*, or *ry*. The *R* and *r* alleles for seed shape are not associated with the *Y* or *y* alleles.

The important conclusion from experiments of this type has been generalized as Mendel's second law, or the **law of independent assortment**:

> During the formation of gametes, the segregation of information for one characteristic is independent of the segregation of information for others.

It is now known that many characteristics in all organisms behave similarly to those that Mendel studied, that is, as single, discrete units of information (genes). Such genes function in pairs to determine traits and are randomly distributed during inheritance. Characteristics that behave according to Mendel's laws are said to follow Mendelian genetics. Mendel's model of inheritance was the first predictive description of inheritance, and it is still very useful today.

Mendel's work went unnoticed for about fifty years, until several geneticists be-

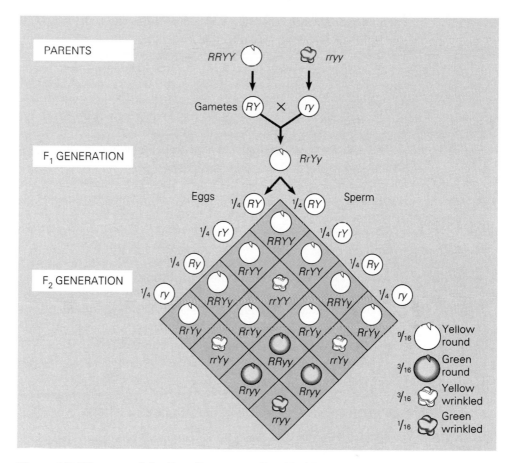

Figure 4.7 Diagram of the F₁ × F₁ cross outlined in Figure 4.6.

came aware of his experiments and realized the impact his statistical predictions would have on the study of inheritance. His two laws allowed vague notions about traits being inherited in some fashion to be turned into scientifically explainable phenomena. Inheritance became predictable.

ESSAY: SCIENCE IS MORE THAN FACTS

Science is often viewed by the general public as a collection of facts that explain the way things work. While this is certainly a part of what makes up science, science is more than just facts. It is a process that humans undertake and thus embodies many of the same difficulties and rewards as any other human endeavor. Mistakes are made, beliefs are held even in the face of incontrovertible evidence to the contrary, and underhanded or perhaps even illegal activities can take place. At the same time, a new discovery can result in levels of excitement usually reserved for children with new toys.

Science reveals itself to be a human undertaking in many ways, a number of which we will explore at various places in this book. One we mention at this point is scientific prediction.

Scientific predictions lend themselves to use and abuse outside the realm of science, particularly when a prediction involves something of social consequence. A certain fatalism or determinism is attached to scientific predictions simply because they are based on perceived factual scientific analysis. Consider an example from economics. An accepted, well-publicized prediction about the state of the economy might cause many people to change their spending or saving habits almost instantly. These changes often tend to move the economy in the predicted direction. Thus, the prediction, because it is believed, becomes self-fulfilling. In the view of many, science deals only in the realm of fact, so predictions based on science are perceived to be already true. Thus, science has a societal impact. The prediction of inheritance based on Mendel's laws is not immune to such misperception. We have indicated that many traits are inherited in Mendelian fashion and thus should follow certain statistical rules. But, as we will see, not all traits are Mendelian, and thus not all can be predicted in the same fashion. In Chapter 6, we will see to what degree society can be influenced by the supposedly fact-based, unbiased predictions of genetics.

EXTENSIONS OF MENDELIAN GENETICS

We now understand how patterns of inheritance are established and transmitted from generation to generation. It is tempting to think that we have a fairly complete picture of heredity. However, the real world does not conform to such simple rules, and years of study of different organisms and different traits have revealed many complexities in genetic systems.

Incomplete Dominance

The seven traits described by Mendel showed virtually perfect dominance and recessiveness—there was no evidence to suggest that purple flower color was ever not dominant over white flower color. But remember that Mendel chose these particular seven traits in part because of the very fact that there were no mixed phenotypes—round peas were always clearly distinguishable from wrinkled peas, and so on. He also determined that a number of traits were not appropriate for further study in part because the phenotypes were not as distinct.

We can understand the problem better by considering one of the same characteristics Mendel studied, flower color, but in a different plant. If pure-breeding white-flowered snapdragons are mated with pure-breeding red-flowered ones, the F_1 generation is all pink (Figure 4.8)! The trait exhibited in the offspring is a mixture or blending of the parental traits and does not exhibit the usual dominance and reces-

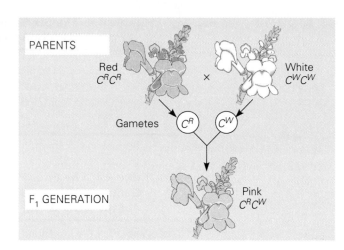

PARENTS

Red
$C^R C^R$

×

White
$C^W C^W$

Gametes C^R C^W

F₁ GENERATION

Pink
$C^R C^W$

Figure 4.8 Incomplete dominance. When pure-breeding red snapdragons are crossed with pure-breeding white snapdragons, the F₁ offspring have pink flowers. This inheritance pattern demonstrates incomplete dominance, in which neither trait is completely dominant over the other. The resulting offspring thus display a phenotype that is intermediate to those displayed by the parents.

siveness that Mendel described. The pink-flower phenotype demonstrates **incomplete dominance**.

Why is one particular allele sometimes dominant over another? Dominance is defined by the observation of phenotypes; nothing inherent in a particular gene causes the gene itself to "dominate." Only through the action of the product of the gene is dominance established. Thus, we must search for biochemical explanations of dominance. Unfortunately, there is no general cause of dominance. It can be biochemically manifested in numerous different ways. Consider the flower color of snapdragons. The genes in question code for enzymes involved in the production of specific pigments. The "red" gene causes the production of an enzyme that results in a red pigment. A "white" gene codes for the production of an enzyme that produces white pigment. A heterozygote, containing one copy each of these genes, produces both enzymes, resulting in the production of both pigments. The net result in the flower is a mixing of the two pigments, red and white, causing a pink color (Figure 4.8). Neither gene is dominant. Obviously, the situation is different in peas, where purple flower color is dominant.

Let's examine the dominance relationships in a more complex trait, that of the ABO blood groups in humans. Individuals can be classified as being in one of four groups: blood types O, A, B, and AB. This classification is based on the presence or absence of either of two types of molecules, A or B, on the surface of red blood cells. If only molecule A is present, this signifies blood type A. If only B is present, the blood type is B. If both A and B are present, the individual is type AB. And if neither A nor B is present, the individual is type O.

A single pair of genes controls ABO blood type, but these genes have three different alleles. The I^A allele causes the production of type A molecules. The I^B allele causes the production of type B molecules. The third allele, designated i, causes neither type of molecule to be produced. I^A and I^B are both dominant to i (hence the standard upper- and lowercase designations). Thus, the homozygote ii does not produce either the A or B molecule, resulting in type O blood. The heterozygote $I^A i$ will result in type A blood, since only the A molecule will be produced. Similarly, $I^B i$

Table 4.2 The Six Genotypes of the ABO Blood Groups

Alleles	i	I^A	I^B
i	ii	iI^A	iI^B
I^A	$I^A i$	$I^A I^A$	$I^A I^B$
I^B	$I^B i$	$I^B I^A$	$I^B I^B$

Table 4.3 The Four Phenotypes Generated by the Genotypes of the ABO Blood Groups

Genotype	Blood Type
ii	O
$I^A I^A$ or $I^A i$	A
$I^B I^B$ or $I^B i$	B
$I^A I^B$	AB

results in type B blood. $I^A I^B$ causes the production of both molecules, resulting in type AB blood.

Alleles I^A and I^B are said to be codominant with respect to each other; if both are present, both will be expressed. **Codominance** is a particular kind of incomplete dominance in which both alleles are equally expressed. Thus, while there are six possible genotype combinations of these alleles, there are only four phenotypic possibilities (Tables 4.2 and 4.3). Dominance in this case is caused simply by the nature of the recessive allele, i, which does not produce a surface molecule. Either I^A or I^B produces surface molecules. If either of these alleles is paired with i, the allele that produces a surface molecule will dominate; i will always be recessive, since it produces no molecule.

Dominance can be manifested in a variety of ways. The point is that there is no easy way to understand all possibilities. While many human characteristics, including many diseases, do follow strict rules of dominance, many other characteristics do not exhibit clear Mendelian dominance and instead show incomplete dominance. Predicting the inheritance of such traits becomes more complex as the dominance relationships become less clear.

Genetic Linkage

Is it reasonable to expect all gene pairs to segregate independently, as suggested by Mendel's second law? Let's put the question another way: If the genes for two different traits are located very close to each other on the same chromosome, why would they be inherited independently? Doesn't it seem likely that a particular trait will be inherited along with the other traits on the same chromosome?

This phenomenon, called **genetic linkage**, is a direct consequence of the fact that each gene is not a physically separate piece of DNA. Many different genes are carried together on the same chromosome. They are physically attached to each other and are therefore carried together during cell division. If every gene were a distinct DNA molecule and thus capable of independent movement, then there could be completely independent assortment. Traits whose genes are on different chromosomes do segregate independently, as described by Mendel. Different chromosomes move independently and so are randomly divided into reproductive cells.

Thus, some degree of independent assortment still exists, even with chromosomal limitations. Remember that recombination can occur during meiosis, in which homologous regions of different chromosome pairs can be interchanged. In essence, recombination serves to overcome linkage. It is a shuffling mechanism that allows for some measure of independent assortment by genes on the same chromosome. Since recombination is essentially a random occurrence, the farther apart any two genes are on a chromosome, the more likely it is that they will be exchanged (Figure 4.9). Genes very close to each other are not likely to be shuffled, whereas genes on opposite ends of the chromosomes are much more likely to be exchanged and therefore segregated independently. If two genes are nearly always inherited together, it follows that they must be located close to each other on the same chromosome. Genes very close to each other are said to be tightly linked.

Linkage is analogous to a measure of physical distance between genes and can actually be used to determine the locations of genes on a chromosome. Such locations are called genetic *loci* (singular, **locus**). The identification and isolation of a specific gene may require knowledge of its location on the chromosome. By examining the percentage of offspring that inherit the gene of interest along with other genes whose locations are already known, the relative position of the gene of interest can be determined. A specific genetic locus of interest, say, for a disease, can often be determined fairly precisely by using a set of human genetic markers in a

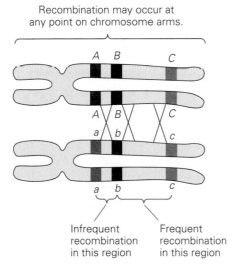

Recombination may occur at any point on chromosome arms.

Infrequent recombination in this region

Frequent recombination in this region

Figure 4.9 Model of a homologous pair of chromosomes showing the likelihood of recombination. The farther apart two genes are on a chromosome, the greater the number of possible sites for recombination. Genes B and C are separated by a large distance. Thus, random recombination events occur frequently between these two genes. Genes A and C are very close together, so it is rare for random recombination to occur between these two genes.

mapping experiment. Currently, a massive genetic mapping effort (the Human Genome Project) is under way that will identify useful marker genes at closely spaced intervals on all chromosomes. These would then be used for mapping the many genes that will continue to be isolated and identified in the future, particularly those involved in causing disease.

Genetic linkage is a critical extension of Mendelian genetics. The value of Mendel's laws is not diminished by the discovery of linkage; rather, the discovery of exceptions to these laws results in a much more complete understanding of heredity.

ISSUES: GENERALIZATIONS AND EXCEPTIONS

Why didn't Mendel come across examples of traits that exhibit genetic linkage? He probably did at some time in the course of his experiments. Remember that he looked at more than seven traits. He chose to ignore some, perhaps because they did not show complete dominance or because they did not yield the same inheritance ratios as the seven traits on which he reported. Genetic linkage would alter such ratios, since multiple genes would not segregate independently.

Scientists sometimes selectively emphasize certain of their observations. While this may not be ideal, in some cases it may be the only way to reach generalizations. Mendel's laws, while they do not explain everything about inheritance, are exceedingly important generalizations. Emphasizing certain traits that fit the statistical patterns allowed him to make these generalizations.

On the other hand, exceptions are often equally important. As we have seen, Mendel's laws do not explain everything about inheritance. Some ideas and concepts, such as genetic linkage, could be explored and articulated only because of exceptions to Mendel's laws. Our understanding of nature requires both incomplete generalizations and detailed attention to exceptions. In the study of inheritance, Mendel provided the first significant generalizations. It was then left to others to incorporate the exceptions into extended theories of Mendelian genetics.

SUMMARY

Our first general understanding of genetics came about through the observation of inheritance in pea plants. Mendel's laws allow a degree of prediction in inheritance. Characteristics that follow simple Mendelian rules can be tracked statistically throughout successive generations. Despite the beauty and simplicity of Mendelian genetics, there are many exceptions to Mendel's laws. Continued exploration of these exceptions has refined our understanding of inheritance and has resulted in a more complete understanding of how genetic characteristics are passed on from parents to offspring.

REVIEW QUESTIONS

1. State Mendel's law of equal segregation. What experiments (matings) gave substance to this generalization?

2. Why did Mendel initiate his work with pure-breeding peas?

3. Distinguish between genotype and phenotype using several of Mendel's pea characteristics. Why can't we determine an individual's genotype based on its phenotype?

4. State Mendel's law of independent assortment. What assumptions are behind the diagrammatic mating depicted in Figure 4.4?

5. Show diagrammatically how Mendel was able to reach the conclusion dictated by the law of independent assortment.

6. Describe what is meant by (a) incomplete dominance and (b) genetic linkage. Why are these complexities not considered part of Mendelian genetics?

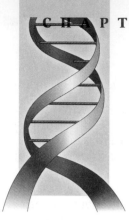

Mutations and Genetic Disease

Imagine knowing that you have genes that *might* cause violent behavior. How would that knowledge affect you? Would you be more wary of confrontation, trying to avoid it in case your genes would cause you to behave violently? Or would you ignore it and assume that you are normal and that any conflict in which you might engage is similar to that experienced by others? At least one particular disease is sometimes described in exactly this way—it *may* result in an increased tendency toward violence. This disease (XYY syndrome) is one of over 4,000 identified genetic diseases affecting humans.

Genetic diseases result from detrimental changes in the genetic material, ranging from the smallest possible change—an alteration of a single base in a single gene—to abnormalities in whole sets of chromosomes. For instance, sickle-cell anemia, a disease that affects the delivery of oxygen to body tissues because of altered red blood cells, results from a single base alteration, whereas Down syndrome, a disease affecting development, appearance, and mental function, results from the inheritance of an extra chromosome. Other diseases, such as cancer, can arise from many different kinds of changes in the DNA.

The common thread among all genetic diseases is an imperfection in the information storage system. Cells are capable of maintaining the integrity of information as long as DNA is accurately replicated and distributed to daughter cells and offspring. But mistakes are made; the replication and transfer of information, while very precise, is not foolproof. To make matters worse, the DNA itself is susceptible to environmental influences that can directly alter the stored information. Permanent changes that occur within the information storage system, that is, within the DNA, are called mutations.

Mutation has come to have a strong negative connotation, when in fact a mutation refers simply to a heritable genetic change, either for better or for worse. It is probably true that the majority of mutations are harmful, or at least not advantageous. On the other hand, not all mutations are harmful; evolutionary changes that arise from mutations may result in some advantage for individual organisms. For example, a mutation that allowed a predator to run faster would increase the chances that it would survive and reproduce and thus increase the likelihood that this advantageous mutation would be passed on to future generations. Strangely, some mutations can be both beneficial and detrimental at the same time. For example, the mutation causing sickle-cell anemia also provides protection against malaria.

This chapter is concerned with **genetic disease**, which can be defined as any heritable disease resulting from an alteration in the information content of the organism. It is estimated that among humans some 5 percent of children have some sort of genetic disease. We will examine the nature of various mutations and their effects in order to understand how changes in normal cellular functions can result in disease. We will characterize two broad groups of mutations: those affecting only one or a few bases, which we refer to as **gene-level mutations**, and those affecting large parts of chromosomes or entire sets of chromosomes, which we call **chromosome-level mutations**. For each type of mutation, we'll examine how they arise and what diseases they cause.

Much of the biotechnology industry is geared toward furthering our understanding of genetic disease, with the aim of devising better diagnostic tools, treatments, and cures. We will return to the treatment of disease in later chapters, after we have examined genetic engineering and immunology.

GENE-LEVEL MUTATIONS

Recall from Chapter 2 that spontaneous mutations can arise in several ways, such as through inaccurate DNA replication. Nearly all mutations involve changes in individual bases. Many mutations can be repaired by cellular enzymes and so do not become permanent. Some, however, manage to escape repair and result in permanent mutations.

Another way mutations develop in DNA is through exposure to chemicals or some types of radiation. Many substances interact chemically with DNA to create mutations. Such substances are called **mutagens** and are said to be *mutagenic*. Most mutagens are also **carcinogenic**, meaning that they cause cancer. Some mutagens cause base changes, while others interact with the DNA helix and cause extra bases to be inserted or removed during DNA replication, causing **insertions** or **deletions**.

Environmental Mutagens

It is clear that certain chemicals, for example, some of those in cigarette smoke, are mutagenic. However, it is not always obvious to what degree mutagens are present in our environment, as either synthetic or naturally occurring chemicals. Some plants in our diet, such as alfalfa sprouts, contain low levels of mutagens. Some molds that grow on corn and peanuts produce compounds known as aflatoxins, which are among the most potent carcinogens known. Charred or burned foods, such as barbecued hamburgers and hot dogs or even toast, contain a wide variety of mutagens.

In some cases, compounds known to be mutagenic are still used industrially because they have other, desirable properties. Nitrite, for example, is widely used as a meat preservative even though animal tests have shown it to be mutagenic. Opponents of using nitrites argue that its consumption is unhealthy. Proponents of nitrite preservatives argue that without such treatment, many people would die from botulism, a generally fatal form of bacterial food poisoning.

The question is: How mutagenic is nitrite? In order to answer this question we need an accurate and efficient means of testing. Since humans cannot be experimented on, we rely on animal tests in which rats, mice, or other animals are exposed to high levels of a test chemical to determine whether it is mutagenic. However, animal testing can take up to three years to complete. Another issue is that, even when testing is completed, there is no guarantee that humans will respond in the same manner as the laboratory animals. A quick, inexpensive test for mutagenicity (the Ames test) is in use. The Ames test employs a specific strain of bacteria to detect low levels of mutation following exposure to a chemical (Figure 5.1). Rates of mutagenesis are established for a range of chemical concentrations. Most importantly, we have learned that about 80 percent of chemicals found to be mutagenic in the Ames test are also found to be carcinogenic when further analyzed in animal tests. Therefore, the Ames test is accepted as a good predictor of carcinogenicity and is used by the U.S. Food and Drug Administration as well as many private testing laboratories.

In addition to chemical mutagens, some types of radiation are highly mutagenic. For instance, X rays and gamma radiation can penetrate cells and damage DNA. One only has to read about the survivors of the atomic bombing in Japan or of the nuclear accident in Chernobyl to realize the harm that results from exposure to high levels of radiation.

Ultraviolet light is another form of radiation known to be mutagenic. Specific wavelengths of ultraviolet light can penetrate our skin cells and interact with DNA, causing significant alterations. Overexposure to sunlight is widely understood to lead to a high incidence of melanoma, or skin cancer. Fortunately, some sunscreens are highly UV absorbent and prevent much of the ultraviolet light from reaching the skin.

Mutagens are present in sunlight and in our food, air, and water. Additionally, radiation is present all around us; in fact, those who live at higher elevations are subjected to more radiation than those living at lower elevations (less is absorbed by the atmosphere). Completely removing environmental sources of mutation is impossible. However, despite the inevitability of mutation from these sources, we can greatly minimize the overall effects by exercising appropriate cautions. As we will see in Chapter 14, most cancers develop only after several mutations have accumulated within a single cell. Lowering the rate of mutation thus has a significant effect on lowering the incidence of cancer.

Even if we could completely insulate ourselves from our environment, the natural instability of DNA and the errors that occur during its replication will eventually cause mutations. Our conclusion is straightforward: There is absolutely no way to avoid the potential for mutation; being alive means being vulnerable to mutations from sources both within and without.

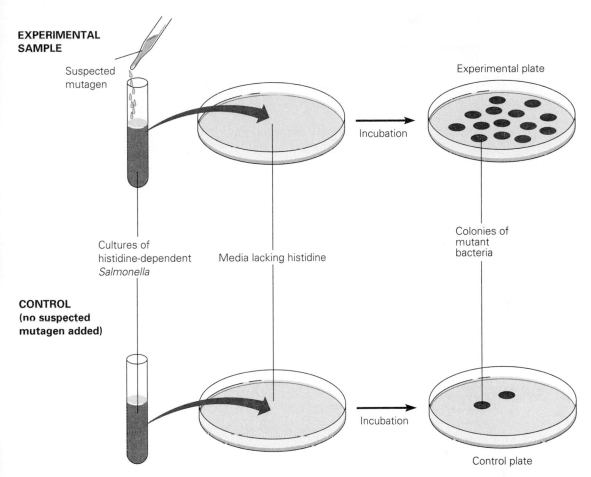

Figure 5.1 The Ames test. The Ames test measures the rate of mutation in a population of a particular strain of bacteria. To perform the test, bacterial cells are grown on dishes containing a mix of nutrients and different amounts of the substance to be tested. The nutrient mix is such that the normal bacterial cells are not able to grow—a particular nutrient, histidine, is missing. However, should a specific mutation occur within the bacterial cell, the cell will be able to grow without that nutrient. Thus, the more mutagenic a substance is, the more bacterial cells will form colonies. Simply counting the colonies gives a good measure of how mutagenic the compound is.

Mutations and the Flow of Information

Now that we've seen how gene-level mutations arise, we can examine exactly how these mutations cause further changes. First, we'll consider single base changes. The results of such mutations depend on what specific change is made and what role that particular base plays in the DNA. Let's consider several examples in which single base changes occur in the coding region for a protein, as shown in Figure 5.2. Figure 5.2a shows a change in a codon for the amino acid alanine, from C to G (see arrow).

Figure 5.2 Some possible effects of single base mutations. (**a**) The change of a single base, from C to G, has no effect, since the new codon, GCG, codes for the same amino acid (alanine) as the original GCC codon. (**b**) A different change in that same codon, from GCC to GAC, does change the amino acid from alanine to aspartic acid. (**c**) In this case, the original TAT codon mutates to a TAA stop codon. This causes premature termination of protein synthesis. (**d**) Mutations within the promoter regions can have various effects. Illustrated here is a mutation that prevents RNA polymerase from binding to the mutated promoter. Since binding is required for the production of mRNA from the gene, the mutation prevents this gene from being expressed.

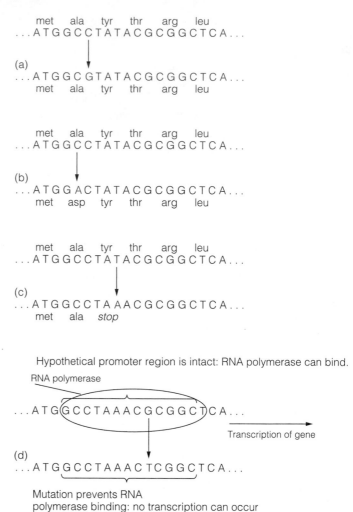

Reference to the genetic code allows us to predict the consequences of this change for the encoded protein. In this case there is no change. The same amino acid will be inserted even though the codon now reads GCG instead of GCC, since both codons specify alanine (see Figure 2.5). The *redundancy* of the genetic code means that some base changes will have no effect on the protein product.

In Figure 5.2b, we see another base change (GCC to GAC) within a codon. In this case, reference to the genetic code reveals that the changed base results in a different codon, causing aspartic acid to be inserted into the protein chain instead of alanine. Thus, depending on the particular protein and the specific amino acid change, a wide range of effects may result, from no measurable change to complete disruption of protein function—presumably with serious consequences.

In Figure 5.2c, we consider yet another base change in a coding region, from T to A (TAT to TAA). This results in the codon for the amino acid tyrosine being changed into a stop codon. What effect will this mutation have on the protein encoded by that gene? A stop codon that appears before the end of the protein coding sequence causes protein synthesis to stop earlier than it should, resulting in an incomplete protein. In general, incomplete proteins are nonfunctional and have potentially serious consequences for the organism.

Figure 5.2d illustrates what could happen if the mutation occurs in a *control sequence* instead of in a coding region. A change in the promoter could weaken or destroy the promoter's ability to bind RNA polymerase, resulting in decreased production of that mRNA. It is also possible to increase expression of a gene through a promoter mutation. As we will discover when we examine cancer, both underexpression and overexpression of a gene can be detrimental to an organism.

Figure 5.3 illustrates what might happen as a result of small insertions or deletions in the base sequence. The insertion of a single base shifts the remaining region of the coding sequence, offsetting it from the original reading frame. This shift results in the production of a protein that bears little resemblance to the intended protein and that, in all likelihood, has no function. Such a mutation is referred to as a **frameshift mutation**, because it shifts the reading frame of the ribosome from the original series of codons to a new series. The deletion of a single base has similar consequences.

In summary, we have seen that gene-level mutations can arise from a number of sources. While many changes are repaired by cellular mechanisms, not all are. Those that are not repaired become mutations. The effects of mutations can range from no effect to the production of completely dysfunctional proteins.

As we become more aware of the molecular effects of chemicals currently being produced and developed, concern has risen about the long-term consequences for humans and other animals. This concern has led to improved methods of chemical treatment and cleanup called **bioremediation**. In bioremediation, microorganisms metabolize a wide range of chemicals that are toxic to humans, converting them to less harmful and perhaps valuable by-products. Oil-eating bacteria, of potential use on oil spills, are a good example of developments in this area. We will say more about bioremediation in Chapter 9.

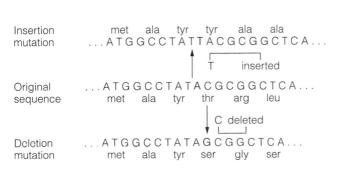

Insertion mutation

Original sequence

Deletion mutation

Figure 5.3 Effects of single-base insertion and deletion (frameshift) mutations. The original gene sequence and the sequence of amino acids for which it codes are shown in the middle. If T is inserted at the position shown, the amino acids coded for change as indicated. If C is deleted from the original sequence, the amino acid sequence also changes. Both changes result in the production of vastly different proteins in the cell, which probably will not function properly.

CHROMOSOME-LEVEL MUTATIONS

In addition to the types of mutations described above, which generally cause a change in a single base or a few bases, more extensive mutations can occur, causing large portions of chromosomes or even entire sets of chromosomes to be altered. We will consider three types of chromosomal mutations: deletions (loss of a region of a chromosome), **translocations** (exchanges of material between nonhomologous chromosomes), and **aneuploidies** (abnormal numbers of chromosomes).

Deletions

While chromosomal deletions in humans are rare, they generally have serious consequences. How they arise is unclear, but some event causes a complete break in the DNA within the chromosome, which results in a loss of genes. The homologous chromosome, therefore, provides the cell with the only copy of these genes. In some instances, this loss can have serious consequences; single genes may not be able to direct the synthesis of sufficient quantities of proteins, a situation called **haploid insufficiency**. In addition, any recessive disease gene in the homologous chromosome is now expressed, since it is the sole copy of the gene. A number of diseases result from specific chromosomal deletions, including retinoblastoma (a rare inherited childhood cancer of the eye) and Wilms' tumor (a childhood kidney cancer).

Translocations

Chromosomal translocations occur when there are breaks in nonhomologous chromosomes and segments are exchanged (Figure 5.4). If the breaks occur in noncoding regions of the DNA, no genetic information is destroyed. In this case, no new or altered phenotypes result from the translocation. However, if the translocations interrupt a coding region, then the information contained in those genes will be dis-

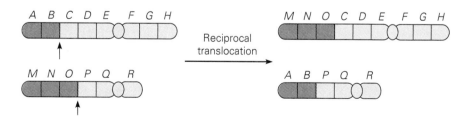

Figure 5.4 Chromosomal translocation. A translocation moves a DNA segment from one chromosome to another. Shown is the most common type of translocation, a reciprocal translocation, in which segments are exchanged between nonhomologous chromosomes. Arrows designate sites of breakage on the chromosomes. Nonreciprocal translocations can also occur. In a nonreciprocal translocation, a chromosome transfers a segment without receiving a segment in return.

rupted. Various cancerous diseases are associated with translocations, including one form of leukemia (cancer of blood cells) and Burkitt's lymphoma, a common form of cancer in parts of Africa.

Aneuploidy

Aneuploidy refers to the presence of an abnormal number of chromosomes within cells. The most common aneuploid condition is **trisomy**, in which three copies of a chromosome are present in a cell instead of the usual two. The most common way that trisomy occurs is through nondisjunction. During meiosis, at either the first or second cell division, the normal mechanisms for separating chromosome pairs fail to function properly, resulting in nondisjunction (Figure 5.5). If this occurs at the first division, two of the four gametes produced will contain an extra copy of this chromosome and two will contain no copies. If nondisjunction occurs at the second division two of the resulting gametes will be abnormal: One will contain two copies, and one will contain none. The other two reproductive cells will be normal. If a cell that contains two chromosomal copies fuses with a normal gamete, a trisomy will result.

The most familiar trisomy, and one of the most common, is Down syndrome, a trisomy of chromosome 21. It occurs about once in every 600–700 human births. There is a strong positive correlation between the age of a female giving birth and the incidence of Down syndrome; women who are forty-five to fifty years old when bearing children have a fifty to sixty times higher risk for having a child with Down syndrome.

Many trisomies involve the sex chromosomes. Klinefelter's syndrome (in which there are two X chromosomes and one Y, making the person XXY) occurs about once per 700 human births. Those who suffer from this disease are male but possess some female features. Other physical and mental effects can be quite variable; some individuals never know they have the defect, while others may suffer from mild retardation and other associated emotional problems.

The XYY trisomy is not only of biological interest but is also an example of how genetic diseases are perceived in a social context. A now famous study, published in the late 1960s, examined the occurrence of this trisomy in the population of a high-security mental institution. The study reported that the incidence of XYY trisomy in the institutional population was twenty times higher than in the general population. The authors attributed the traits of mental illness, mental retardation, tallness, and aggressiveness to this trisomy. Their study generated considerable interest and spawned numerous studies of the incidence of XYY trisomy in individuals in mental hospitals and prisons. Most of these studies reported higher-than-average incidence of XYY and associated it with aggressive behavior. The media reported the results as studies of "criminal genes," and a social stigma based on "factual" predictive genetics arose: The XYY condition produces males with aggressive behaviors who are very likely to commit violent crimes. Ultimately, further studies negated the conclusions of the original research. A thorough examination of the general population revealed that most persons with XYY trisomy live normal lives without ever knowing they

Figure 5.5 Nondisjunction leads to trisomy. Nondisjunction can occur during meiosis I and meiosis II. Following nondisjunction, at least one of the reproductive cells contains an extra chromosome. When that cell combines with a normal reproductive cell, the resulting fertilized cell has three copies of that chromosome—a trisomy.

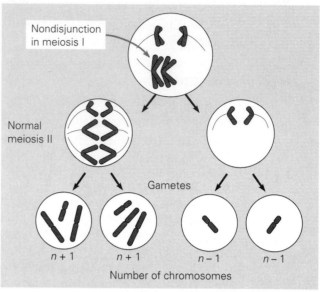

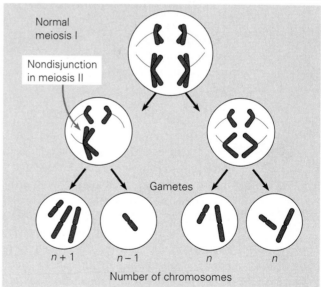

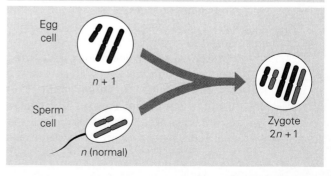

have the extra Y chromosome. In addition, most incarcerated inmates who were XYY had committed nonviolent crimes. Thus, there is no apparent basis for the association of an XYY genotype with violent criminal behavior.

....

ESSAY: CAN KNOWLEDGE OF GENETIC MAKEUP CREATE STIGMAS?

Because our ability to determine genetic makeup for any individual is increasing, we need to be more cautious in using this information than we have been in the past. Such knowledge has a huge potential for stigmatization.

The XYY trisomy is a good example of our need to be as socially responsible as possible. This includes parents who are raising an XYY child as well as teachers and others who will affect the child's development. Consider parents or teachers who believe that XYY syndrome results in more aggressive behavior. Would they be able to distinguish between normal pushing and shoving as the child grows and learns to play and socialize and aggression that might be caused by the XYY syndrome? Probably not. The child might be treated differently from other children, and overly protective treatment might cause rebelliousness, fulfilling the expectations of the parents or teachers.

It is entirely possible that our belief in the predictive abilities of genetics oversteps reality. Genetic tests can now be performed on fetuses (Chapter 10) to determine their genetic makeup even before birth. If we believe that a certain genetic makeup will result in particular behaviors, whether this is true or not, our actions may fulfill our expectations.

....

GENETIC DISEASES

This chapter is concerned primarily with genetic diseases caused by mutations in a single gene. These mutations can lead to metabolic changes, such as the lack of an enzyme necessary for the production of a required product (as in hemophilia), the production of a substance that turns out to be harmful to cells (as in sickle-cell anemia), or a change in the normal regulatory properties of a cell such that growth or metabolism is altered in some way (as in cancer).

Genetic diseases can be manifested in two ways. *Inherited diseases* stem from defects in the genetic information contained within reproductive cells (in most cases meaning that they are contained in all cells of the affected individual) and thus are passed from parent to offspring. Table 5.1 lists inherited genetic diseases, their prevalence in populations, and their characteristic symptoms. **Somatic disorders** are caused by mutations in nonreproductive cells and are passed only to cellular progeny (never to offspring) through mitosis. Cancer is primarily a somatic cell genetic disease, and we will discuss it more fully in Chapter 14.

Changes in genes that cause genetic diseases exhibit the same patterns of inheritance as other genes. Inherited diseases can be either recessive or dominant, and the

Table 5.1 Incidence of Genetic Disorders

Condition	Estimated Incidence	Major Symptoms
cystic fibrosis	1 in 2,500 Caucasians	
diabetes	1 in 80 individuals	loss of control of blood-sugar levels
Down syndrome	1 in 1,050 individuals	mental retardation
Duchenne muscular dystrophy	1 in 7,000 males	
hemophilia	1 in 10,000 males	inability of blood to clot
Huntington's disease	1 in 2,500 individuals	nervous disorder
phenylketonuria	1 in 12,000 Caucasians	brain damage, if not treated
Rh incompatibility	1 in 100 individuals	
sickle-cell anemia	1 in 625 African Americans	severe, life-threatening anemia
Tay-Sachs disease	1 in 3,000 Ashkenazi Jews	nervous disorder; death at an early age
β-thalassemia	1 in 2,500 individuals of Mediterranean descent	blood disorder

genes associated with disease can be found on either autosomes or sex chromosomes. We will explore specific examples of each pattern of inheritance to gain a better understanding of what kinds of effects can be caused by mutations. We will also consider genetic predispositions to disease. Other diseases, termed **multifactorial** or **polygenic**, involve multiple genes and environmental influences. The special complications involved in analyzing multifactorial traits are considered in Chapter 6.

Autosomal Recessive Disorders

Approximately one thousand known genetic diseases are caused by recessive alleles on *autosomes*. This means that the disease is manifested only when two copies of the mutated allele are inherited by an individual. Individuals who carry only a single copy of the mutated allele are called **carriers**. In many cases, carriers exhibit no symptoms of the disease; however, for certain genetic diseases, carriers are affected to some degree. A mutation in one allele usually causes that gene to encode a recessive nonfunctional product. In a carrier, another allele encodes a normal product. Metabolism suffers from the absence or abnormality of the protein only when both copies are mutated.

ADA Deficiency Adenosine deaminase (ADA) deficiency is an example of an autosomal recessive disease. It is caused by a lack of the enzyme adenosine deaminase, which is normally involved in the breakdown of excess adenosine. Without this enzyme, high levels of deoxyadenosine triphosphate (dATP) accumulate in the blood. Excessive levels of dATP are selectively toxic to dividing **immune system** cells (B cells and T cells; see Chapter 11). Thus, patients with ADA deficiency have markedly reduced counts of mature B and T cells, resulting in a condition known as **se-**

vere combined immunodeficiency disease (SCID). They have no immune defenses and are subject to a multitude of infections early in life. Until recently, children with this disease were kept in sterile conditions (in a plastic bubble) to prevent infections. Weekly injections of ADA-coated beads now allow patients to leave the hospital. Although fewer than one hundred cases of this disease have been reported worldwide, it is the target for the first approved gene therapy experiments (Chapter 10).

PKU Phenylketonuria (PKU) is another classic autosomal recessive disease. Afflicted individuals are deficient in the enzyme phenylalanine hydroxylase, which is necessary for breaking down the amino acid phenylalanine. Without this enzyme, levels of phenylalanine in the blood increase, causing severe mental retardation within a few months of birth.

In the United States, all babies are now routinely tested for high phenylalanine levels at birth. Treatment consists of modifying the diet to include virtually no phenylalanine, making the disease manageable. After about ten years, the adverse effects seem to disappear, and most patients can partake of a normal diet.

PKU has been newsworthy because of the widespread use of aspartame (Nutrasweet) as an artificial sweetener. Aspartame contains phenylalanine; phenylketonurics who drink large quantities of diet soda, for example, will accumulate phenylalanine when their bodies metabolize aspartame. High dosages of aspartame have proved fatal in test animals with PKU, although dosages used in these tests are much higher than would occur in humans as a result of normal aspartame consumption.

Sickle-cell Anemia Not all recessive disorders are caused by enzyme deficiencies. Sickle-cell anemia is a disease in which red blood cells assume an elongated, sickle shape instead of the normal disk shape, in capillaries where there is typically a low concentration of oxygen (Figure 5.6). These sickled cells clog the capillaries and prevent the flow of blood to tissues, resulting in severe anemia and tissue damage. Until recently, victims of this disease rarely survived to adulthood. Some treatments now decrease the extent of sickling, providing some relief; however, the treatments have significant side effects.

The disease is caused by a single base change in one of the two genes encoding hemoglobin, the oxygen-carrying protein of red blood cells. A change from A to T results in the change from a GAG (glutamic acid) codon to a GTG (valine) codon. The substitution of this single amino acid in hemoglobin results in a profound change in the structure of the protein, causing it to bind to other hemoglobin molecules under oxygen-deprived conditions, forming long fibers. These fibers change the shape of the cell, resulting in sickling.

It is well known that certain populations (African Americans, Africans, and people of Mediterranean descent) have a much higher frequency of sickle-cell alleles than other populations. Why? Carriers of the disease, in whom about 40 percent of red blood cells sickle, have a greater resistance to malaria. Malaria results from infection of red blood cells by any of four protozoan species in the genus *Plasmodium*. Sickled red blood cells provide a less hospitable environment for the protozoan, and therefore carriers are more resistant to malaria than the rest of the population. (See

Figure 5.6 Sickle-cell anemia.
(**a**) Normal red blood cells have a concave, disklike shape. They are very pliable and fit through relatively small passages in the circulatory system. (**b**) Sickle cells are elongated and spiky. They are inflexible and often cannot pass through small constrictions in the circulatory system.

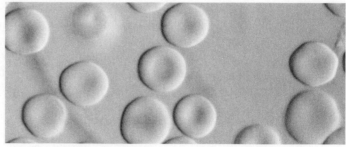

(**a**) Normal red blood cells   10 μm

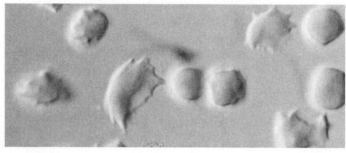

(**b**) Sickle cells   10 μm

Chapter 11 for the role of red blood cells in malaria.) Populations with higher frequencies of the sickle-cell gene coincide with populations in which malaria is a major threat (Figure 5.7). Thus, being heterozygous for sickle-cell anemia—that is, being a carrier—is an advantage in parts of the world where malaria is prevalent. The selective pressure of the environment—the prevalence of malaria—has served to maintain a higher frequency of sickle-cell alleles, even though it also results in a higher incidence of sickle-cell anemia. Apparently, a balance has been struck between the advantage of greater resistance to malaria and the disadvantage of sickle-cell anemia.

Autosomal Dominant Disorders

Dominant traits are expressed regardless of the allele on the homologous chromosome. Over 1,500 **autosomal dominant diseases** are known, but most are exceedingly rare. In humans, there are two primary mechanisms by which the dominance of a disease can be expressed: by the production of a product that compromises the survival of the cell and by haploid insufficiency, the inability of the single normal allele to provide sufficient product. We will now examine three examples of autosomal dominant diseases.

Thalassemia We have already discussed hemoglobin as the oxygen carrier in red blood cells. Hemoglobin is a complex protein consisting of four separate polypeptide chains and an extra prosthetic group (a portion of a protein not composed of amino

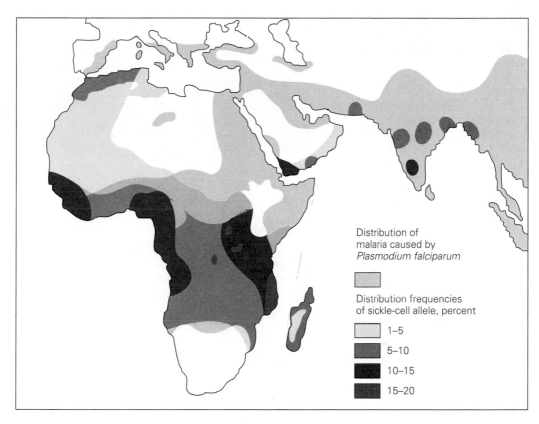

Distribution of
malaria caused by
Plasmodium falciparum

Distribution frequencies
of sickle-cell allele, percent

1–5

5–10

10–15

15–20

Figure 5.7 Sickle-cell anemia and malaria. The sickle-cell allele is found at high frequencies in certain parts of Africa, the Mediterranean, and the Middle East. This map also highlights regions where the protozoans that cause malaria live. The overlapping distribution of the sickle-cell allele and malaria is quite remarkable. We now know that sickle cells provide a less hospitable environment for the malaria parasite, allowing individuals with at least one copy of the sickle-cell allele to better withstand the disease. (Adapted from A. C. Allison, "Abnormal Hemoglobin and Erythrocyte Enzyme-Deficiency Traits" in *Generic Variation in Human Population* by G. A. Harrison, ed., 1961, Oxford: Elsevier Science.)

acids) called heme (Figure 5.8). The polypeptides of functional hemoglobin are of two types: two α-globin chains and two β-globin chains. The heme group contains an iron atom, which alternately binds and releases oxygen. The iron must have a +2 charge on it to function properly. Sometimes, the iron is oxidized to a +3 charge, forming methemoglobin, which does not bind oxygen. To regenerate the +2 charge on the iron, a specific enzyme called methemoglobin reductase is required. Certain mutations in the α- or β-globin polypeptide chains make it very difficult for the enzyme to act on the iron atom. If a hemoglobin molecule contains even a single defective chain, the entire protein is prevented from functioning normally.

Consider what happens in an individual who is heterozygous for a defective β-globin gene. Half of the β-globin chains produced by this gene pair will be defective. Thus, the probability of a hemoglobin molecule having a normal β-chain at the first position is 1/2. The probability of having a normal β-chain at the second position is also 1/2. Therefore, only one-fourth of the assembled hemoglobins (1/2 × 1/2) will

Figure 5.8 The structure of hemoglobin. Hemoglobin consists of four protein chains assembled into a compact structure. Four heme groups, which contain iron atoms, are also part of the structure. As the hemoglobin moves to the lungs, the iron atoms bind oxygen. This "loaded" hemoglobin then travels through the blood to tissues, where the oxygen is unloaded. The iron is absolutely essential to the function of the hemoglobin—it must have a charge of +2 to bind oxygen.

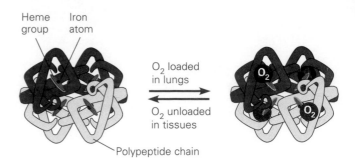

Heme group Iron atom

O_2 loaded in lungs

O_2 unloaded in tissues

Polypeptide chain

contain normal β-chains at both positions; the rest will have at least one defective chain. This means that only one-fourth of the hemoglobin will be fully functional, too little to sustain normal life. Infants with such mutations are called "blue bloods" because the high level of iron in the +3 state causes their blood to be blue instead of red. This is a good example of an autosomal dominant disorder in which a mutant allele can generate a product that can compromise the organism's survival.

Familial Hypercholesterolemia Other autosomal dominant diseases may result from haploid insufficiency, in which the production of a normal protein from a single allele is simply insufficient for metabolism. A good example is **familial hypercholesterolemia**, which is caused by a mutant allele of the gene for the low-density lipoprotein receptor. Low-density lipoproteins (LDLs) transport cholesterol in the bloodstream. The LDL receptor, a protein on the surface of cells, is responsible for removing low-density lipoproteins from the bloodstream and transporting them into cells, where the cholesterol is extracted and used for various purposes. Cells obtain a significant portion of their required cholesterol in this fashion. If one LDL receptor allele is mutated, cells produce only half as much receptor, and a large fraction of LDLs remain in the bloodstream. As a result, high levels of cholesterol accumulate and contribute to the early development of atherosclerosis, increasing the risk of heart disease. Another way of saying this is that the amount of receptor produced by a single normal allele is insufficient to maintain normal cholesterol transport into the cells, and high serum cholesterol levels result. Remember that dominance is defined by phenotypes. Thus, the mutant allele is dominant simply because the single normal allele in a heterozygote cannot produce enough protein to make up for the loss of the second normal allele.

Huntington's Disease The last autosomal dominant disease we will consider is Huntington's disease. Sufferers of this disease develop involuntary movements, and eventually neurological degeneration leads to a loss of motor control. These symptoms do not usually appear until afflicted individuals are in their forties or fifties. Thus, many people who will develop the disease have already reproduced before they know they have it. Since it is a dominant disease, any offspring will have a fifty-fifty chance of having the disease if either parent is afflicted.

A way to detect the presence of mutations in the gene responsible for Huntington's disease now exists, allowing us to predict whether individuals will develop the disease later in life. The test can be performed before reproductive age; in fact, it can be performed on a fetus. Needless to say, the emotional and moral questions raised by this technological advance can be tremendous for individuals. We will consider Huntington's disease again in several other chapters, including our discussion of genetic screening (Chapter 10).

X-linked Diseases

A special class of genetic diseases consists of diseases whose genes are part of the X chromosome. The X chromosome is the larger of the two sex chromosomes; females have two X chromosomes, and males have one X and one Y chromosome. A small amount of genetic information is common to both chromosomes, but a large region of the X chromosome is not duplicated on the Y (Figure 5.9). This effectively means that males are haploid for information on the X chromosome, while females are diploid, just as with the autosomes. The term applied to this genetic feature of males is **hemizygous**, and genes found on the X chromosome are said to be **X-linked**.

The hemizygous nature of the male X chromosome creates an unusual inheritance pattern. What if a male has a mutation in an X-linked gene? Since that mutated gene is the only one for that trait in the cell, the allele will be expressed, regardless of whether it behaves in a dominant or recessive manner. A mutated allele on the X chromosome results in males having a significantly higher incidence of disease than females. Females would have to have two recessive alleles, one on each X chromosome, to be afflicted with a recessive X-linked disease.

Color Blindness There are about two hundred known X-linked diseases, most of which are quite rare. The most common is color blindness. The genes for the visual pigments necessary for color vision are on the X chromosome. Red-green color blindness is caused by defects in the genes for these pigments. Males suffer from color blindness to a much greater extent than females (approximately 8 percent of males).

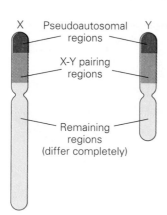

Figure 5.9 The X and Y chromosomes. These two chromosomes contain two regions of similarity. The top regions, called the pseudoautosomal regions, are identical in both the X and the Y chromosome, and they behave as if they were autosomes. The middle regions, called the X-Y pairing regions, are similar but far from identical. Recombination can occur within these regions, but it is rare. The remaining regions of the chromosomes are completely different.

This makes sense, knowing what we do about X recessive genes. If a male inherits an X chromosome with a mutation in this gene, he will be red-green color-blind. A female, on the other hand, must inherit *two* copies of each mutated allele to be color-blind, a less likely possibility.

Hemophilia Hemophilia refers to a group of diseases that affect blood clotting. This mechanism involves a great number of blood and cellular proteins, as well as cells and specialized cell fragments (platelets), that respond quickly to a wound by creating a fibrin clot. The clot is built around long fibrous proteins (fibrin) that assemble at the site of the wound (Figure 5.10). The regulation of this process involves many different proteins. If any of the proteins involved are defective, the result is an inability to clot properly, or hemophilia.

Two of the genes for proteins involved in clotting are carried on the X chromosome, those for factor VIII and factor IX. This means that males will be particularly susceptible to the effects of mutations in these genes, in typical X-linked fashion. He-

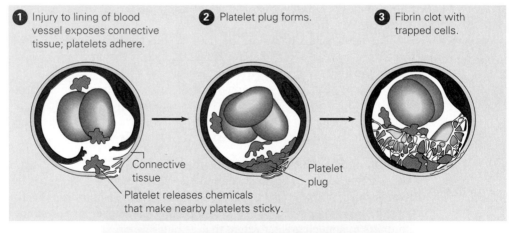

1 Injury to lining of blood vessel exposes connective tissue; platelets adhere.

2 Platelet plug forms.

3 Fibrin clot with trapped cells.

Connective tissue

Platelet releases chemicals that make nearby platelets sticky.

Platelet plug

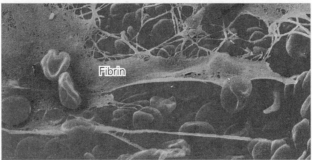

Fibrin

Figure 5.10 Blood clotting. This sequence of events leads to the formation of a blood clot. A wound triggers a cascade of reactions, resulting in the formation of fibrin, the long, sticky, fibrous protein that is largely responsible for trapping cells and particles at the site of the wound, filling the hole. Each reaction in the cascade is catalyzed by an enzyme. If any of the enzymes (or other factors) involved is missing, proper clotting cannot occur. Hemophilias are diseases in which at least one factor is missing.

mophilia A, which results from mutations in the gene for factor VIII, is the most common form of the disease, affecting approximately one in 10,000 males. Only a few cases of female hemophilia A have been reported, as would be predicted based on a statistical analysis of the mutant allele frequency. Mutations in factor IX lead to hemophilia B, which is less common. Treatment for all forms of hemophilia consists of injecting the missing blood proteins into the patient. These proteins can be purified from blood or produced through biotechnological methods that we will describe in Chapters 8 and 9.

GENETIC PREDISPOSITIONS

A person who is *predisposed* to a disease has a higher risk of getting that disease than the general population. Genes correlated with incidence of disease can be identified, even though not everyone with that gene may develop the disease. The gene is said to predispose the individual or, in other words, put the individual at higher risk. The interesting question is why some individuals with the gene develop the disease and others don't. Clearly, other genes and environmental effects must be acting together with this predisposition gene to actually cause the disease.

There have been recent reports of the identification of genes that predispose people to manic depression, schizophrenia, alcoholism, dyslexia, many forms of cancer, lung disease, and so forth. As we will see in Chapter 10 and elsewhere, the identification of a specific gene for a disease or a predisposition has many implications, including two of particular medical and social importance. It may allow the development of a diagnostic test that can determine whether the gene responsible for the disease or predisposition is present in an individual, even an unborn fetus. It also allows, in theory, the development of potential genetic treatments or cures through repair or replacement of the predisposing gene using gene therapy (Chapter 10).

We are now at the stage of medical technology in which certain predispositions can be diagnosed. Such predictions might be helpful; an individual diagnosed as having a predisposition to skin cancer presumably would make an effort to avoid overexposure to the sun. The behavior of a child diagnosed as having a predisposition to schizophrenia would be closely monitored and probably modified by concerned parents, similar to the situation described for children diagnosed with XYY syndrome. However, we can imagine that knowledge of a predisposition might also lead to expectations that have detrimental effects.

SUMMARY

We have seen that different types of mutations can lead to nonfunctional gene products, depending on how the protein (the gene product) normally functions and how it has been disrupted. We have concentrated on diseases caused by single-gene defects; to a large degree, the inheritance of such diseases is predictable through the use of Mendelian genetics. One of the most important medical applications of biotechnology involves the diagnosis and treatment of genetic diseases. Our later discussion of genetic screening, genetic counseling, and gene therapy will refer back to the inheritance patterns of many of the diseases we have noted here.

REVIEW QUESTIONS

1. Many chromosomal translocations and deletions are associated with cancers. Why might that be?

2. What is the difference between gene-level and chromosome-level mutations?

3. Briefly outline how biotechnology and disease interact today and what is expected in the near future.

4. List several examples of chemical mutagens and environmental mutagens.

5. Select specific mRNA codons that are known to encode specific amino acids (the universal genetic code) and show how the amino acids encoded by the codons would be affected if bases were changed, removed, or inserted. Refer to Figure 2.5 for the genetic code.

6. Define and provide at least one example of a disease resulting from these chromosomal mutations: deletion, translocation, and aneuploidy.

7. Explain the meaning of the statement "Cancer is primarily a somatic cell genetic disease."

8. Define or describe ADA deficiency, PKU, sickle-cell anemia, thalassemia, familial hypercholesterolemia, and Huntington's disease.

9. What is considered unique about X-linked conditions such as color blindness and hemophilia?

10. Offer your most convincing arguments, first on behalf of one extreme and then on behalf of the opposite one, regarding this position: Health insurance companies have the right to know if an individual they insure carries a gene for Huntington's disease (or one that predisposes him or her to cancer).

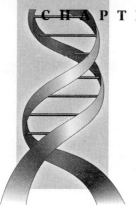

CHAPTER 6

Complexities of Genetics: Polygenes, Behavioral Traits, and Eugenics

This chapter deals with the subject of complex characteristics and variation. Many characteristics, in fact probably most human characteristics, are affected by more than one gene. Further, some characteristics are affected by external, nongenetic influences. For example, a person's weight, while determined to some extent by genes, is subject to how much food the person eats and how much he or she exercises. Is a particular characteristic determined by the individual's genetic makeup (nature) or by the individual's environment (nurture)? The "nature versus nurture" debate has always created controversy and continues to affect our views of why we are the way we are. Nowhere is this debate more germane than in the area of human behavior. Unfortunately, deciphering behavior is more difficult than studying traits such as pea shape or flower color, and it can easily be tainted by prejudices and preconceptions. In this chapter, we will examine the eugenics movement, an early-twentieth-century program of prejudicial and oppressive social reforms that was based in large part on the genetics of behavioral traits.

POLYGENIC TRAITS

The characteristics on which we focused in Chapters 4 and 5 are discrete, which means they are all-or-nothing traits: A pea is either round or wrinkled. **Polygenic traits**, traits conferred by more than one gene, are exceedingly important in plant and animal breeding. From the stock market to the futures market to dinner tables around the world, economic and nutritional interests center on grape, soybean, corn, and tomato yields. For years, animal breeders have been concerned with milk production in dairy cattle, frequency of egg laying in various poultry strains, the weight of wool sheared from sheep, and litter size and carcass quality of hogs. Many human characteristics as well are not all-or-nothing but vary continuously—for example, height (Figure 6.1), infant growth rate, adult weight, skin color, blood pressure, serum cholesterol, musical ability, aggressiveness, academic performance, and length of life.

Figure 6.1 Height of humans. Height is an example of a continuously varying trait or characteristic.

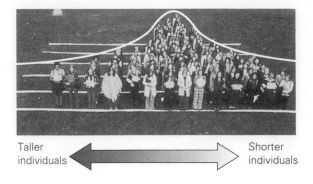

Taller
individuals

Shorter
individuals

Variation in traits is due to two general factors: the involvement of multiple genes and environmental influences. Here we consider some implications of more than one gene influencing a given trait. Polygenic traits, sometimes called *quantitative traits*, can display a continuum of phenotypes. Kitten coat color is a good example because the genes involved have been well characterized. In this example, an uppercase letter signifies a set of alleles, or a genetic locus. Italicized letters identify individual alleles, uppercase for dominant and lowercase for recessive.

Five primary sets of genes determine kitten coat color, and they are identified as A, B, C, D, and S (Figure 6.2). The A gene exists in two allelic forms. The dominant allele, *A*, produces hair with the color called agouti, grayish with a band of yellow in the middle. In a homozygous recessive cat, *aa*, the yellow band is missing, so the color is solid gray.

There are also two allelic forms of the B gene. The *B* allele yields normal agouti only if at least one *A* allele is present; if the animal is *aa*, *B* yields solid black. The recessive *bb* also yields two different colors: *bb* with *AA* or *Aa* gives cinnamon, while *aabb* produces solid brown. The A and B genes for coat color provide a classic example of **interacting genes**. The alleles present at one genetic locus interact somehow with the alleles at the other locus, yielding many more possible phenotypes than the individual genes would allow.

When we say that genes "interact," we do not mean that the two regions of DNA come together physically. Rather, the products of the genes interact in some fashion. Consider a protein that functions in a particular metabolic pathway. Another protein (a product of a different gene) can function as a regulator of the first; that is, it can bind to the first protein and modify its activity, perhaps by increasing or decreasing its output or by making it more or less responsive to other signals. Because their products interact, we describe the genes as interacting. Changes (mutations) in either gene will then affect the same trait. For example, mutations in either A or B genes can affect a kitten's final coat color.

Returning to our example, let's consider the C gene. It functions differently from the A and B genes in that it either permits or prevents the expression of the color determined by A and B. The *C* allele permits color expression, whatever that color may

(a) Well-Known Interacting Genes with the Corresponding Coat Color

(c) A Calico Cat

Alleles	Hair Color
A_	Gray with midband of yellow (agouti)
aa	Gray
A_ + B_	Gray with midband of yellow (agouti)
aa + B_	Black
A_ + bb	Cinnamon
aa + bb	Brown
C_	Coat color expressed
cc	No coat color expressed; albino (white coat)
D_	Allowance of full coat-color expressions
dd	Washed-out coat color (not full expression)
S_	No spots
ss	Piebald or spotted (different-colored patches)

(b) Combined Individual Genotypes and the Actual Coat Colors Produced

Genotypic Individuals	Actual Coat Color
AaBBCcDDSs	Agouti
aabbCcddss	Washed-out brown with spots
AaBbccDDSs	Pure white (albino)

Figure 6.2 Kitten coat color.

be, and cc prevents color expression. A cc animal is an albino, which lacks pigment and thus is white. The D gene controls the intensity of the color determined by the other genes. The D allele allows full expression of color, but dd yields a washed-out color. This kind of gene is sometimes referred to as a *modifier gene*, since it functions only to modify what is established by other genes. The fifth gene in the coat color system, the S gene, controls the presence or absence of spots. The S allele produces no spots, but ss results in what is called the piebald pattern, which yields spots or, more precisely, patches or blots (Figure 6.2).

The complexity of polygenic traits makes phenotypic predictions difficult. For example, can we predict the ratios of phenotypes resulting from a cross between two cats, one of which is *AabbCcDDSs* and the other of which is *aaBbCcDdSs*? Since there are five gene pairs, in general there will be 32 (2^5) possible combinations of alleles in each haploid reproductive cell. The other haploid reproductive cell will also have 32 possible combinations. That means 1,024 theoretically possible genotypes (32 × 32) in a matrix-type analysis. Sorting through all of these possibilities is a daunting task, although it can be done given that we know the genes involved and their relationships and interactions. It took tremendous effort to unravel the workings of this single trait, coat color, which is determined solely by genetics and is easily observed and measured. Imagine trying to dissect the genetics of musical ability or aggressiveness in humans!

VARIABILITY AND THE ENVIRONMENT

Asking about a trait such as musical ability is unfair, because not only is this characteristic polygenic, it is further complicated by the second way variability can be introduced: through the environment. Environment influences the final outcome on many different levels. The products of genes exist within cells, which are constantly bathed in an environment that provides hormones, nutrients, toxins, and so forth. Further, an organism exists in an environment that constantly supplies signals and sensations, germs and viruses, heat and cold. Many organisms can also learn and thus modify their behavior such that environments can be altered or avoided if desired. For example, an animal living in a cold climate might seek shelter, which might alter its need for a dense fur coat to keep warm. It seems probable that many characteristics or traits are continuously influenced by environmental signals.

Familial and Heritable Traits

Some traits are determined entirely by environment, with no basis in genetics. The social traits that show the highest correlation between parents and offspring are political and religious affiliation. Neither of these traits is genetically determined, yet both appear to be inherited if we measure simple correlations among relatives. It is easy to understand how these particular traits are environmentally determined—parents expose their children to their own religious and political beliefs.

A distinction must be made between two types of traits. **Familial traits** are shared by relatives for whatever reason, including social, behavioral, and environmental influences as well as heredity. **Heritable traits** are shared by relatives because they have inherited the same genes. Establishing that a trait is familial is fairly easy, since we only need to measure correlations among related individuals and compare them to correlations among unrelated individuals. However, establishing that a trait is heritable is much more difficult, because this involves being able to exclude all external, nongenetic influences. We can do this only by studying individuals who have absolutely no environmental differences. We can study some animals in laboratory settings in this manner, but it is impossible to do so with humans. Therefore, it is impossible to establish that a complex trait in humans (one that does not obey simple Mendelian genetics) is due solely to the action of genes.

For most traits, the question is not *which* of these, genes or environment, determines the trait, but rather to what *degree* each contributes to the final trait. Genes play a role in most characteristics that can be defined, and many traits are primarily determined by genes. The diseases we considered in Chapter 5 are good examples of this. Yet the role of the environment can be important even here; for example, phenylketonurics suffer no symptoms as long as their environment is controlled so that they do not eat phenylalanine.

Measuring Heritability: Twin Studies

Is it possible to estimate what relative contribution genes and environment make to a particular trait? The only hope is to measure how variable a trait is in situations in which either the influence of genes or the influence of the environment is controlled. The extent of variability caused by nature or nurture would then indicate how important each is in establishing the trait. In theory, we can measure the variability in a trait among relatives and compare it to the variation measured in unrelated individuals. If the trait has a genetic component, then relatives should exhibit less variation than unrelated individuals. However, this type of analysis ignores subtleties of environmental influences. Relatives usually share more similar environments than unrelated people. Does the decreased variability in a trait among relatives derive from the similarity in genes or from the similarity in environments?

Studies of twins, particularly those raised in separate environments, shed some light on the relative contribution of genes and environment. Identical twins arise from a single fertilized egg that divides very early in development into two cell masses, each of which goes on to form a separate fetus. Identical twins thus have identical genetic information, so the study of identical twins raised in separate environments should provide information specifically related to the role of the environment in the development of traits. However, studies of separated twins are difficult to conduct because it is hard to find sets of twins whose environments have no correlation. Even twins separated by adoption tend to share similar environments, since adoptive families tend to be of like social class. Interpretations based on twin studies are exceedingly difficult and usually unconvincing, since all of the variables cannot ever be controlled. Complicating the interpretation is the fact that the numbers of subjects are frequently very low, making statistical analysis less reliable.

In spite of these difficulties, however, there is a legacy of using twins in studies of many human traits, particularly behavioral traits such as homosexuality, alcoholism, intelligence, and schizophrenia. These studies usually attempt to assign a numerical value to the relative contribution of genes to the variability of the trait and therefore to the trait itself. Such a measurement is called the **heritability** of a trait. For example, many studies and textbooks report values corresponding to the heritability of intelligence, that is, the degree to which intelligence is determined by genes. However, none of these studies ever accounts completely for all environmental influences, making interpretations risky and heritability measures of these complex behavioral traits of questionable value.

Even if we could make a heritability measurement that completely accounted for the environment, it could not be applied to an entire population. Particular groups may experience drastically different environments that alter traits in ways that don't affect other groups at all. For example, Americans traveling in foreign countries often experience illness as a result of drinking water. The environment—in this case, a region and its water—has a tremendous effect on health. The native population, however, experiences no ill effects from drinking its own water. Thus, a measurement of heritability for one set of individuals under a certain set of circumstances does not correlate with the heritability of that same trait for a different group or a different set of circumstances.

Behavioral Traits

What exactly is a behavioral trait? Any type of behavior that can be identified and somehow measured can be considered a **behavioral trait**. Behaviors can be instinctive or learned, genetically or environmentally induced. Discerning the cause of a behavioral trait is difficult because many genes and environmental factors operate simultaneously. Further, the interpretation of behaviors is subject to the state of knowledge at any given time. Consider a boy with a habit of compulsive nail biting. Sometimes he gets so worked up that he injures his lips and the ends of his fingers in a biting frenzy. Without further information, his behavior might be thought of as a reflection of some mental illness or deficiency. However, it turns out that this behavior is characteristic of Lesch-Nyhan syndrome, an X-linked recessive disease in which nucleotide metabolism is disrupted. Males who inherit the recessive mutant allele (remember that, since it is X-linked, only a single copy is present in males) are compulsive nail biters. How the lack of a particular enzyme results in this behavior is not known.

Another example of a disease with behavioral consequences that is caused by a defect in a single gene is Huntington's disease, a neurological disorder brought on by a dominant autosomal mutation (Chapter 5). The progression of the disease results in marked changes in behavior, including muscle spasms and personality disorders. For many years, Huntington's disease was characterized as a behavioral disease, a form of mental illness. Around 1910, it was found to be an inherited dominant trait that follows simple Mendelian laws, indicating that a single gene was responsible for the disease. This gene was found to be on chromosome 4 (Figure 6.3), and further analysis revealed that the molecular basis for the disease is a triplet mutation—patients have forty-two to sixty-six copies of a CAG (codon) repeat rather than the eleven to thirty-four copies found in normal individuals. We still don't understand the impact of the additional repeats. Even so, the isolation of this gene and the develop-

Figure 6.3 Huntington's disease. The causal mutation, excess CAG repeats, is located on chromosome 4.

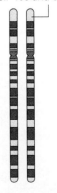

Huntington's disease
Neurodegenerative disorder tending to strike people in their 40s and 50s

ment of diagnostic tests are excellent illustrations of the application of molecular biology, as well as of its social implications.

These examples illustrate that behaviors are interpreted within a changing climate of scientific knowledge and social acceptability. What is now perceived as a behavioral disorder caused by a genetic defect may well turn out to be greatly influenced by environmental factors, or vice versa. A complete understanding of behavioral traits requires a great deal of study in order to identify both genetic and environmental influences. It is interesting to note, however, that behaviors that are deemed acceptable at a given time are generally not subjected to analysis. It is only traits deemed unacceptable for which we seek explanations.

Huntington's disease illustrates one additional point concerning genetic causation. We know that mutations in a single gene are responsible for this disease. However, signs and symptoms of the disease differ: onset occurs at different times in different individuals, and the severity and the time it takes for the disease to run its course both differ. In other words, even though the disease is caused by a single mutated gene, there is still great variability in the phenotype, the behavior of a victim. How can we explain this variability? Different sorts of mutations in the gene can cause differing effects, but even patients with exactly the same mutation show great variability. Do other genes affect the development of this disease in ways we do not understand? Do environmental factors affect the time of onset of the disease? Even though a genetic explanation for Huntington's disease can now be used to diagnose this defect, a great deal about it is still not understood. In time, a more complete understanding may affect the way we view this disease.

ESSAY: GENETIC DETERMINISM

It is interesting that, in scientific parlance, a disease or trait that is known to be derived from any amount of genetic influence, no matter how little, is usually called an inherited disease. Very few noninfectious diseases are thought of as "environmental diseases," with the exception of cancers—and, as we will see in Chapter 14, the primary focus even of cancers is turning toward inherited predispositions. This manner of identifying traits and diseases as genetic gives no indication of the influence that other factors may have and contributes to the growing expectation that genes ultimately will be found to be responsible for all aspects of every individual. This view even affects the type of research that is financially supported to further our understanding of genetic traits. Once a disease becomes identified as an "inherited disease," more emphasis is placed on understanding its underlying genetics, with less emphasis placed on understanding the role that any other contributory factors may play.

Studies aimed at identifying genetic correlations with complex traits such as alcoholism, homosexuality, schizophrenia, dyslexia, intelligence, and a host of others are proliferating. To the extent that these studies attempt to identify genes involved in disease and provide the groundwork for potential diagnoses and treatments, their value is inestimable. However, as the increasing number of popular media reports focusing on genetic causation shows, acceptance of genetic causation is becoming more widespread. In some ways, the current popular acceptance of a kind of "genetic determinism"—the notion that most human characteristics and behaviors are predetermined by one's genes—is similar to that which existed early in the twentieth century, when great advances in genetics fueled the eugenics movement.

EUGENICS—A SOCIAL MOVEMENT BASED ON THE INHERITANCE OF COMPLEX TRAITS

Eugenics was defined in 1883 by Sir Francis Galton, an eminent British scientist and a cousin of Charles Darwin who was obsessed with measurements of all kinds. He was particularly interested in human abilities and the measurement of human traits. His book *Hereditary Genius* first espoused his eugenic ideas. Galton defined **eugenics** as "the science of improving human stock by giving the more suitable races or strains of blood a better chance of prevailing speedily over the less suitable." Bear in mind that the breeding of crops and animals to obtain more and better food had been occurring for centuries and, in fact, practices that we might now identify as eugenic have been a part of every social group that forbade marriage across class, ethnic, or religious lines. But before eugenic ideas surfaced, no one had put forth specific suggestions for improving the human race based on ideas of inheritance, measurement of traits, and judgments about the desirability of those traits. The development of eugenics is the most dramatic example to date of the misuse and misapplication of genetics.

The eugenics movement of the early twentieth century was based largely on misunderstandings and incorrect or prejudiced interpretations of Mendelian genetics. Unsubstantiated inferences drawn from supposed scientific fact led to the rise of eugenics as a popular social and political force, with widespread impact in both the United States and Europe. Although the eugenics movement itself was not responsible for the Holocaust in Nazi Germany, it did provide a purported scientific basis for racism in Germany long before Hitler came to power.

According to its proponents, eugenic improvement could be gained in two ways. What came to be called **positive eugenics** emphasized reproduction by "good" people—that is, those who exhibit desirable traits. Such people should reproduce with each other and do so often so that "good blood" will become more prevalent. **Negative eugenics** emphasized the necessity of limiting reproduction of those with "bad blood" so that undesirable traits would be lost from the population. In general, advocates of positive eugenics did not try to force such a reproductive policy on society but rather hoped that all "enlightened" people would voluntarily perform this duty to the human race. However, advocates of negative eugenics assumed that **degenerates**—those with "bad blood"—would not voluntarily choose to limit their reproduction for the good of the human race. Thus, they thought it essential for the state to legislate a negative eugenics policy to limit reproduction among those deemed unfit—meaning people who were alcoholics, insane, epileptic, criminals, or just plain poor.

Who were the eugenicists? During the formative years of eugenics, as its scientific basis was being established, eugenicists were prominent scientists and psychologists. The scientists included Charles Davenport, who in 1910 became the head of the Eugenics Record Office, a laboratory supported by donations from Mrs. E. H. Harriman, wife of a railroad tycoon; Alexander Graham Bell, inventor of the telephone; and Luther Burbank, the great plant breeder. Other famous people allied with the eugenics movement included authors George Bernard Shaw and H. G. Wells; the first president of Stanford University, David Starr Jordan; and Theodore Roosevelt. The American Eugenics Society was formed in 1923 to promote the cause of eugenics; much of its annual income was contributed by John D. Rockefeller, Jr., and George Eastman of Eastman-Kodak. In 1926, this organization published *A Eugenics Catechism*, a book designed to explain to laypersons the benefits and methods of eugenics. Consider a typical statement from the book:

> Q: Does eugenics mean less sympathy for the unfortunate?
> A: It means a much better understanding of them, and a more concerted attempt to alleviate their suffering, by seeing to it that everything possible is done to have fewer hereditary defectives.

The *Catechism* was replaced and expanded in 1935 by a book produced under the direction of the Directors of the American Eugenics Society entitled *Tomorrow's Children: The Goal of Eugenics*. In its preface the author, Ellsworth Huntington, noted that everyone who contributed to the publication effort agreed that "everything possible should be done to encourage large families in the right kinds of homes and to discourage them in undesirable homes." In our time one wonders how "right" and "undesirable" homes were distinguished.

How widespread was the acceptance of eugenics? Garland Allen, in his 1975 essay *Genetics, Eugenics and Class Struggle*, argues that the eugenics movement was primarily a social program of the upper class, designed and implemented in order to maintain the desirable class structure they commanded. On the other hand, Elof Axel Carlson, in his review *Human Imperfection: Unresolved Responses*, suggests that the eugenics movement had extensive grass-roots support and became a popular cause of all but the most destitute. While such debate by historians may ultimately change the views we have of the social acceptance of eugenics, it is safe to suggest that genetics was used in horribly mistaken ways to support a social program for the betterment of the human race that simply could not, and did not, work.

The eugenics movement grew out of a unique combination of social, economic, political, and scientific influences. Exploring the origins of these influences allows us to more completely understand the early-twentieth-century attitude toward the unfit. We will explore three components, each of which contributed to the development, acceptance, and popularization of eugenics. First, we will uncover the origins of the "degeneracy" concept; next we'll encounter attempts to quantify degeneracy through intelligence testing; finally, we'll examine some of the political and social aspects of eugenics.

The Origin of "Degeneracy"

Because the concept of degeneracy plays a fundamental role in eugenics, we should examine it in some detail and define what it meant in the early twentieth century. There have always been people characterized as "degenerate": sick, poor, handicapped, uneducated and illiterate, mentally ill, socially ostracized, homeless, and so on. The perception of "degenerates" includes a belief that all of these symptoms are interrelated. As we study eugenics, we must remember that degeneracy was a very broad category, including all manner of "undesirable" characteristics exhibited by the less fortunate.

What causes degenerates to be the way they are? When did the idea arise that degeneracy had a biological basis? Part of the underlying explanation of "degeneracy" or "unfitness" that existed in the nineteenth and early twentieth centuries derives from a biblical story. *Genesis* 38 tells the story of Onan, who, following the death of his brother, is commanded by God to take his brother's wife and give her children. Onan, recognizing that these children will not be considered his, chooses not to impregnate his new wife, instead "spilling his seed upon the ground." This angers the Lord, and Onan is killed for his sin. The lesson is not only that "seed" is not to be spilled but also that sexual pleasure is to be avoided. Several religious groups reinforced this view, and many Christians enforced a sexual repression that endured for centuries. During the Inquisition, all possible means were used to gain evidence of heresy against the church in order to eliminate religious competition. Sins of sexual passion were used as evidence and quickly became synonymous with witchcraft, demons, and the devil; "spilling seed," that is, masturbation, was one of these sins.

In 1720, the book *Onania* was published. It voiced the popular view that "loss of strength" (that is, disease) would result from the spilling of seed. This loss of strength could take many forms, including distempers, ulcers, and madness—all degenerate traits. Such ideas were given medical credence in the 1750s by a Swiss physician, Samuel Tissot, who developed an entire "theory of degeneracy" based on loss of sperm. Masturbation, thought to be a sexual vice, through some unknown biological mechanism resulted in degeneracy. Lacking any real medical knowledge of infectious disease or mental illness, people agreed that vices were at least as good a cause of degeneracy as anything else. Incidentally, there is now little doubt that many of the symptoms attributed by Tissot to masturbation were actually caused by advanced venereal disease. The concept of infectious disease was still a century away.

While it is obvious how degeneracy itself, as a description or measure of unfitness, fit into a eugenic theory, it is perhaps less apparent why masturbation became such a focal point for eugenicists. Its importance was twofold. First, masturbation was viewed as a degenerate trait. From its initial status as a cause of degeneracy, the theories of Tissot and others quickly established masturbation itself as a degenerate behavior. Second, castration was proposed as a method of treating degeneracy, probably based on observations that castrated animals become more docile. Patients suffering from a wide variety of "degenerate" traits—hysteria, insanity, and masturbation, to name a few—were experimentally castrated. Public outcry was sufficient to prevent wide acceptance of castration, so vasectomy was tried as a means to the same end. Harry Clay Sharp, head physician at the Indiana State reformatory, was the first to try vasectomy as a cure for the specific purported degenerate behavior of masturbation, in 1899. Patients appeared to be cured—they no longer masturbated. Very quickly, Sharp began to campaign for the widespread use of sterilization to help cure all forms of degeneracy. Following his advice, the state of Indiana enacted a compulsory sterilization law in 1907 that called for the mandatory sterilization of habitual criminals, rapists, epileptics, the insane, and the institutionalized mentally ill. It is easy to see how sterilization, first proposed as a way of treating disease, was also viewed as a way to accomplish the negative eugenic goal of preventing the reproduction of degenerates.

Society's Responsibility to Degenerates Prior to the early twentieth century, Western notions of charity were primarily church-based. It was a Christian's duty to give alms to the needy. The English Poor Laws of 1601 were a secular extension of the church's responsibility, placing a burden of charity on all individuals through taxation. In the midst of research into how best to deliver this government charity to the needy, it was found that investments of time and effort were more productive than monetary investments in helping the poor. This idea formed the basis of a New Charity movement, in which volunteers donated time and help instead of money. However, the vast majority of people simply interpreted the New Charity to mean that giving money to the poor was counterproductive, encouraging people to remain poor, become beggars, commit crimes, and become even more degenerate. Not donating money became virtuous.

Of course, such ideas only exaggerated differences between social classes. The very wealthy class, developing as a result of industrialization, became even more distrustful of the poor; hatred, fear, and disgust followed quickly. The prevailing belief was summarized nicely by William Graham Sumner, a prominent Yale political scientist who was firm in his belief that the rich deserved to be rich. His 1873 book *What Social Classes Owe to Each Other* argued that the rich owed nothing to the poor, and vice versa. The rich were intelligent and could plan and save, but the poor and degenerate were extravagant and foolish. In other words, the rich and the poor had characteristics that made them behave as they did. Degeneracy was caused by these individual characteristics, and degenerates could not and should not be helped by charitable social programs.

Degeneracy and Inheritance What light did science shed on the nature of degeneracy? Two scientific advances were important to the development of eugenics. In 1887, August Weismann proposed his germ theory of inheritance, which postulated that each individual possesses a germ line, or a set of cells that contain hereditary information, and that this set of cells is shielded from the environment by the body such that hereditary information is unchanged—and, in fact, unchangable—from generation to generation. Weismann's theory seriously challenged the prevailing notion of heredity, that characteristics acquired by individuals during their lifetime were passed on to offspring, and it became widely accepted because it seemed to be the best explanation for inherited characteristics. However, when the germ theory of inheritance—unchanging hereditary information—was coupled with a second important scientific advance, the rediscovery and popularization of Mendelian genetics, the basis for eugenics was strengthened. Combining these two theories, science was able to explain both what was responsible for inheritance (units of hereditary information passed on to offspring) and the unchanging nature of this information. The inheritance of all characteristics was then seen as being completely predictable, completely scientific.

Support for this view came from the discovery that diseases can be inherited and explained by Mendelian laws. In 1902, alkaptonuria, a disease in which patients excrete urine that darkens as it contacts air, became the first disease to be attributed to a genetic defect. In 1909, Charles Davenport, an influential geneticist and leader of the eugenics movement, discovered that Huntington's disease was hereditary. Until that time, Huntington's disease had been characterized as a behavioral disorder. It seemed logical to assume that all behaviors, degenerate or not, were inherited. Thus, the unchangeable hereditary nature of degeneracy was "proved."

The alleged hereditary basis for degeneracy was used as a justification for the Indiana sterilization law, which begins "Whereas heredity plays a most important part in the transmission of crime, idiocy, and imbecility. . . ." Other sterilization laws incorporated similar statements that blamed heredity for virtually every imaginable degeneracy. Not only was science able to explain the origin of degeneracy, but, more important, it was poised to provide solutions to the problems posed by the degenerates of society.

Intelligence Testing and Eugenics

Some scientists used their studies of characteristics to help establish eugenic principles. More than any other characteristic, eugenicists used measurements of intelligence to develop and support their social agenda. To a geneticist, a characteristic must be observable and measurable so that it can be followed through generations. How could intelligence be measured?

The formal development of intelligence testing began in 1904, when Alfred Binet was commissioned by the French government to develop a method for identifying children who needed special attention or extra help in their schooling. He developed a series of tests based on the performance of age-related tasks. His goal was to pinpoint children who were unable to perform up to norms so that they could be given remedial help. Binet explicitly stated that his tests did not measure intelligence or innate ability. Inherent in his testing program was the belief that performance would improve as a child was given extra help.

Binet's tests were translated into English by Henry Goddard, a psychologist and the director of a laboratory for the study of mental deficiency at the Training School for Feeble-Minded Boys and Girls in Vineland, New Jersey. Note the name of the institution. Feeblemindedness was a catchall categorization synonymous with degeneracy. Inherent in the use of this categorization was a link between intelligence and behavior. Many members of the upper class generally believed that they behaved in a much more civilized, "better" manner than degenerates. They were people of a social type that exhibited intelligence, foresight, thrift, and self-control. They attributed their overall superiority to higher levels of innate intelligence, in addition to the absence of degenerate traits. Lower native intelligence was attributed to degenerates. Thus, the "feebleminded" included mental defectives, people with behavioral problems, and diseased people, and all of these "traits" were interrelated, hereditary, and unchangeable.

Goddard made many contributions to the development of eugenics, including the following:

1. He imported Binet's tests and began using them in the United States, first at his New Jersey school and then elsewhere.

2. He took a preexisting classification scheme that was based on functional testing and provided intelligence ratings for the categories: *idiots*, who could not develop full speech, were assigned a mental age of less than three; *imbeciles*, who could not master writing, were assigned a mental age of three to seven; and *morons* (a category he identified and named), or "high-grade defectives," had a mental age of eight to twelve.

3. He championed the idea that Binet's tests measured innate, inheritable intelligence, which could not be changed. Feeblemindedness, Goddard said, was "a condition of mind or brain which is transmitted as regularly and surely as color of hair or eyes."

Goddard was most concerned about the effects of morons on society; idiots and imbeciles were already "dealt with" by institutionalization and therefore presented a minimal threat. But morons could not be easily identified by their language difficulties and so were part of the degenerate class that he saw as threatening the advance of humanity. He warned that morons might appear normal but are not. Therefore, much of his work was designed to develop the ability, through testing or other means, to identify morons. In a 1919 book, *Dwellers in the Vale of Siddem*, the eugenically inclined author wrote:

> It is not the idiot or, to any great extent, the low grade imbecile, who is dangerous to society. In his own deplorable condition and its customarily accompanying stigmata, he is sufficiently anti-social to protect both himself and society from the results of that condition. But from the high grade feeble-minded, the morons, are recruited the ne'er-do-wells, who . . . drift from failure to failure, spending a winter in the poor house, moving from shack to hovel and succeeding only in the reproduction of ill-nurtured, ill-kempt gutter brats to carry on the family traditions of dirt, disease, and degeneracy.

Thus, it was not difficult to establish a distinct linkage between inherited characteristics, behavioral traits, poverty, criminality, and disease.

Goddard translated his beliefs into a social program by proposing negative eugenic solutions to the problem of morons. The chief methods of negative eugenics, aimed at limiting the number of children born with an *undesirable inheritance* and directed mostly against morons, were limiting family size, sterilization or segregation of those unfit for parenthood, and the postponement of marriage. Two specific approaches became popular. First, morons should not be allowed to reproduce. This would prevent the spread of their degenerate genes. Compulsory sterilization was the method of choice for accomplishing this. People who were congenitally feeble-minded, epileptic, afflicted with certain types of insanity, or subject to dangerous emotional instability, or certain other socially inadequate persons who have inherited their defects—that is, the morons or "moronic-related defectives"—should have a vasectomy or have their fallopian tubes severed and tied. Second, the immigration of morons or related defectives to the United States should be prevented in order to preserve "our" stock. In an effort to meet the second goal, Goddard helped establish a program in 1913 at Ellis Island, New York (the receiving point for immigrants at that time), with the goal of identifying feebleminded immigrants and returning them to their home countries. Stephen Jay Gould, in his book *The Mismeasure of Man*, describes the methods used to determine whether or not an immigrant was feeble-minded. Often, immigration officials made decisions after simply looking at a person. Goddard described the activities of those making the determinations:

> After a person has had considerable experience in this work, he almost gets a sense of what a feebleminded person is so that he can tell one afar off. The people who are best at this work, and who I believe should do this work, are women. Women seem to have closer observation than men. It was quite impossible for others to see how these two young women could pick out the feebleminded without the aid of the Binet test at all.

The results of Goddard's program were staggering. In 1913 deportations increased 370 percent over the previous five-year average, and in 1914 they increased 514 percent! Goddard's program was considered "successful" because it was able to identify so many feebleminded individuals. Thus, the necessity for such programs was firmly established in the minds of those who believed in the degeneracy of immigrants.

Other psychologists became involved in intelligence testing, including Stanford psychologist Louis Terman, who believed that social classes were established by innate intelligence. Therefore, people could be categorized through intelligence testing.

> The evolution of modern industrial organization together with the mechanization of processes by machinery is making possible the larger and larger utilization of inferior mentality. One man with ability to think and plan guides the labor of ten or twenty laborers, who do what they are told to do and have little need for resourcefulness or initiative.

Apparently, the more intelligent individuals, in other words, those who could perform well on intelligence tests, would be the thinkers and planners. Degenerates, identified by poor performance on the tests, would be the laborers. Terman believed that people could be channeled into acceptable professions suited to their intelligence: Persons having IQs below 75 were most suited for unskilled labor, and "anything above 85 IQ in the case of a barber probably represents so much dead waste." Terman also believed that sociopaths, because they were degenerates, could be identified and removed from society before they committed crimes.

Terman concerned himself with one of the major difficulties in administering the Binet tests: They were labor-intensive and required a trained examiner to spend several hours with one person. He developed a variant of the test, the Stanford-Binet test, that could be administered to large groups. He also introduced the term IQ, or intelligence quotient, defined as an individual's mental age divided by his or her chrono-logical age, times 100.

Terman's purpose in intelligence testing showed clearly when he wrote the following in 1916:

> It is safe to predict that in the near future intelligence tests will bring tens of thousands of these high-grade defectives under the surveillance and protection of society. This will ultimately result in curtailing the reproduction of feeblemindedness and in the elimination of an enormous amount of crime, pauperism, and industrial inefficiency.

In his attempts to institute universal testing in order to identify the "merely inferior" class, Terman took the ever-popular approach of considering the economic savings that would result from identifying inferiors and somehow preventing them from committing crimes.

> Considering the tremendous cost of vice and crime, which in all probability amounts to not less than $500,000,000 per year in the United States alone, it is evident that psychological testing has found here one of its richest applications.

The next development in intelligence testing came from Robert Yerkes, a psychologist who declared that "man is just as measurable as a bar of steel." He wanted to quantify psychology in order to give it the firmest possible scientific foundation. He recognized a fantastic opportunity for the collection of massive amounts of data (which, of course, would provide the raw material for innumerable quantitative studies) by testing all Army recruits and draftees in World War I. His selling point to the military was that individuals could be routed into appropriate positions within the Army based on their measured intelligence. Yerkes, along with other testing experts such as Goddard and Terman, developed a series of streamlined IQ tests that could be delivered en masse to warehouses full of draftees. All told, almost 2 million tests were administered between 1916 and 1918. Test 8 of the U.S. Army mental tests, examination alpha, is shown in Figure 6.4. Take the time to read over the test (essentially forty multiple-choice questions and fill-in or completion statements). How do you think you would have done? Was your innate intelligence tested, or was something else really the focus of the exam?

The results of Army tests remained classified until 1921, at which time this tremendous store of information—including not only the IQ test results, but also other personal information such as educational background and medical history—was declassified and subjected to intense analysis. As often happens, mostly brief snippets—those most likely to be shocking—were reported to and remembered by the general population:

1. The average mental age of a white, American male adult was 13 (just above that of a moron). This differed from that determined in a previous study by Terman, who had reported an average mental age of 16. Apparently, average intelligence was getting worse!

2. European immigrants could be graded and ranked by their country of origin; Southern and Eastern Europeans were intellectually inferior to Western and Northern Europeans.

TEST 8

Notice the sample sentence:

People hear with the   eyes   <u>ears</u>   nose   mouth

The correct word is ears, because it makes the truest sentence.

In each of the sentences below you have four choices for the last word. Only one of them is correct. In each sentence draw a line under the one of these four words which makes the truest sentence. If you can not be sure, guess. The two samples are already marked as they should be.

SAMPLE {People hear with the   eyes   <u>ears</u>   nose   mouth
France is in   <u>Europe</u>   Asia   Africa   Australia

1. America was discovered by   Drake   Hudson   <u>Columbus</u>   Cabot 1
2. Pinochle is played with   rackets   <u>cards</u>   pins   dice.. 2
3. The most prominent industry of Detroit is   <u>automobiles</u>   brewing   flour   packing.................... 3
4. The Wyandotte is a kind of   horse   <u>fowl</u>   cattle   granite 4
5. The U.S. School for Army Officers is at   Annapolis   <u>West Point</u>   New Haven   Ithaca............... 5
6. Food products are made by   Smith & Wesson   <u>Swift & Co.</u>   W.L. Douglas   B.T. Babbitt 6
7. Bud Fisher is famous as an   actor   author   baseball player   <u>comic artist</u> 7
8. The Guernsey is a kind of   horse   goat   sheep   <u>cow</u> 8
9. Marguerite Clark is known as a   suffragist   singer   <u>movie actress</u>   writer............................... 9
10. "Hasn't scratched yet" is used in advertising a   duster   flour   brush   <u>cleanser</u> 10
11. Salsify is a kind of   snake   fish   lizard   <u>vegetable</u> 11
12. Coral is obtained from   mines   elephants   oysters   <u>reefs</u> 12
13. Rosa Bonheur is famous as a   poet   <u>painter</u>   composer   sculptor.. 13
14. The tuna is a kind of   <u>fish</u>   bird   reptile   insect 14
15. Emeralds are usually   red   blue   <u>green</u>   yellow... 15
16. Maize is a kind of   <u>corn</u>   hay   oats   rice 16
17. Nabisco is a   patent medicine   disinfectant   <u>food product</u>   tooth paste.............................. 17
18. Velvet Joe appears in advertisements of   tooth powder   dry goods   <u>tobacco</u>   soap................. 18
19. Cypress is a kind of   machine   food   <u>tree</u>   fabric 19
20. Bombay is a city in   China   Egypt   <u>India</u>   Japan 20
21. The dictaphone is a kind of   typewriter   multigraph   <u>phonograph</u>   adding machine 21
22. The pancreas is in the   <u>abdomen</u>   head   shoulder   neck 22
23. Cheviot is the name of a   <u>fabric</u>   drink   dance   food.. 23
24. Larceny is a term used in   medicine   theology   <u>law</u>   pedagogy.. 24
25. The Battle of Gettysburg was fought in   <u>1863</u>   1813   1776   1812 25
26. The bassoon is used in   <u>music</u>   stenography   book-binding   lithography 26
27. Turpentine comes from   petroleum   ore   hides   <u>trees</u> 27
28. The number of a Zulu's legs is   <u>two</u>   four   six   eight... 28
29. The scimitar is a kind of   musket   cannon   pistol   <u>sword</u> 29
30. The Knight engine is used in the   Packard   Lozier   <u>Stearns</u>   Pierce Arrow........................... 30
31. The author of "The Raven" is   Stevenson   Kipling   Hawthorne   <u>Poe</u>.................................. 31
32. Spare is a term used in   <u>bowling</u>   football   tennis   hockey 32
33. A six-sided figure is called a   scholium   parallelogram   <u>hexagon</u>   trapezium........................ 33
34. Isaac Pitman was most famous in   physics   <u>shorthand</u>   railroading   electricity...................... 34
35. The ampere is used in measuring   wind power   <u>electricity</u>   water power   rainfall.................. 35
36. The Overland car is made in   Buffalo   Detroit   Flint   <u>Toledo</u> 36
37. Mauve is the name of a   drink   <u>color</u>   fabric   food 37
38. The stanchion is used in   fishing   hunting   <u>farming</u>   motoring 38
39. Mica is a   vegetable   <u>mineral</u>   gas   liquid... 39
40. Scrooge appears in   Vanity Fair   <u>The Christmas Carol</u>   Romola   Henry IV 40

Figure 6.4 A U.S. Army mental test. U.S. Army mental test 8, examination alpha (for those who could read), which was administered to almost 2 million men about the time of World War I. (Correct answers are marked here.)

3. Blacks, as a race, scored at the bottom of the scale, with an average mental
    age of 10.4.

Volumes have been written describing the peculiarities of analysis to which these
data were subjected. We'll discuss just one example that illustrates the kind of bias
that can exist in the name of science. The Army IQ tests found that Northern blacks
scored higher than Southern blacks. They also found that Northern blacks attended
school longer than Southern blacks. Such a correlation might be interpreted today as
indicative of the effect that schooling has on test performance: More education results in better performance. However, since the tests were believed to measure "innate" intelligence, not experience or learning, other explanations were offered. First,
Northern blacks must have a greater mixture of white blood. Second, "the operation
of economic and social forces, such as higher wages, better living conditions, identical school privileges, and a less complete social ostracism, tend to draw the more intelligent Negro to the North" (Yerkes). Unmistakably, cultural and racial prejudice
played an important role in the "unbiased interpretation" of the test results.

However, biased interpretations merely followed in the footsteps of the tests
themselves. The IQ tests administered during this time period were highly biased.
People with better language skills and more education scored higher. The tests included questions that reflected cultural knowledge, which obviously does not reflect
innate intelligence.

Any numbers that rank people can be used in at least two ways. First, they can
identify inequalities, allowing social programs to be designed to alleviate the differences. Alternatively, the ranking can support the belief that inequalities are inherent
in society and that both upper and lower classes deserve their position. In hindsight,
we can see how intelligence testing was used by eugenicists.

Political Effects of Eugenics

What were the political effects of this emphasis on innate intelligence? The primary
effect was in providing a scientific basis for the Immigration Restriction Act of 1924,
which severely curtailed the number of immigrants who could come from countries
of "bad stock," primarily Southern and Eastern Europeans. President Calvin Coo-
lidge, when signing the bill into law, summarized the eugenic rationale of this legislation: "America must be kept American. Biological laws show . . . that Nordics
deteriorate when mixed with other races." During the 1930s, many Jews who wished
to emigrate from Eastern European countries could not do so because of this act and
were therefore abandoned to endure the ensuing Holocaust.

Eugenicists won another political victory with the passage of compulsory sterilization legislation. The conclusions drawn from the Army IQ data, along with the
growing sentiment among the populace that eugenics could offer a solution to social
problems, worked to support the passage of compulsory sterilization laws in twenty-
eight states by 1934. These laws targeted incarcerated criminals or individuals with
"degenerate diseases," who were then sterilized without their consent.

Many of these laws were soon challenged in court on the grounds that they were
cruel and inhumane. Advocates of compulsory sterilization wanted a test case to take
to the Supreme Court. Therefore, the most influential eugenicists gathered to propose

a new sterilization law for Virginia that, upon passage, could immediately provide such a test case.

A seventeen-year old woman named Carrie Buck, who was committed to the Virginia Colony for Epileptics and Feebleminded, had just delivered a baby out of wedlock. Her mother had already been committed to the same institution. IQ tests were given to Carrie and her mother, both of whom scored on the lower end of the moron scale. (Carrie had a mental age of nine, her mother eight.) Eugenic experts were called in by the State of Virginia to determine if the feeblemindedness could be hereditary. Without examining or even talking with the two women, the experts determined by pedigree analysis and other, more questionable approaches that Carrie's supposed feeblemindedness was indeed hereditary. Some of the "evidence" included testimony by a Red Cross worker that Carrie's seven-month-old daughter had "a look" about her that was "not quite normal."

Eventually, the case was heard before the Supreme Court, where an eight to one majority upheld the order for sterilizing Carrie Buck. Oliver Wendell Holmes, Chief Justice, wrote the majority opinion:

> We have seen more than once that the public welfare may call upon its best citizens for their lives. It would be strange if it could not call upon those who already sap the strength of the State for these lesser sacrifices . . . in order to prevent our being swamped with incompetence. . . . The principle that sustains compulsory vaccination is broad enough to cover cutting the Fallopian tubes. . . . Three generations of imbeciles are enough.

American eugenics reached its high point with this decision. However, at least in one country, eugenic policies would continue for another twenty years.

Eugenics in Nazi Germany

We should not end our exploration of eugenics without considering some aspects of the policies of Nazi Germany during the 1930s and 1940s. While Adolf Hitler's ideal of racial purity was an extreme eugenic position and was not directly supported by any scientific research, it must be recognized that Germany, like other European countries, was developing its own eugenics programs long before Hitler came to power. In the 1920s, an extensive system of Hereditary Health Courts was established to pass judgment on the suitability of marriages between people of "questionable" stock. In 1933, a eugenic sterilization law was passed, compulsory for all people—institutionalized or not—who suffered from (alleged) hereditary disabilities, including feeblemindedness, schizophrenia, epilepsy, blindness, drug or alcohol addiction, or physical deformities. Physicians were to report all "unfit" persons to the Hereditary Health Courts. Within three years of the passage of this law, about 250,000 people were sterilized, half of whom were classified as feebleminded.

Incidentally, when the German laws were passed in 1933, they were heralded by the American Eugenics Society as showing great courage and statesmanship. As we noted at the beginning of our review of eugenics, this society claimed as members many prominent scientists and politicians. Many of these eugenicists felt that Germany was far ahead of the United States in eugenics.

Of course, the German government went on to commit atrocities in the name of racial purity. The efforts were far beyond any ideas implemented by eugenicists in other countries and did not have the direct support of the scientific dogma of the day.

The Decline of Eugenics

The eugenics movement in the United States began to die in the mid-1930s. Three events were prominent in its decline. First, naive understandings of inheritance were replaced by more informed knowledge. Complex traits could be analyzed in greater detail, and the improved understanding of genetics did not support the claims of eugenicists. The same was true of psychological advances. "Feeblemindedness" was replaced by a host of other, more specific characterizations, some of which have been identified as genetic diseases, some as social disorders, and some as acceptable behaviors. Eugenicists oversimplified traits and used these traits to identify, characterize, and determine the fate of entire social classes. When it became obvious that the "trait" of feeblemindedness was either too simplistic or not real, hopes for a eugenic solution to undesirable behaviors evaporated. Second, the Great Depression of 1929 forced many people into poverty. Almost overnight, the "degenerate" class grew to contain many people who had previously held higher social status. The realization that circumstance, chance, and environment did in fact play a role in determining social class quickly followed. It became very difficult to blame poverty on the inheritance of "bad blood" during the Depression! Third, people responded with total revulsion against the practices of the Nazis before and during World War II.

ISSUES: THE BALANCE OF SOCIAL AND SCIENTIFIC BELIEFS

During the time of the eugenics movement, the accepted scientific dogma (germ theory and Mendelian genetics, leading to genetic determinism) supported the popular social belief that class structure was predetermined and unchangeable. Following the eugenics movement, in the post–World War II era, behavioral science, not genetics, was thought to provide the more complete explanation of human traits. During this period of "free will," seemingly everything about a person could be explained as a consequence of upbringing or social milieu. Nurture, not nature, was credited with being the main ingredient in the determination of human behavior.

During each of these two periods, prevailing scientific beliefs and popular social attitudes supported each other. But which came first, the social attitude or the scientific evidence to support it?

A Final Question

This brief look at eugenics helps point out that science is not simply a factual progression of knowledge. Social, political, and economic attitudes determine not only what scientific questions are pursued, but also how the results of these studies are interpreted and integrated into society. The ultimate power of genetics, as harnessed by the biotechnological revolution, is too great to ignore or leave to the experts alone.

SUMMARY

We have addressed the polygenic or quantitative traits of animals, especially humans, for several cautionary reasons: (1) We need to understand that genetic complexity goes far beyond Mendelian laws before we tackle genetic engineering and the nuts and bolts of biotechnology, (2) we need to see that human behavior is probably never determined solely by genetic makeup since environmental factors play a role, and (3) we should examine the historical fact that the idea championed by eugenicists that we would be better off without "degenerates" was based on very incomplete genetic information. In human society, the heritability of a trait is seldom straightforward. Genetic influences clash with environmental forces, forming some very complex problems—most of them currently unsolved.

As we arrive at the millennium, we might be a little smug about our extensive scientific knowledge, especially in genetics, compared to that of sixty to a hundred years ago. We should remind ourselves that the elite, the upper class, the wealthy, the educated, and, yes, even the university professors from about 1900 to the mid-1930s were the strongest supporters of eugenics and the social programs used to "squash" degenerates.

REVIEW QUESTIONS

1. Briefly outline the "nature versus nurture" debate, which has been viewed as a conflict between genetics and the environment. Why is it important to understand the interactions between genes and the environment?

2. Why is cat hair color characterized as polygenic?

3. Distinguish familial traits from heritable traits. Why, practically speaking, is it difficult to decide between the two?

4. Outline evidence that both multiple genes and environmental factors operate simultaneously to produce specific behavioral traits.

5. What is genetic determinism?

6. Define eugenics. What served as the basis for this movement? What kinds of people were termed "degenerates" by eugenic proponents? What are the dangers of applying eugenics?

7. What did the U.S. Army, World War I, and IQ testing have to do with eugenics?

8. List three reasons for the decline of the eugenics movement.

9. How were masturbation, castration, negative eugenics, and prisoners interrelated in the forceful establishment of eugenics?

10. List several agriculturally related polygenic traits (first of plant crops and then of domestic animals) that have strong economic ties. Does the relationship suggest anything concerning the attention paid to applied science versus basic research?

Biological Control

Previous chapters introduced the basic ideas of genetics, the science that describes the inheritance of characteristics. We learned much about how cells operate and the things that cells must do in order to grow and divide. Now we will explore the details of how cells accomplish their tasks.

The wonder is that all the cells of our body manage to function so smoothly. Of course, we do become ill occasionally, and we often experience muscle soreness following a strenuous workout. However, we are generally oblivious to our own cellular activity. We seldom even begin to appreciate the intricate system of controls that constantly operates in all our living cells, tissues, and organs.

This chapter introduces some of the mechanisms that control biological systems. In many cases, biotechnology involves our attempts to understand and gain control of specific cellular activities. Control is at the heart of biotechnology. Recall that biotechnology is the use of biological systems for practical purposes. To accomplish this, we must usually wrest control of some biological system away from the cell or organism.

REGULATION OF METABOLISM

In Chapter 1, we noted some aspects of metabolism, the chemical reactions that occur within cells. Here we will concentrate on the regulation of metabolism. Recall that we examined metabolism in terms of burning fuel in a controlled fashion so as to capture some of the energy. Three interrelated groups of reactions make up metabolism. First are the reactions geared toward releasing chemically useful energy. Second are reactions that break up large, complex molecules into smaller, simpler ones for use as building blocks. Third are reactions that assemble building-block compounds into useful biological macromolecules by various biosynthetic pathways. These groups of reactions might be going on at the same time in the same cell. Thus, cells must be able to precisely control metabolism in order to maintain the proper balance between energy extraction and macromolecular synthesis.

No simple diagram can explain how cells precisely control metabolic processes, but you already know that enzymes, or biological catalysts, facilitate virtually every reaction that occurs within cells (Chapter 1). Metabolism is controlled in large part through a precise tailoring of enzyme availability and activity within living cells.

Metabolic control becomes more complex when we consider large animals or plants composed of multiple cells, tissues, and organs. One organizational principle of complex creatures is that various types of cells make up functional tissues, which in turn make up organs. Specialized cells, tissues, and organs perform specific functions. For instance, skin cells offer protective, waterproofing functions, while some cells of the liver are engaged in activities such as food storage and enzyme synthesis. Humans have many different cells, tissues, and organs, each with specific functions that require exacting metabolic control.

For example, when you examine the human design you will note that humans do not need to eat continuously. Instead, we eat periodically and store some of the energy and the chemical building blocks for later use. The advantages are that we eat when we want and we can be active when it suits us. However, flexibility in eating and physical activity has some associated costs.

One cost is that we must have ways of storing fuel (energy or building blocks). One of the major means of storage is through the synthesis of **glycogen** (a large polysaccharide composed of glucose molecules). Remember that glucose is a major primary carbohydrate (fuel molecule) for us. When you eat fruit, cookies, bread, or cereal, large quantities of glucose are released within your digestive tract. The glucose travels through your blood to various tissues and organs. In response to rising blood-sugar levels, the pancreas releases the hormone insulin, which causes the liver to remove the excess glucose from the blood and store it as glycogen (Figure 7.1).

During a late afternoon workout, hours after lunch, you need extra energy. Where do your muscles get it? The glycogen that is stored in the liver is hydrolyzed to glucose, released into the blood, and then taken up by the muscles for use as fuel.

Your liver and muscles are intimately involved in the blood-glucose cycle, and each is designed to perform particular functions. Muscle cells are designed to contract and move our bodies. For that movement, they need huge amounts of energy, which they get from glucose molecules obtained from the breakdown of stored glycogen by the liver. As the muscle cells remove glucose from the blood, blood-glucose levels begin to fall. What happens when the level of blood glucose gets low? The liver regulates the level of blood glucose so that it does not become exceedingly low. When blood glucose drops below a certain level, the pancreas releases another hormone, glucagon, which causes the liver to break down stored glycogen to glucose, which can pass into the blood and be carried to working muscles. Notice that the previously stored glycogen is used as the fuel source directly, without your having to eat another meal. Later, following dinner, your glycogen stores will be replenished, preparing you for any energy requirement that may occur between then and breakfast.

Each living cell, tissue, or organ has specific metabolic functions that differ substantially in the same organism. What causes liver cells to behave differently from muscle cells? What allows the liver to export glucose to the blood, while muscle cells use it? Ultimately, the responsibility lies with different sets of enzymes. Liver cells

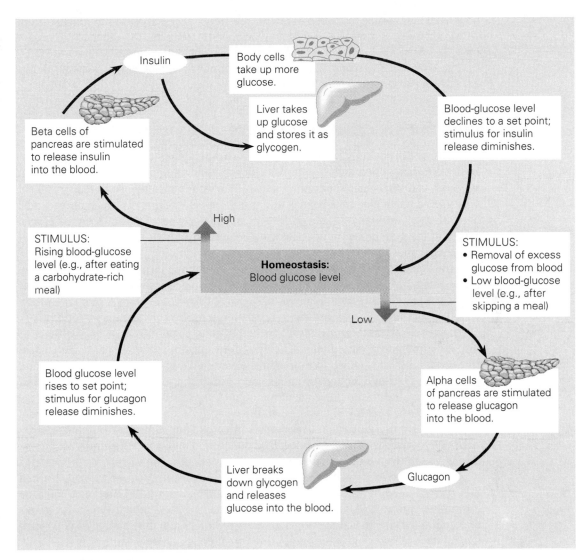

Figure 7.1 Regulation of blood glucose. Blood-glucose levels are regulated primarily by insulin and glucagon. When blood glucose is too high, the pancreas secretes insulin. This signals liver and muscle cells to take up the excess glucose and store it as glycogen until glucose is needed. Should the blood-glucose level drop below a certain point, the pancreas secretes glucagon. In response, the liver breaks down the stored glycogen and secretes the resulting glucose into the blood, raising blood glucose to a normal level.

have enzymes that can catalyze all the reactions necessary to take glucose from within the cells and move it into the blood. Muscle cells, on the other hand, do not have that capability, since it would be metabolically unfavorable for them to transport glucose into the blood. The existence of different sets of enzymes allows different tissues and organs to function uniquely.

CONTROL OF GENE EXPRESSION

Enzymes are products of gene expression. That is, genes coding for particular enzymes are transcribed and then translated into enzymatic proteins. We understand, however, that the cells of our bodies generally have identical genes. If a liver cell expresses one particular set of genes and a muscle cell a different set, then each cell must be using different subsets of the available genetic information. This is a crucial aspect of the genetics of multicellular organisms—different cell types use different subsets of genes.

The regulation of gene expression (gene regulation) in living cells can occur at many points before, during, or following transcription or translation. As an example of gene regulation, we will focus on bacterial (prokaryotic) transcriptional control, which is not only a common form of regulation but also the best understood.

Bacteria have interesting mechanisms for gene regulation that enable them to respond rapidly to environmental changes. Consider, for example, the bacterial species *Escherichia coli*, which lives in your intestine. When you drink a glass of milk, you introduce a large amount of milk sugar (a disaccharide called **lactose**) into your body. In response to the presence of abundant lactose, *E. coli* produces the enzyme β-galactosidase, which will break down the disaccharide into glucose and galactose. *E. coli* can use the glucose as an energy source and for molecular building blocks. How does *E. coli* produce β-galactosidase only when needed, that is, only when lactose is present in the immediate environment?

Years ago, two French scientists, François Jacob and Jacques Monod, learned that there were different groups of bacterial genes: **structural genes** that could be transcribed into mRNA for later translation into enzymatic protein, **operator genes** that controlled the structural genes, and **regulator genes** that indirectly controlled the operator. They termed a group of three such genes an **operon**.

In their experiments, Jacob and Monod found a protein, the ***lac* repressor protein**, in *E. coli* that could detect lactose, the substrate for β-galactosidase. When there was no lactose in the vicinity of *E. coli*, the *lac* repressor protein was continuously produced and actively inhibited transcription of the β-galactosidase gene. However, when lactose was present, β-galactosidase mRNA was produced. Jacob and Monod showed that the *lac* repressor protein inhibited transcription of the β-galactosidase gene by binding to the operator (Figure 7.2). When the *lac* repressor protein attached to the operator, RNA polymerase could not bind; thus, there was no transcription of the β-galactosidase gene and no β-galactosidase synthesis.

What was the role of the lactose substrate? The lactose acted as an **inducer molecule**. When lactose was abundant, it bound to the *lac* repressor protein, and that connection caused a change in the *lac* repressor protein's shape. The altered *lac* re-

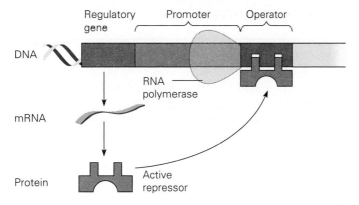

(a) Lactose absent, repressor active, operon off

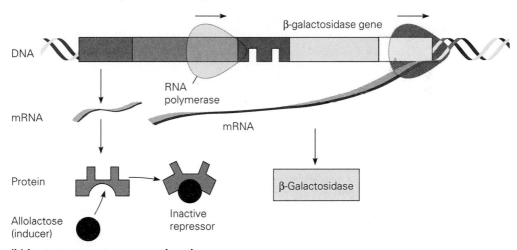

(b) Lactose present, repressor inactive, operon on

Figure 7.2 The *lac* operon in *Escherichia coli*. (**a**) When lactose is not present, the *lac* repressor protein binds to the operator and prevents RNA polymerase from transcribing the structural gene for β-galactosidase. (**b**) When lactose is present, it binds to the *lac* repressor protein, causing it to change shape and be unable to bind to the operator. RNA polymerase can then bind to the promoter and transcribe the β-galactosidase gene. The resulting mRNA is subsequently translated into β-galactosidase.

pressor protein then either was unable to bind to the operator or, if previously bound, released from the operator. As the operator was uncovered or unblocked, RNA polymerase could bind and transcribe β-galactosidase mRNA, which ultimately would be translated into the lactose-digesting enzyme, β-galactosidase. Once the lactose was broken down (removed from the bacterial environment), the newly formed *lac* repressor protein would reattach to the operator, once again shutting off β-galactosidase synthesis. Thus, we can see that by using operons bacteria efficiently control gene expression; that is, they synthesize specific proteins only when they are needed.

CELLULAR COMMUNICATION

There are several types of communication in which cellular needs are communicated to molecules (for example, enzymatic proteins) that, in turn, affect specific processes. No matter what signals might have caused the initial response, these signals must be communicated in order to have an effect. Cellular communication thus boils down to the interactions of particular molecules. Many interactions essentially operate as on/off switches—if the interaction occurs, then a process gets turned on (or off). If the interaction does not occur, then the process remains off (or on), as we saw with the *lac* operon. Other interactions are designed to operate like dimmer switches, generating different levels of response. What controls molecular interactions?

Specificity of Molecular Interactions

The specificity of many different biological interactions is immensely important. Enzymes are quite specific about the reactions they catalyze, the nitrogenous bases of DNA are specific about their pairings (A–T, C–G), and transcription factors interact with RNA polymerase to either initiate or, in some cases, inhibit transcription. How are such specific interactions created? What is it about the molecules involved that makes them interact specifically and selectively?

All molecules have a particular shape, determined by the chemical building blocks and the order in which they are linked. The different types of chemical bonds noted in Chapter 1 act in concert to determine the final shape of each molecule. Each shape is unique, and, most distinctly, molecular shape effectively controls the specificity of molecular interaction.

When we say that molecules interact, we really mean that they bind or attach to each other. Consider two bar magnets. When they are brought close to each other in the proper orientation, they will attract each other; if you are holding them, you can feel the magnetic attraction. That force is attempting to bring the two magnets as close together as possible. The same principle operates with molecules. For example, consider a molecule with a convex surface coated with positive charges. Another molecule might have a concave surface coated with negative charges. If these molecules come very close to each other with these surfaces in the proper orientation, strong attractive forces will draw them together, and the two molecules will bind to each other (Figure 7.3a).

The required conditions for two molecules to bind can be thought of as complementary-shaped surfaces with enough shared forces to hold them together. The binding of molecules can be strong, weak, or in between; the net total of all the individual interactions between the two surfaces will determine how tightly the two molecules bind.

How does molecular interaction result in a signal being transmitted? Imagine a pair of molecules with shapes that are not exactly complementary (Figure 7.3b). If these two molecules come near each other, they will tend to be drawn together due to the attractive forces. If the forces are sufficiently strong, the molecules will bind even though their shapes might not fit perfectly. The strength of the forces will "pull"

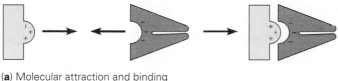

(**a**) Molecular attraction and binding

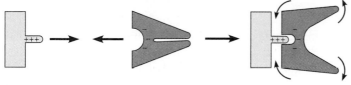

(**b**) Molecular attraction leading to changes in configuration

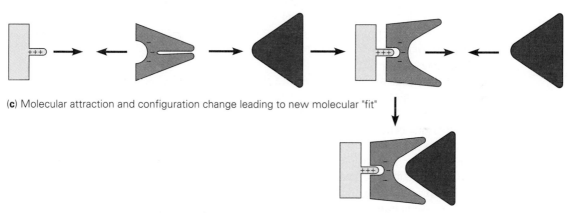

(**c**) Molecular attraction and configuration change leading to new molecular "fit"

Figure 7.3 Binding and communication. (**a**) Two molecules with complementary surfaces and forces can interact and bind to each other. (**b**) A molecule can be induced to change its shape when it binds to certain other molecules. (**c**) The resulting change in shape can lead to another binding event, which could not have happened before the first two molecules bound to each other. Such changes in shape and chains of binding can affect the activity of enzymes, altering the metabolism in a cell.

or "deform" the misshapen molecule so that it precisely fits the other molecule. Thus, the actual shapes of the molecules can be altered by binding to each other. Shape changes can have consequences elsewhere in the altered molecule. For example, as the molecular parts are connected, a change in one end of the molecule can sometimes cause a change at the opposite end. Further, if the shape of the opposite end changes, the resulting configuration might have a surface that is more likely to interact or bind to yet another protein (Figure 7.3c). A second binding event could cause yet another change and result in yet another molecular interaction. One of these molecules might be an enzyme, and the change caused by the interaction with another molecule might be enough to cause it to lose its activity or to become increasingly active. That is, alterations in molecular shape may be reflected in changes in an enzyme's ability to do its job.

This is a general picture of our current understanding of how signals are physically transmitted from one molecule to another. Shape changes, sometimes subtle, lead to a change in enzymatic activity. A series of interconnected shape changes is "transmitted" from one molecule to another, resulting in the alteration of a particular molecular target.

Signal Transduction

We have seen how signals move from one molecule to another, but how is a signal transmitted between cells that are far away from each other? How can chemical signals move from one location in an organism to a distant one? One answer is **signal transduction**.

Signal transduction can be accomplished in two ways. One way is through nerve impulses. Electrical signals travel along nerve cells and cause changes in other cells, ultimately producing a response through changes in molecular interactions. A second long-distance communication mechanism found within organisms involves hormones.

Hormones are macromolecules that are produced in one place in the body and travel through body fluids (usually blood) to other locations, where they cause a particular set of responses in the target tissues. Many hormones (such as insulin, testosterone, and estrogen) act within the human body and exert their effects by altering the activity of specific enzymes or by altering the expression of genes. Thus, hormones often serve as signals that set into motion the interactions that result in metabolic regulation. To examine how a particular hormone actually causes its effects, let's consider glucagon.

As we mentioned previously, glucagon is one of the hormones involved in regulating blood-sugar levels. It works in conjunction with insulin (generally, in an opposite way) to maintain an appropriate level of blood sugar (Figure 7.1). Insulin serves as a hormonal indicator that blood sugar is high. After a meal, insulin is made and secreted into the bloodstream. It tells cells, particularly muscle and liver cells, to take up and store the excess sugar as glycogen. Glucagon works in an opposite manner: It is made and secreted by cells of the pancreas in response to low blood sugar. Extremely low blood sugar impairs the functioning of the brain, with potentially disastrous effects. Thus, it is important that glucagon be able to signal when blood sugar is running low.

How does glucagon exert its effects? Glucagon's target cells in the liver have specific **cell surface receptors** on their cell membranes that can bind glucagon. The binding of glucagon causes a change in receptor shape inside the cell membrane (Figure 7.4). The shape change initiates a series of signaling events. A particular protein, called a **G-protein** becomes activated, is released, and binds to another enzymatic membrane protein called adenylate cyclase. Binding turns on this enzyme's activity, resulting in the conversion of ATP to a signal molecule called **cyclic adenosine monophosphate** (**cAMP**), which moves within the cytoplasm of the cell and binds to and activates another enzyme called a **kinase**. Kinases modify a whole set of other proteins by attaching phosphate groups to them. Once phosphates are attached, the proteins become either active or inactive. In the case of glucagon, the

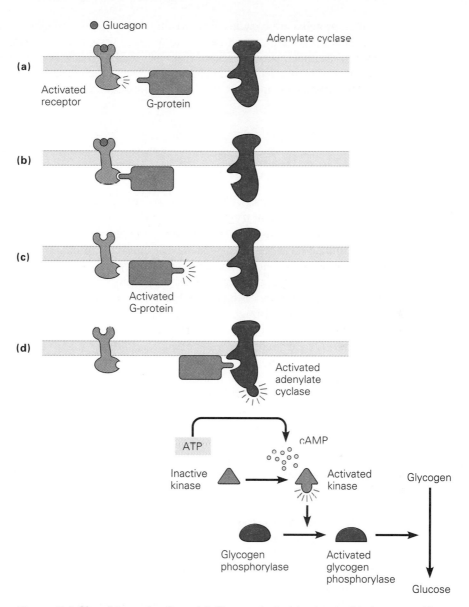

Figure 7.4 Signal transduction. (**a**) Glucagon in the bloodstream binds to a specific receptor in the membrane of a liver cell. (**b**, **c**) The binding of glucagon triggers a change in receptor shape, causing it to bind and activate the G-protein. (d) The activated G-protein then binds to and activates the enzyme adenylate cyclase, which in turn catalyzes the formation of cAMP from ATP. Cyclic AMP acts as a signal molecule and activates a kinase. The activated kinase then activates the enzyme glycogen phosphorylase, which finally breaks down glycogen into glucose.

modified proteins are enzymes involved in the metabolism of glucose. Enzymes that break down glycogen are activated, while those responsible for storing glucose as glycogen are inactivated. Thus, in response to the glucagon message, cells of the liver break down glycogen and export the resulting glucose to the bloodstream to raise the level of blood sugar. This signal transduction mechanism illustrates how hormones act.

DISEASE: THE LOSS OF BIOLOGICAL CONTROL

Metabolism that occurs within a cell is tightly controlled. Different metabolic schemes or sequences of reactions occur in the same cell at different times or in different cells at the same time. Since many of the metabolic activities are vital for continued cellular existence, precise metabolic control is of the utmost importance.

Do these metabolic processes ever go awry? Certainly, and the result can be **disease**. While there are many types of human disease, we will concentrate on two categories relevant to biotechnology. One set of diseases is caused by infectious microorganisms (fungi, bacteria, and protozoans) or the larger animal parasites such as tapeworms and flukes. Fungal diseases include histoplasmosis (*Histoplasma capsulatum*), coccidioidomycosis (*Coccidioides immitis*), and blastomycosis (*Blastomyces dermatitidis*). Some well-known bacterial diseases are tetanus *(Clostridium tetani)*, cholera (*Vibrio cholerae*), and tuberculosis (*Mycobacterium tuberculosis*). Two radically different but important protozoan diseases are malaria (four species of *Plasmodium*) and toxoplasmosis (*Toxoplasma gondii*). One specific condition caused by a larger animal parasite is taeniasis (*Taenia saginata*). These infectious organisms cause changes in the normal anatomy and physiology of cells or tissues and, in the extreme, cause inflammation or actually kill cells and destroy tissues. Microbial agents such as viruses act somewhat more subtly. For instance, papilloma viruses cause warts by entering specific living cells and affecting processes such as DNA replication and protein synthesis.

Another set of diseases, which we will consider in Chapter 10, includes those that arise from alterations (mutations) in the genetic material. Mutations may result in the production of defective gene products (proteins), as in sickle-cell anemia. Other consequences of mutation include the inhibition of protein synthesis (for example, in hemophilia) or an increase in protein synthesis (as in Burkitt's lymphoma).

In some genetic diseases, such as muscular dystrophy and Huntington's disease, the defects are present within reproductive cells (egg or sperm) and thus may be inherited. Mutations can also occur in nonreproductive cells, which sometimes leads to cancer. In a mature human, most cells behave like nerve cells, which generally do not divide once they have reached their mature state. In general, cell division occurs only when cells need to be replaced. For example, skin cells are regularly reproduced to replace those surface cells that normally die and fall off. However, in cells that have become cancerous, the normal checks and balances that control cell growth and division have been altered. Regardless of the cause, cancerous cells have lost their normal function and exhibit rapid, uncontrolled cell division. Orderly development and maintenance are crucial to all normal individual cells; when either of these activities is disturbed by infectious agents or genetic defects, disease may result.

BIOTECHNOLOGY'S CONCERN: USEFUL CHANGES IN CONTROL

Much of biotechnology is concerned with diseases such as cancer. New and quicker diagnostic tools are being developed using biotechnology, as are means of disease prevention (vaccines) and chemotherapeutic agents. Ultimately it is hoped that, through basic research, a more complete understanding of disease processes, including cancers, will provide the basis for new vaccines and more successful chemical treatment.

Remember that a loss of biological control is an underlying cause of disease. Generally, control is lost as a result of an infectious agent or a change in genetic information. In large part, biotechnology exists as an agent of designed changes in, rather than the loss of, metabolic control. Many biotechnologists are currently engaged in the research and development of products, processes, and sometimes whole organisms that will carry out specific metabolic functions. Usually, this means that normal control mechanisms can be bypassed or inactivated or that a new metabolic function can be added. As we will see in the next chapter, bacterial cells are routinely used to make products, notably pharmaceuticals, that cannot easily or inexpensively be made in any other way. For such production to take place, the normal bacterial metabolism must frequently be altered to allow the "new" metabolic reactions to occur. Often, this is accomplished by changing or adding genetic information in or to living cells.

Keep in mind as you read the rest of this book that biotechnology, as a science designed to produce products or processes, ultimately revolves around the ability to manipulate the normal biological control mechanisms operating within cells. The more we know about these control mechanisms, the more likely it is that we will be able to manipulate them in desirable ways.

ISSUES: ETHICS IN BIOTECHNOLOGY

Like all new technologies, biotechnology brings with it a host of ethical questions and dilemmas. We will discuss many of these ethical considerations as we discover applications of biotechnology throughout the rest of this book. But even now we can ask a general philosophical question: Are we playing God by developing ways to alter and control living things? For those who believe in the sanctity of life, any alteration of the "natural state" may seem to be immoral. Yet most of us readily submit to medical procedures and treatments that at least temporarily alter the normal functioning of our bodies. The question seems to be one of degree. People have been "controlling" living things for centuries. We plant what we want where we want. We breed all sorts of animals and plants, trying to obtain the most desirable traits, even if that means simply a particular appearance. Just look at the wide variety of pure-bred dogs and cats. In essence, we already exert a tremendous amount of control over many living things. Is our newfound biotechnological capability any different?

Some people would argue that biotechnology allows control through unnatural means. Traditionally, selective breeding uses natural methods to produce new breeds (varieties) of animals and plants. The offspring result from reproduction involving members of the same species. However, genetic engineering can create combinations that simply could never exist in nature; for instance, a cow's DNA would not combine with bacterial DNA through natural means. Some people believe we are tampering with evolution.

To think of ourselves as being able to sit outside evolution and affect it one way or the other is to presume that we are not ourselves a part of evolution. Frankly, everything we do is part of evolution. Survival of the fittest applies to us as well as to every other living thing. Is logging to produce lumber for house construction affecting evolution by not allowing the trees to grow as they would naturally? Perhaps not; after all, beavers cut down trees, and birds break off twigs to build their nests. We all participate in shaping and molding the world and its inhabitants. Simply developing a new way to reshape the world is not antievolutionary. Even so, that does not mean that it is necessarily moral or ethical.

There are no easy answers to questions involving ethical and moral choices. And while there are different ways of viewing the world, the development of biotechnology is radically different from most other human endeavors. It has tremendous potential to both benefit humankind and cause us colossal problems.

SUMMARY

We have gained an important understanding—that regulation or control of cellular metabolism is of the utmost importance for all life processes. The genetic information carried on chromosomes contains all the possibilities for what a cell can be and do. To that end, we have examined some of the ways that gene expression (transcription and translation) is regulated, including the details of how cellular communication is involved in regulatory events and how signal transduction facilitates communication. Having been reminded that on numerous occasions disease results from a loss of biological control, we saw that our attempts to manipulate control mechanisms are, in fact, the basis of biotechnology.

REVIEW QUESTIONS

1. Outline three general types of reactions that make up cellular metabolism. Indicate what seems to be the main purpose of each reaction type.

2. Compare the metabolic control functions of liver and muscle cells.

3. Diagram how DNA, mRNA, and enzymatic proteins are related in cellular metabolism.

4. Using a labeled diagram, explain how the following are interrelated in the transcriptional control of β-galactosidase production by *E. coli*: *lac* operon, lactose, glucose, galactose, β-galactosidase, *lac* repressor protein, operator, messenger RNA.

5. Describe how the following function together in a normal adult: glucagon, insulin, liver cells, muscle cells, and abnormally high, abnormally low, or normal levels of blood glucose.

6. How is modern biotechnology related in a practical way to infectious human diseases and cancer? (*Hint*: Consider prevention, diagnosis, and treatment.)

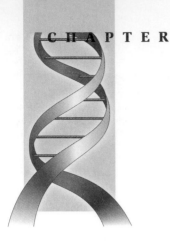

CHAPTER 8

Genetic Engineering: Tools and Techniques

We are now ready to move beyond descriptions of the way biomolecules, cells, and organisms work and begin to consider one way biology can be applied to technology. Scientific progress usually occurs much faster than most nonscientists realize. **Genetic engineering**, the ability to alter the genetic makeup of an organism, is no exception. Technological advances take place in the relative privacy of academic research laboratories and the secrecy of the corporate world. By the time new technologies are made known to the public, they are usually already being applied. Is this always in the best interest of society? Are there times when it would be better not to develop a particular technology or, at least, not to apply it in a certain way?

This chapter introduces the basic tools and techniques of genetic engineering. We have chosen a specific medical example to illustrate the technology, whose power and potential we will begin to see in this chapter. We'll also explore ways in which the development of this new and extremely powerful technology might be creating new problems even as it solves others.

DIABETES AND THE ROLE OF INSULIN

Most of us know someone who is diabetic. *Diabetes mellitus*, derived from the Greek for "siphon" and the Latin for "honey," is characterized by the excretion of large amounts of sugar in the urine. In its most serious form, the disease results from the body's inability to make sufficient **insulin**, one of the hormones involved in the regulation of blood sugar (glucose). As we have already seen, if the level of glucose is not controlled within certain limits, potentially fatal physiological complications can result. In response to high levels of blood glucose, insulin is secreted into the bloodstream from cells within the pancreas. Once insulin enters the blood, it travels throughout the body, signaling the appropriate tissues (primarily liver and muscle cells) to remove the excess glucose.

There are two major types of diabetes. Insulin-dependent, or juvenile-onset, diabetes affects children and is caused by a lack of insulin. Non–insulin-dependent, or adult-onset, diabetes results from a deficiency of insulin receptors. Diabetes is a manifestation of a malfunctioning communication system within the body. Recall from Chapter 7 how hormones are involved in communication. Insulin works in a manner similar to that of most other hormones. Insulin in the bloodstream recognizes specific insulin receptors on particular cells and initiates a cascade of reactions that results in the uptake of glucose. Regardless of the type of diabetes, the same principle is in effect: Despite high levels of glucose in the bloodstream, the proper signal does not trigger its uptake. Thus, individual cells begin to starve even though plenty of glucose is available.

In the absence of glucose uptake, cells begin to use fats as a primary source of energy. The catabolism of fats results in the synthesis of compounds known as **ketone bodies**, which are secreted into the bloodstream and function as an alternative source of energy for the brain (which cannot utilize fat directly). However, large amounts of ketone bodies are harmful. Not only do they cause the blood to become more acidic than normal (acidosis), but they also are toxic at high levels. One type of ketone body is acetone, which smells somewhat fruity. "Acetone breath" is a symptom of diabetes, because some of the excess acetone is transferred out of the blood in the lungs and exhaled. The large excesses of glucose and ketone bodies in the blood are excreted in urine. When they are excreted, they take with them huge amounts of water and salt, which can result in severe dehydration.

Muscle cells, which require large amounts of glucose for ATP synthesis, react to the perceived starvation caused by the inability to take up glucose by metabolizing protein. When proteins are used as fuels, a large amount of ammonia is produced, which must be processed by the body because it is toxic to humans. Normally, small amounts of ammonia are removed by its conversion to urea and subsequent excretion in the urine. However, under diabetic conditions, ammonia can rise to toxic levels. Thus, a simple breakdown in communication results in greatly altered metabolism in many cells. The long-term effects of these changes can include kidney failure, heart disease, cataract formation, brain damage, and ultimately death.

Non–insulin-dependent diabetes is treated primarily through diet and weight reduction. Thus, for our purposes, we will focus on the treatment of insulin-dependent diabetes, which involves genetic engineering. Insulin-dependent diabetes is treated by regular injections of carefully controlled amounts of insulin that serve to bring insulin levels to normal. As long as the dosage is carefully controlled and diet and exercise are carefully monitored, a relatively normal life-style can be maintained. Such a treatment requires a ready supply of insulin. Until recently, the needed insulin was obtained from the pancreas of cows or pigs. Massive numbers of this organ, which produces insulin, were obtained from slaughterhouses, and insulin was extracted from them.

A number of interesting observations can be made concerning the production of insulin by pharmaceutical companies. First, the incidence of juvenile-onset diabetes is increasing steadily each year, requiring an increased supply of insulin. Second, the availability of animal pancreases from slaughterhouses is decreasing due to reduced consumption of red meats. Third, the production of insulin is a large, lucrative market, particularly in the United States.

....

ISSUES: TREAT THE SYMPTOMS OR PREVENT THE DISEASE?

An affluent society such as the United States often has the luxury of developing technology that can mask the symptoms of medical problems without curing the underlying disease. Of course, this is often the only approach possible. Since there is no cure for the common cold, having drugs available that can alleviate the symptoms and allow us to function relatively normally is a wonderful development. On the other hand, sometimes new treatments are developed at great expense when different approaches could have been chosen that might reduce the incidence of the disease instead of simply treating the symptoms. Diabetes might be a case in point.

There is overwhelming evidence that many facets of our affluent life-style contribute to the incidence and seriousness of diabetes. Sedentary life-styles, diets high in sugar and fat, and obesity are all known to be risk factors for developing diabetes. One way to approach the medical problem of increasing incidence of diabetes might be to greatly increase efforts to modify our behavior in ways that would reduce obesity, improve our diets, and increase our exercise levels. However, such educational efforts are time-consuming and expensive to implement and generally have met with little success. Consider how difficult it has been to decrease the incidence of smoking, despite the fact that it is clearly harmful.

An alternative approach to treating diabetes is to develop better, less expensive drugs that allow diabetics to maintain at least some aspects of their unhealthy life-style. The development of a new way of producing insulin is one example. There is no doubt that the economics of the situation demanded a new source of insulin. At the same time, having a more readily available insulin source might influence personal choices concerning healthy life-styles. It certainly would seem less important to alter an unhealthy practice if drugs can alleviate the symptoms of the disease. How many people would stop smoking if there were effective treatments for the associated diseases, such as emphysema, bronchitis, and lung cancer? A significant number of people might choose to continue smoking or return to smoking.

Genetic engineering affords us remarkable opportunities to develop new medical procedures, both to treat symptoms and to eliminate disease. The choice of which path to pursue is governed primarily by economic forces, and any attempts to reform the fabric of health care in our country and reduce the cost of medical care will have to face questions about such choices.

....

In 1976, during a scientific conference to discuss the future needs of diabetics, it became clear that an alternative source of insulin was desirable, if not absolutely necessary. At about the same time, the technology called genetic engineering was developing. Could this new technology help produce insulin? Genetically altered bacterial cells that produce human insulin could be created, negating the need for animal pancreases. The intent was to create "bacterial factories" that would provide cheap, readily available sources of the hormone and would not be subject to the economic forces that determine the availability of livestock.

Thus, one of the first commercial genetic engineering ventures was conceived. If it was successful, a number of commercial difficulties would be avoided. First, genetic engineering would provide a ready source of product, since bacterial cells, in contrast to pancreases, are easy to grow in mass quantities. Second, a small but significant portion of the population has an allergic reaction to insulin derived from non-human sources, and producing human insulin from bacterial cells would alleviate this problem. Such a feat would have been considered pure fantasy twenty years earlier, but the rapid development of genetic engineering opened a multitude of new possibilities. Genetic engineers have learned how to manipulate genes by altering them, replacing them, or moving them into different organisms. The potential of this technology is just starting to be realized.

GENERAL CONSIDERATIONS

From this point on, the primary focus of the text will be to explore genetic engineering. In earlier chapters, we paved the way by describing how cells store and use information and how genes function to control the properties of cells. Now we will consider how genetic information can be manipulated. Genes can be manipulated for a variety of purposes, such as the production of pharmaceuticals, increases in crop yield, the creation of organisms that can help clean up toxic waste, and the conviction or acquittal of suspected criminals through DNA testing. Before we consider the specific techniques involved in moving a gene, or a specific piece of DNA, from one organism to another, we should focus on three general considerations that will be important for the rest of this chapter. After we have laid the groundwork, we will then be able to describe the cloning of the gene for insulin.

Biological Reactions Inside and Outside the Cell

First, we need to understand that scientists can cause some biological reactions to take place outside the cells in which they normally occur. For example, we know that transcription is accomplished through the action of RNA polymerase and other proteins. Through a variety of chemical manipulations, the RNA polymerase can be isolated from all the other cell components and purified. If the isolation is performed properly and the enzyme is stored and maintained under proper conditions, it will remain active when it is added to an appropriate solution containing DNA and RNA building blocks. In other words, we can actually perform a transcription reaction outside the cell, in a test tube. The ability to do this depends on the nature of the enzyme—it is not possible to do it for every enzyme involved in metabolism—but as time goes on more enzymes are becoming available as purified components ready for use in experiments outside the cell. The great advantage to performing such reactions outside the cell is that they can be performed with no interference or contamination from other biological molecules. The ability to work outside the cell is crucial to genetic engineering.

On the other hand, many processes cannot be performed outside the cell. One of the primary sets of reactions in this category is the replication of DNA. We can perform replication in a test tube, but it is somewhat inefficient compared to what the cell can accomplish. If we want to obtain copies of a particular DNA molecule, in many instances it is easiest to allow an intact cell to perform this function and then isolate the copied DNA. Under the proper conditions, cells can generate many identical copies of a DNA molecule. In fact, one common use of the word *cloning* is to denote the production of many identical copies, whether of a particular gene or an entire organism. To clone a gene generally means to use organisms to generate, through genetic engineering techniques, many copies of the specific gene in question.

Growing Bacteria in the Laboratory

The second major consideration is that scientists must use an intact organism as a host for the gene they are cloning. The most often used host is the common bacterium *Escherichia coli*, which is an inhabitant of our intestinal tract. One of the reasons *E. coli* is so popular as a host is that it is probably the most well-studied organism in nature. Years ago, scientists pursued studies of this bacterium because it is easy to maintain, grows quickly, and is quite amenable to study in the laboratory. Now, after many years of study, we know more about the inner workings of *E. coli* than about any other organism.

An understanding of how bacterial cells are grown and maintained in the laboratory will help in understanding later presentations of specific techniques. Bacterial cells can easily be grown by providing them with a suitable mixture of metabolic fuels and other nutrients. Such a mixture is called a **culture medium**. Commonly, a culture medium can be prepared as a liquid, or **broth**, into which bacterial cells can be placed for growth. The cells will grow in this culture medium, with some dividing as quickly as every fifteen to thirty minutes, until they produce too many waste products and/or run out of suitable nutrients. Broth cultures are usually used to rapidly produce large quantities of bacterial cells.

Bacteria can also be grown on a solid medium containing nutrients and the solidifying agent **agar**, an indigestible polysaccharide derived from red algae. Agar is hydrated, heated to 100°C, and sterilized. Just before the agar solidifies, it is poured into petri plates and allowed to harden. Petri plates containing solidified culture media are referred to simply as **plates**.

To grow bacteria on plates, a small number of cells are spread onto the surface of the plate. The cells will rest on the surface of the agar. Bacteria can extract nutrients from the culture medium and grow, but since they are not in broth they cannot move. Therefore, a particular cell will grow and divide wherever it happens to end up after being spread onto the plate. The progeny will also be unable to move, so the reproducing cells will pile up on and around one another at this spot. After twelve to twenty-four hours of growth, the piles of cells will be large enough (the minimum required number is about 1 million cells) to be visible to the naked eye (Figure 8.1a, b). Such a pile of cells is called a colony. Every cell in an iso-

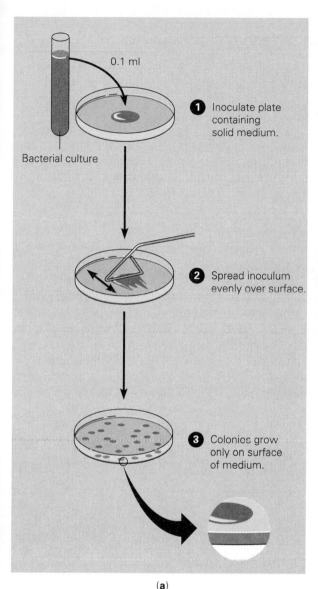

(a)

Figure 8.1 Growing bacteria. Bacterial cells can be grown in either a liquid medium or a solid medium. (**a**) On a solid medium, as individual cells grow without being able to move, they ultimately give rise to many genetically identical cells that pile on top of one another, forming a colony. (**b**) Colonies can be seen with the naked eye. Each colony represents what was originally a single viable cell.

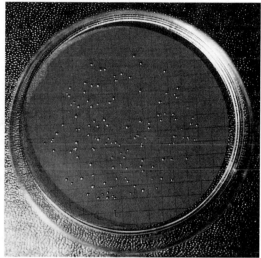

0.1 ml

1 Inoculate plate containing solid medium.

2 Spread inoculum evenly over surface.

3 Colonies grow only on surface of medium.

Bacterial culture

lated colony is a descendant of one single cell—thus, all the cells within a single colony are identical.

Cells are usually grown on plates when separating individual cells is important. Usually, we will be looking for differences between the cells in one colony and those in another. Growing the cells on a solid medium allows us to separate the individual cells and examine their unique properties.

Detecting What Happens to Individual Molecules

The third consideration is really a problem that faces most scientists when they attempt to describe the world in which we live. As we know, DNA is a very small molecule, and individual genes within DNA are smaller yet. How can we even contemplate moving specific pieces of DNA from one organism to a test tube and into a bacterial cell? How will we know whether we have DNA at all, let alone a specific gene? Much of science ultimately deals with difficulties involving unseen forces or minute molecules that cannot be seen directly. Therefore, the actual work of scientists can become focused on developing indirect ways of seeing forces and molecules. In this case, we need to know when we have a particular gene. Much of the rest of the chapter deals with techniques designed to show whether or not a particular piece of DNA is present.

THE CLONING AND EXPRESSION OF GENES

Any attempt to clone a gene can be reduced to a number of smaller operations. We will consider five primary steps that must be carried out in order to successfully obtain insulin from genetically engineered bacteria:

1. Obtain the gene for insulin from human DNA.
2. Insert the gene into bacterial cells.
3. Select cells that have the desired gene.
4. Induce the bacterial cells to express this "foreign" gene in order to produce insulin.
5. Collect and purify the final product, insulin.

Obtaining the Insulin Gene

In the simplest terms, our task is to try to find the small piece of DNA that codes for insulin from among the rest of the DNA that makes up the human genome. The insulin gene is only a tiny fraction of the DNA present within a cell. How can we find this genetic needle in the DNA haystack?

The most common method begins by isolating mRNA rather than DNA. One reason for this is that the mRNA that is transcribed from a particular gene is more abundant than the gene itself. Insulin production is the primary function of the **β-cells** of the pancreas. If you examined the mRNAs present within β-cells, you would find a large amount of insulin-encoding mRNA. The insulin mRNA is much more abundant within the mRNA population than is the insulin gene within the DNA and contains the same genetic information as the original gene, in a slightly different chemical form. If we could turn the insulin-encoding mRNA molecules of β-cells into DNA, then we would have large amounts of the insulin gene available.

Our approach, then, is to use pancreatic β-cells as our starting material for the isolation of mRNA. Eukaryotic mRNAs contain a particular chemical modification, a

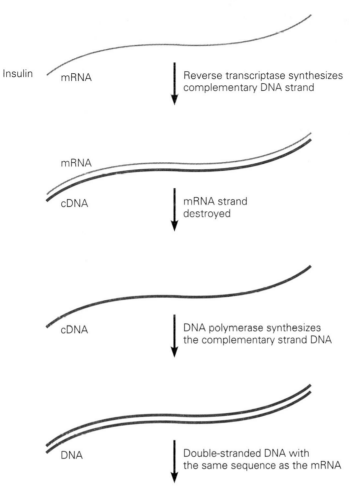

Insulin mRNA

Reverse transcriptase synthesizes
complementary DNA strand

mRNA

cDNA

mRNA strand
destroyed

cDNA

DNA polymerase synthesizes
the complementary strand DNA

DNA

Double-stranded DNA with
the same sequence as the mRNA

Figure 8.2 Obtaining genes using mRNA. Isolated mRNA serves as the template for reverse transcriptase, which synthesizes a strand of complementary DNA (cDNA). The mRNA template is then destroyed, leaving a single-stranded cDNA molecule. The cDNA is then used as a template by DNA polymerase to synthesize a complementary strand. The net result is the formation of a double-stranded DNA molecule that contains the same genetic information as the mRNA. In effect, a gene has been made using the information from the mRNA.

series of adenosines called a poly A tail, attached to the end of each mRNA. This chemical modification allows for the separation of mRNAs from all other nucleic acids once cells are broken open. Thus, we can obtain a population of mRNAs from a sample of cells.

After this population of mRNAs is isolated, they are converted into RNA-DNA hybrids by an enzyme called **reverse transcriptase**. This enzyme uses the mRNA as a template and synthesizes a complementary strand of DNA that base-pairs with it (Figure 8.2). The RNA strands are then chemically degraded, leaving single-stranded DNA molecules. These DNA molecules are the exact complements of the information present in the original mRNA and hence are referred to as **cDNA** (copy DNA or complementary DNA). The cDNAs are then converted into double-stranded DNA molecules through the action of the DNA polymerase. The resultant double-stranded DNA molecules contain the coding sequences found in the genes from which the mRNAs were transcribed. The more abundant a particular mRNA was to begin with, the more abundant the cDNA for that product will be. Since the insulin mRNA is abundant in β-cells, by performing the described procedure we have, in effect, multiplied the insulin gene.

At this point, we face the first of several selection problems. Ultimately, we need to be able to identify and select a particular cDNA that encodes insulin. Even though we have, in effect, amplified the insulin gene, we still face the problem of determining which of the double-stranded DNA molecules carry the information for insulin. One way to perform such a selection is to see which DNA molecules can actually function as insulin genes. This requires that the DNA be inside a cell, which leads us to the next step.

Inserting Genes into Bacterial Cells

Unfortunately, it is virtually impossible to take small pieces of DNA and insert them into bacterial cells so that they will direct the production of protein products. First, DNA does not enter a cell easily. Second, bacterial cells do not tolerate DNA that does not form a circular structure; linear pieces of DNA are rapidly destroyed. Third, even if we could get such DNA into cells intact, it will not necessarily contain the proper signals for transcription, translation, and replication. In other words, the DNA might not be used by the bacterial transcription, translation, and replication systems. Thus, we must solve a number of problems in order to successfully insert DNA into bacterial cells.

Let us first address a number of difficulties related to the nature of linear DNA itself. With few exceptions, bacteria will rapidly degrade DNA that is not circular. Genes obtained using cDNA molecules are linear, not circular. Therefore, if they were inserted into bacterial cells, they would be rapidly degraded.

However, if the genes were incorporated within other DNA molecules that can exist within bacteria, they could then be safely introduced into bacteria. Such a "carrier" DNA is called a **vector**. DNA vectors are used to carry genes of interest as they are moved between test tube and cell. There are many types of vectors, but the most common type is a **plasmid**. Plasmids are circular pieces of DNA, found in many different kinds of microorganisms that are replicated by the cell but exist separate from the chromosomes of the organism (Figure 8.3a). The plasmids used as cloning vectors contain only a few genes, sometimes only one. Often these genes encode proteins that allow the organism to grow in the presence of antibiotic drugs. Plasmids possess a number of traits that are relevant to cloning:

- They are small, easily manipulated DNA molecules.
- Genes encoding antibiotic resistance will prove useful during subsequent steps in the cloning procedure.
- Plasmids can be readily transferred into cells and are also easily isolated.
- Plasmids contain signals for independent replication within cells.

Thus, any DNA inserted into a plasmid will be replicated along with the plasmid DNA.

By using a plasmid vector to carry genes of interest, we overcome the difficulties

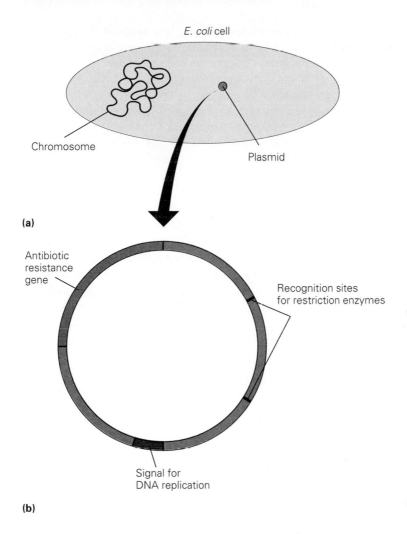

Figure 8.3 Plasmids. (**a**) Plasmids are found in many bacteria. They exist separate from the chromosome and are replicated independently within the cell. (**b**) Those plasmids useful as cloning vectors generally contain one or more genes encoding antibiotic resistance, as well as the signal for replication. In addition, different recognition sequences for restriction enzymes are located throughout the plasmid.

of replication and degradation of linear DNA molecules. We also have a means of determining whether the plasmid is present within cells: testing for antibiotic resistance. A number of other problems can be overcome by careful planning at the stage of vector selection and design.

How are genes of interest inserted into vectors? We need a cutting and joining process that can break open a circular plasmid and allow a linear piece of DNA to become a part of the circle. The discovery of enzymes that can do this is what has led to the rapid development of genetic engineering. This ability to "cut and paste" DNA sequences is integral to the process.

Restriction enzymes are, in essence, DNA scissors that cut double-stranded DNA (Figure 8.3b). The process of cutting DNA with restriction enzymes is called **restriction digestion**. An important feature of this type of enzymatic activity is that each different restriction enzyme recognizes and cuts at only one very specific DNA sequence. Commonly used restriction enzymes cut at specific six-base sequences. Over three hundred different restriction enzymes have been identified.

Another important feature of restriction enzymes is that many of them cut the two DNA strands in a staggered fashion, resulting in small, single-stranded regions that overhang the ends of the DNA (Figure 8.4). The sequence of bases in an overhang is characteristic of the particular restriction enzyme used to cut the DNA. These short, single-stranded sequences are available to base-pair with a complementary sequence should one be available; hence, they are often referred to as **sticky ends**. In many instances, it is these sticky ends that direct the formation of new combinations of DNA.

ISSUES: WHY DO RESTRICTION ENZYMES EXIST IN NATURE?

Restriction enzymes are found in a wide variety of microorganisms and play a vital defense role in bacterial cells. We've already encountered viruses that can infect bacterial cells. What if a cell contained a set of restriction enzymes that cut up "foreign" DNA into pieces and ultimately destroyed it as soon as it entered the cell? The entry of such DNA would be "restricted." That would certainly provide for an effective defense.

But what about the organism's own DNA? Wouldn't it also be restricted by the enzymes? Organisms that contain restriction enzymes also contain special modification systems that chemically alter the host DNA such that it is not recognized by the enzymes and therefore is not attacked. Most of these systems modify, by the addition of methyl groups, one or more bases in the recognition sequences of the restriction enzymes. The modified sequences are not recognized by the restriction enzymes, and thus the cell's own DNA is protected. The action of the restriction-modification system is very important to the life of the bacterium and also provides an essential tool—restriction enzymes—for use in genetic engineering.

Let us now examine the specifics of inserting genes into plasmid vectors. The overall process is outlined in Figure 8.5. Millions of copies of the desired plasmid are first isolated from bacterial cells. These plasmids are digested, or cut, by a particular restriction enzyme, resulting in the creation of linearized plasmids. These molecules are no longer circular. Note that the restriction digestion results in the generation of sticky ends.

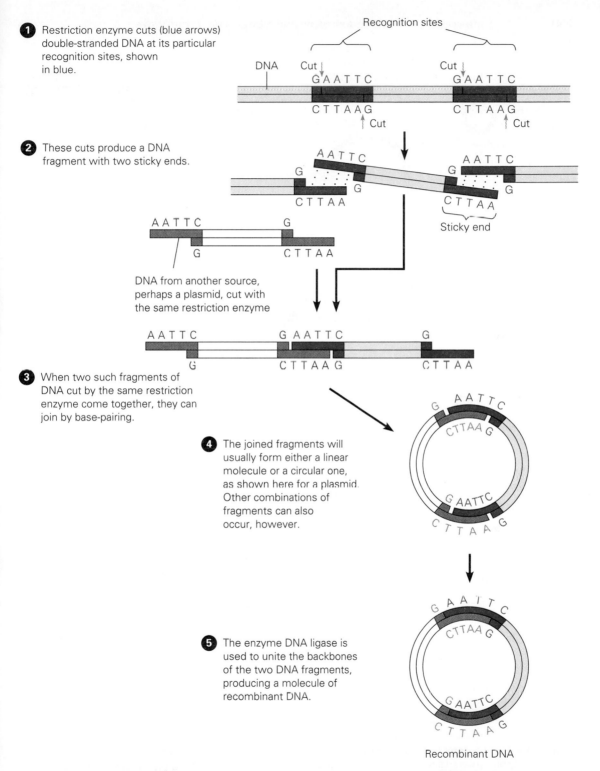

1. Restriction enzyme cuts (blue arrows) double-stranded DNA at its particular recognition sites, shown in blue.

Recognition sites

DNA

Cut ↓
G A A T C
C T T A A G
↑ Cut

Cut ↓
G A A T C
C T T A A G
↑ Cut

2. These cuts produce a DNA fragment with two sticky ends.

A A T T C
G
G
C T T A A

A A T T C
G
G
C T T A A

Sticky end

A A T T C
G
G
C T T A A

DNA from another source, perhaps a plasmid, cut with the same restriction enzyme

A A T T C
G
G A A T T C
C T T A A G
G
C T T A A

3. When two such fragments of DNA cut by the same restriction enzyme come together, they can join by base-pairing.

4. The joined fragments will usually form either a linear molecule or a circular one, as shown here for a plasmid. Other combinations of fragments can also occur, however.

G A A T T C
C T T A A G
G A A T T C
C T T A A G

5. The enzyme DNA ligase is used to unite the backbones of the two DNA fragments, producing a molecule of recombinant DNA.

G A A T T C
C T T A A G
G A A T T C
C T T A A G

Recombinant DNA

Figure 8.4 How restriction enzymes are used in constructing recombinant DNA molecules.
The numbered steps depict how restriction enzymes and DNA ligase can be used to join DNA fragments from two different sources.

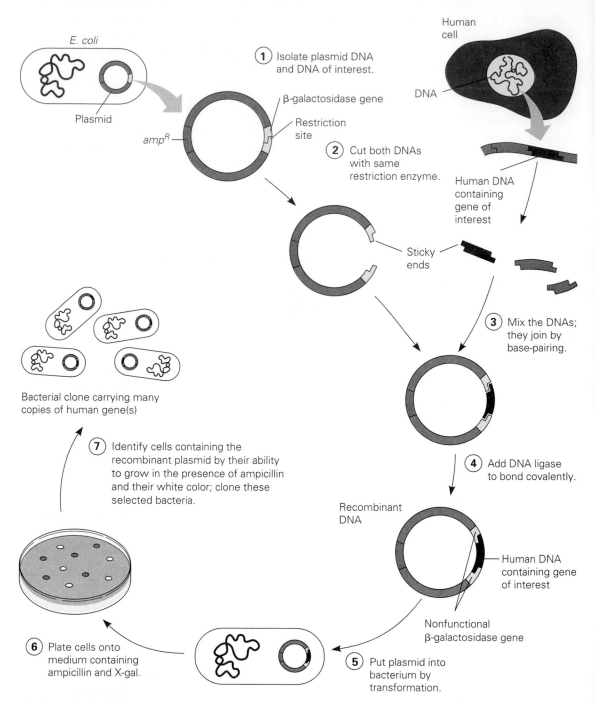

Figure 8.5 How a gene may be cloned in bacterial cells. The numbered steps depict the steps involved in cloning a gene in bacterial cells. The selection of desired bacterial cells, using ampicillin and X-gal, is described in Figure 8.6.

The genes obtained from cDNAs now need to be prepared for cloning. The genes do not have the same sticky ends as the linearized plasmids, so small artificial pieces of DNA that contain the same restriction enzyme recognition sequence as the plasmid vector are added to them. These **linkers** are attached to the genes by using DNA ligase, an enzyme that joins DNA fragments within cells. After the linkers are attached, the genes are then digested by the appropriate restriction enzyme to generate the same sticky ends present on the linearized plasmid vectors.

We now have a population of DNA molecules with sticky ends as well as linear plasmids with the same sticky ends. All of these molecules are now mixed together. As we know, single-stranded DNA will readily base-pair with a complementary sequence, forming a double-stranded DNA molecule. Since the genes and the plasmids have the same sticky ends, some of these ends will base-pair, or **anneal**, as illustrated in Figure 8.5. The base-paired ends can then be connected by DNA ligase, which joins the individual DNA strands by constructing new phosphate-sugar bonds. Why the sticky ends are so important now becomes clear: It is much more difficult to join two pieces of DNA together if they do not readily base-pair to each other.

Several different possible combinations can form during the annealing process. The new combination of DNA pieces is referred to as a **recombinant DNA** molecule, since the two pieces have combined to form a new molecule. Genes can be inserted into plasmids in this way; this is the desired result. Unfortunately, other combinations are possible as well: The plasmid can recircularize, two plasmids can form a single circle, or several genes can join together within a single plasmid. Out of the entire mixed population of molecules, some recombinant and some not, we are interested only in those plasmids that have a gene inserted into them. For simplicity, a gene that is placed into a plasmid is often referred to as an **insert**, and a recombinant plasmid is called a **construct**. Thus, we began with a population of mRNA molecules, a large number of which contained the genetic information for insulin; we created a population of insulin genes; and we now have a collection of recombinant plasmids (constructs), some of which contain inserts but even fewer of which contain a gene for insulin as an insert.

The constructs can now be inserted into bacteria. The easiest way to do this is through a process known as **transformation**. Bacterial cells treated with a solution of calcium chloride somehow allow DNA to enter. Cells treated in this manner are said to be **competent**. Competent cells are mixed with a solution containing the population of plasmids, some of which contain the desired insert, the insulin gene. After a period of time, some of the cells will take up the DNA. Usually, conditions are adjusted so that a single cell rarely takes up more than a single plasmid. In this manner, some of the constructs containing the gene for insulin are inserted into bacterial cells.

Selecting Cells with the Desired Gene

We now begin the challenging task of selecting a particular bacterium containing the desired recombinant plasmid. How can we find a cell that has taken up this particular construct? It will help to consider several selection processes, each of which narrows down the number of candidates.

We can first distinguish cells that have taken up any plasmid DNA, regardless of whether it is a recombinant molecule, from cells that have not. During transformation, not every cell takes up DNA—in fact, the majority do not. Since we are interested only in cells that have taken up DNA, it is useful to eliminate the unwanted bacterial cells from the population. Our ultimate goal is to find a bacterial cell that has taken up a plasmid vector containing the insulin gene. Anything we can do to help reduce the population of potential candidates is helpful.

Recall that the plasmid vectors we used contained at least one gene for antibiotic resistance. Many commonly used plasmids have genes for resistance to ampicillin, a variant of penicillin. Normally, ampicillin kills *E. coli*. But if *E. coli* has a plasmid containing an ampicillin resistance gene (amp^R), it will grow in the presence of ampicillin and form a colony. The gene for ampicillin resistance encodes a protein that is secreted from the bacterial cells and degrades the ampicillin before it can affect the cells. Cells that do not contain this gene cannot degrade the ampicillin and are subject to its toxic effects. All we have to do to obtain the bacteria that have taken up a plasmid is grow them in a medium containing ampicillin. Only those cells that have plasmids can grow and form visible colonies (Figure 8.6).

ISSUES: ANTIBIOTIC-RESISTANT BACTERIA— EVOLUTION IN ACTION

The development of antibiotics has been one of the crowning achievements of medicine. Antibiotics have saved countless lives and have relieved the suffering of almost everyone at one time or another. It is disturbing to note that an increasing number of infectious agents have become resistant to treatment with antibiotics. While this issue has little to do directly with genetic engineering, it does involve plasmids and antibiotic resistance genes, and so discussing it here seems appropriate.

It has become fairly commonplace for hospitals to experience outbreaks of infections that do not respond to antibiotic treatment. Antibiotic resistance arises because of mechanisms that exist in bacterial populations for the exchange of genetic information between different cells. When we considered evolution, we saw that natural selection operates to ensure the survival of the fittest organisms. A bacterial population infecting humans is subjected to a wide variety of selective pressures, including antibiotic treatments. Any individual bacterium that is resistant to the particular drug being used will survive and produce offspring that are also resistant. In addition, bacteria can sometimes share genetic information, effectively allowing non-resistant cells to become resistant. The more frequently a particular antibiotic is used, the greater the selection that is applied to the bacterial population. Eventually, enough cells become resistant to render the drug ineffective.

Our great dependence on antibiotics has recently been called into question for exactly this reason. The greater the use of antibiotics, the greater the potential for the development of resistant bacteria. Limiting the use of antibiotics to situations in which they are absolutely required can help alleviate this problem. Unfortunately,

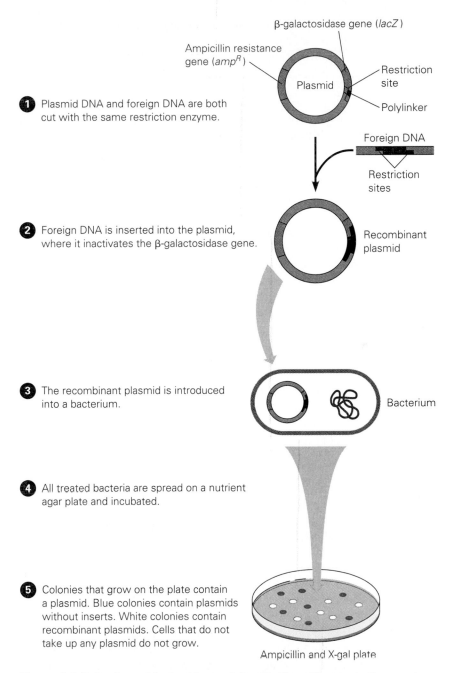

Figure 8.6 Selection of desired bacterial cells. Two different selections can be performed on the same plate: one that will select for cells harboring a plasmid (ampicillin selection) and another that will distinguish between cells containing a recombinant plasmid and those that have no DNA inserted into the plasmid (blue-white screening).

the only way to fight infections by resistant bacteria is to develop new drugs that can still fight the infection. And, as we will see later, developing new drugs is an increasingly difficult, time-consuming, and expensive proposition.

••

Now we want to distinguish between those bacterial cells that have taken up a plasmid with an insert and those that have plasmids with no insert. Recall that a plasmid that has been digested with a restriction enzyme may recircularize without taking in an insert. It would be helpful to identify candidates that have an insert so that we can reduce the population even further before trying to determine whether we have the insulin gene.

There are several ways to accomplish this selection. One is to use what is known as blue-white screening, which utilizes the properties of the bacterial enzyme **β-galactosidase**. This enzyme breaks down the disaccharide lactose into its constituent sugars, glucose and galactose. However, the enzyme will also act upon several artificial substrates, including a compound known as **X-gal** (5-bromo-4-chloro-3-indoylgalactoside). X-gal is normally clear and colorless in solution. However, once it is cleaved by β-galactosidase, it forms a dark blue compound. Thus, X-gal can serve as an indicator of the presence or absence of β-galactosidase. If the enzyme is present within bacterial cells growing on a plate, a blue colony will be formed. If the enzyme is absent, the colony will be white (Figure 8.6).

To take advantage of this system, two things are necessary. First, the host cells to be transformed must be missing the β-galactosidase enzyme. In other words, they must be mutant cells lacking a functional copy of the β-galactosidase gene. Second, we must utilize a plasmid vector that contains a gene for β-galactosidase. Recall that we utilized a particular restriction enzyme to cut the plasmid in order to receive an insert. What if that restriction site were contained within the β-galactosidase gene? The plasmid would then be cut within the gene for the enzyme. If the plasmid simply closes up again during the formation of recombinant molecules, an intact β-galactosidase gene will again be formed. If the β-galactosidase–deficient host is transformed by the plasmid, it will give rise to a blue colony. If, on the other hand, the plasmid takes up an insert, the β-galactosidase gene will be disrupted, causing it to be inactive (Figure 8.6). When β-galactosidase–deficient host cells are transformed by such plasmids, they give rise to colonies that are white. It is the white colonies in which we are interested, since only they have plasmids that contain inserts.

The two selection processes we have described can be applied at the same time. Following transformation of cells with the mixed plasmid population, the bacteria can be plated onto agar containing both ampicillin and X-gal. Of the colonies that grow, only the white ones are of further interest. These colonies must have taken up a plasmid, because they are ampicillin-resistant, and they must contain an insert within the plasmid, because they do not have an intact β-galactosidase gene.

Now it is time to make the final selection, that of selecting bacteria that contain the insulin gene insert. Again, as with many genetic engineering procedures, there are several ways to accomplish this. We will examine one method that introduces some powerful tools.

Antibodies are proteins that function as part of the disease-fighting immune systems of animals. When our bodies are invaded by foreign objects or organisms, a number of responses are elicited, including the production of large amounts of specialized antibody proteins. Antibodies recognize and attach to the foreign invaders, tagging or identifying them as foreign. Other systems within our bodies either destroy the antibody–foreign body complexes or remove them from circulation. We will have much more to say about antibodies and the immune response later. It turns out that antibodies can be purified and used outside the cell. The property that makes them useful for our purpose is that each particular antibody recognizes one particular structure and only that structure. In other words, antibodies bind only to a single kind of molecule. As we will learn later, a very important branch of biotechnology deals with producing mass quantities of specific antibodies for use in medical and research applications. At this point, we only need to know that antibodies can be made that will bind to almost any specific target molecule and that they can be purified.

Thus, we can produce antibodies that will specifically recognize insulin and use these antibodies to select the desired bacterial cells. We will need to label our antibodies somehow in order to be able to visualize them later. One common way to do this is to make them **radioactive**.

Radioactive atoms are unstable atoms that release radiation as they decay. Many different atoms are radioactive, ranging from some forms of uranium and plutonium that release large amounts of energy to forms of hydrogen and carbon that release small amounts of energy as they decay. A number of radioactive atoms are forms of atoms used in biological molecules, such as hydrogen, carbon, sulfur, phosphorus, and iodine. The radioactive forms of these atoms behave exactly like the nonradioactive forms until they decay. As they decay and release energy, they are transformed into different atoms.

The property of radioactive atoms that makes them so useful to biotechnologists is the release of energy. Most of the radioactive atoms used in biology give off radiation that can be detected either by a Geiger counter or by the exposure of X-ray film. The sensitivity of these methods is so great that minute quantities of radioactive materials can easily be detected. Thus, by incorporating some radioactive atoms into biological molecules, the molecules can be more easily detected or "visualized." We will encounter numerous uses of the radioactive labeling of molecules; for now we simply need to understand that it is relatively easy to chemically add radioactive atoms to antibody molecules, thus tagging or labeling them for detection.

Let's return now to our agar plate containing colonies of candidate cells that have inserts. To identify cells that have a plasmid containing the insulin gene, we can use a technique called replica plating (Figure 8.7a). In this procedure, a sterile piece of velvet cloth is pressed face-down on top of the agar plate containing the bacterial colonies. Some bacteria from each colony are transferred onto the velvet. The velvet is then pressed against a fresh agar plate to transfer some of the cells. The transferred cells form colonies that are oriented exactly like those of the original (master) plate. Once the colonies have grown on the replica plate, the cells are broken open by holding the plate over some chloroform. This releases the contents of the cell, including insulin. If we then lay a nylon filter on top of the replica plate and press down on it, the contents of the cells will be transferred to the nylon in the same positions as on

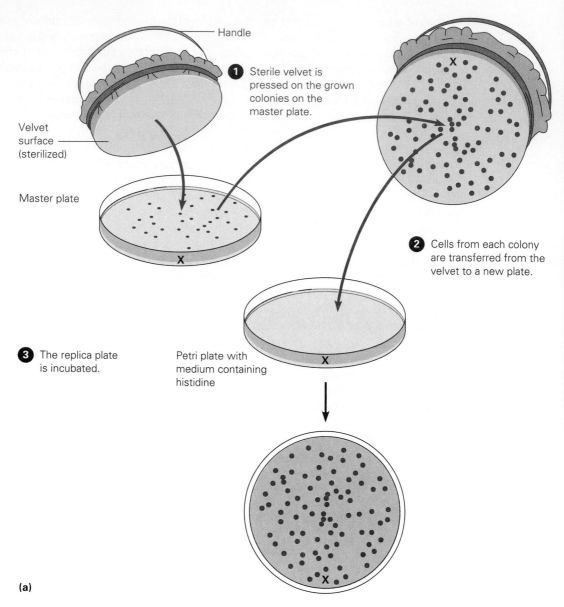

(a)

Figure 8.7 Procedure for identifying cells that produce insulin. (**a**) The first step in this process is replica plating. A piece of sterile velvet cloth is pressed down on an agar plate containing colonies, to pick up cells. The velvet is then pressed against another agar plate (replica plate) to transfer the cells, which then grow and form colonies in the same positions as on the original plate (master plate).

the plate (Figure 8.7b). What we have done in this step is create a replica of the original colonies, except that this replica has the cell contents, including proteins, from each colony attached to it.

The nylon filter is now placed in a dish and coated with radioactively labeled in-

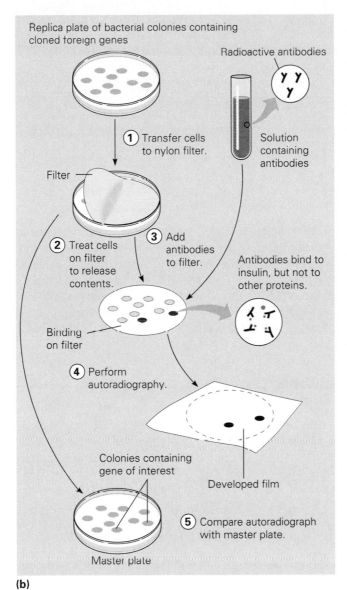

Replica plate of bacterial colonies containing cloned foreign genes

Radioactive antibodies

1 Transfer cells to nylon filter.

Solution containing antibodies

Filter

2 Treat cells on filter to release contents.

3 Add antibodies to filter.

Antibodies bind to insulin, but not to other proteins.

Binding on filter

4 Perform autoradiography.

Colonies containing gene of interest

Developed film

5 Compare autoradiograph with master plate.

Master plate

(b)

Figure 8.7 (b) Cells from the replica plate are transferred to a nylon filter and then broken open to release the cell contents. Insulin-specific antibodies added to the filter will bind only to insulin. Since the antibodies were made radioactive, they will expose X-ray film. Dark spots on the developed film identify colonies containing cells that synthesize insulin. By comparing the developed film to the master plate, we can determine the position of insulin-containing colonies on the plate.

sulin-specific antibodies. The antibodies will bind specifically to insulin and to nothing else. After the passage of a period of time to allow this binding to occur, the unbound antibodies are washed away. The only antibodies left will be those that are bound to insulin. Since the antibodies are radioactive, we can visualize them by exposing the nylon filter to X-ray film. The detection of radioactive materials using X-ray film is known as **autoradiography**. The radioactivity will cause a dark spot on the X-ray film, indicating the presence of insulin. By comparing the exposed X-ray film to the original master plate, we will know exactly which bacterial colonies contain cells that are producing insulin.

What have we accomplished? We began with a population of bacterial cells, some of which contained the gene for insulin. We eliminated those cells with no plasmid at all and those cells with nonrecombinant plasmids. We then utilized antibodies to tell which cells were producing insulin.

Unfortunately, our procedure has a few difficulties that we need to iron out. For one thing, our final selection presumes that bacterial cells containing a gene for insulin will actually synthesize insulin. However, we know that it is unlikely that eukaryotic genes can be expressed in prokaryotic cells, for the following reasons:

- The genes may contain introns, which cannot be removed by prokaryotic cells.
- The genes may contain eukaryotic transcription and translation control signals, which cannot be utilized by prokaryotic cells.

Even if the bacterial cells are producing insulin, there is a possibility that the insulin will be immediately degraded—it is known that some foreign proteins (although not all, by any means) are degraded very rapidly inside bacterial cells. It is also possible that the insulin protein will be toxic to the bacterial cell or harmful in some other way. Predicting the effects of such a foreign protein on a bacterial cell is usually difficult. Thus, for a number of reasons, insulin might not be produced within the bacterial cells, at least given our methods thus far. Solving these technical difficulties leads to our next step.

Inducing the Expression of Insulin in Bacterial Cells

Let's try to develop ways in which the problems we identified above can be avoided. First, we'll address the differences between eukaryotes and prokaryotes in expressing genetic information. These differences can prevent the expression of human insulin in bacteria.

Conceptually, the presence of introns is a severe barrier to the use of eukaryotic information in prokaryotic cells, since prokaryotes lack the cellular machinery to recognize and remove introns. How can we overcome this barrier? Recall that we initially isolated the genetic information for insulin from mRNA. One reason for this approach was that it allowed us to begin with a sample enriched for insulin information. However, we can now recognize another reason. Because the eukaryotic mRNA was processed, that is, the introns were enzymatically removed, the gene obtained from that mRNA will have no introns. Thus, by selecting mRNA as our starting material, we have eliminated the problem of introns altogether! If we had cloned the insulin gene from chromosomal DNA, it would have contained introns, making it impossible to use this gene in prokaryotes. At the present time, using mRNA as a starting material is the only method we have of overcoming the problem presented by introns.

We are still left with the problem of initiating transcription of the inserted insulin gene. Recall that transcription initiation requires a promoter. What if our plasmid vector contained a prokaryotic promoter adjacent to the restriction site used to create the recombinant plasmid? By inserting the insulin gene into the restriction site,

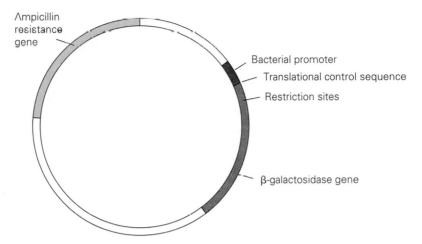

Figure 8.8 A plasmid expression vector. Plasmid expression vectors are used in bacterial cells to drive the production of foreign proteins. The plasmid has been genetically engineered to contain bacterial transcriptional and translational control sequences. Also, recognition sequences for restriction enzymes (restriction sites) are present within the β-galactosidase gene, near the promoter, allowing foreign genes to be inserted in proper position relative to the promoter.

we bring it under the control of an appropriate promoter. Prokaryotic translational control sequences could also be provided in a similar manner. Plasmids that contain such control sequences are referred to as **expression vectors** (Figure 8.8), because no matter what gene is inserted into the plasmid it can be transcribed and translated using these "artificial" signals.

What about the chances of the insulin being degraded? Fortunately, our knowledge of bacteria is such that we know about the systems and enzymes (**proteases**) that degrade proteins, and we can genetically engineer bacterial cells that are deficient in this activity. Foreign proteins have a much higher chance of being synthesized in such deficient host cells than in normal cells. If we suspect that rapid degradation of the expressed protein is a problem, we can use a protease-deficient strain as the host and increase the protein content.

If the foreign protein is toxic or otherwise harmful to the host cells, little can be done other than trying a different host cell. A number of species of bacteria besides *E. coli* can be used as hosts. A variety of eukaryotic cells can also serve as hosts.

In summarizing our efforts to control the expression of the desired gene in the host cells, we find that the necessary actions have all been part of careful planning, primarily involving a specifically designed expression vector that contains necessary DNA control sequences. By initially choosing such a vector, we assure ourselves of a high likelihood of gene expression.

One difficulty has not yet been mentioned. Eukaryotic cells often modify proteins extensively after they are synthesized. Usually, this involves the addition of carbohydrate or lipid groups to the protein or removal of a portion or portions of the polypeptide chain. Unmodified proteins are usually not functional, and prokaryotes cannot perform these modifications.

With regard to insulin, the problem stems from the way in which it is normally produced and stored in the body. It is initially made as a larger, inactive protein that is then converted to a smaller, active form. Think of what must happen in order for insulin to be released in the bloodstream. First, it must be synthesized by the β-cells of the pancreas. Then it must be stored in some manner so it is not released until it is needed by the body. In response to high blood sugar, it must be secreted from the β-cells into the blood. These different steps are accomplished by creating different forms of the protein (Figure 8.9). **Preproinsulin** is the name given to the form synthesized first. It is inactive and thus will not function as insulin. However, it includes the signals necessary to direct the protein to storage. As it is being stored, the part of the protein responsible for signaling storage is specifically removed, resulting in a smaller protein known as **proinsulin**. Proinsulin is the form that is stored; it is also inactive within the body. Active insulin is formed only when the body signals a need for insulin. Then proinsulin is activated by specific processing: Two specific cuts are made in the protein chain that result in the removal of a length of protein. The remainder of the protein, which is now actually two separate protein chains held together by bonds between specific amino acids, is active insulin. It is this form that is secreted into the bloodstream.

What does this have to do with our cloning project? Consider how we obtained the insulin gene. It is the preproinsulin form that is encoded by insulin mRNA. It is also this form that is produced during protein synthesis in β-cells. Thus, if the gene for insulin, which we obtained from mRNA, is cloned into bacterial cells and expressed as protein, the bacterial cells will produce the inactive preproinsulin, rather than active insulin.

Is there a way to get the cells to make active insulin instead of inactive preproinsulin? We know that the active form of insulin is composed of two distinct polypeptide chains joined by disulfide bonds. If each of these chains could be made separately and then properly assembled with the other, active insulin could be made. This is the way recombinant insulin was finally produced. A gene encoding information for one insulin chain was cloned into bacterial cells. The chain was expressed and purified from these cells. The information for the other chain was cloned separately, ultimately providing purified second chains. The two strands were mixed outside the cells and bound to each other in proper fashion to form the active insulin molecule. Thus, cloned insulin was really produced as a mixture of two separately cloned products.

How was the genetic information for each individual chain obtained? There is no "gene for the first chain"—there is only a single gene for preproinsulin, the protein that gets processed into active insulin. Cloning the information for just one chain required a method of obtaining this information separately from the rest of the insulin information.

At the time, the study of the insulin protein was much more advanced than knowledge about its gene. The sequence of amino acids that comprise the two chains was known. This information was used in conjunction with the genetic code to work backward. That is, the known sequence of amino acids was used to determine the

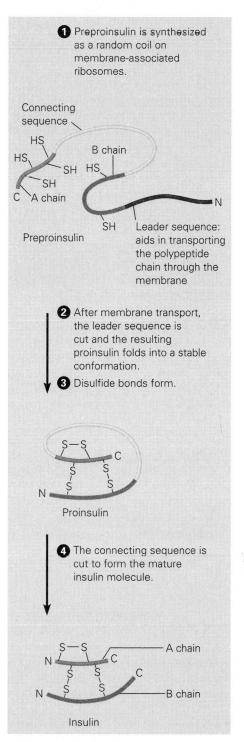

1 Preproinsulin is synthesized as a random coil on membrane-associated ribosomes.

Connecting sequence

HS
HS
SH HS
SH
C A chain
B chain

Preproinsulin

SH Leader sequence: aids in transporting the polypeptide chain through the membrane

N

2 After membrane transport, the leader sequence is cut and the resulting proinsulin folds into a stable conformation.

3 Disulfide bonds form.

S—S
C
S   S
S   S
N   S   S

Proinsulin

4 The connecting sequence is cut to form the mature insulin molecule.

S—S
N   C
S   S
S   S
N   S   S   C

A chain
B chain

Insulin

Figure 8.9 Processing of insulin. Processing of insulin occurs in two steps. Preproinsulin contains a segment that directs the protein to storage compartments. Once there, this segment is removed. Following stimulation of the cells, another internal segment is removed, leaving two small chains bound together to form active insulin.

corresponding DNA sequence of each chain of insulin. The degeneracy of the genetic code actually allows many potential DNA sequences to code for insulin. Another technical advance, the chemical synthesis of genes, finally made it possible to complete the cloning of insulin.

Let's follow the synthesis of the gene for one of the insulin chains (Figure 8.10). We start with a solid support in a small column. We know that C is the first base of the desired sequence, so C residues are chemically attached to the support. A solution containing the next nucleotide, A, is passed through the column. The As will bind to the Cs already attached to the support. The As are chemically blocked to prevent one A from attaching to another A. Excess As are rinsed away. A deblocking solution is then passed through the column, removing the blocking groups from the As. The process is repeated: Attach the next nucleotide to the growing chain and pass successive solutions through that will add the next nucleotide in the sequence to the previous one.

Today, the entire process of synthesizing a piece of DNA has been completely automated. Computer-controlled machines will take the sequence that is given to them and go through the ordered reactions, producing the desired DNA molecule. These **DNA synthesizers**, or "gene machines" as they are often called, are important tools in biotechnology.

The chemistry has been perfected to such a degree that virtually all of the growing chains will have the proper sequence. Although very few improper reactions occur, they ultimately limit the size of a DNA chain that can be produced. If each and every reaction were absolutely perfect, then the process could be continued indefinitely. But the chemistry is not perfect. Let's assume that it is 99 percent accurate. That means that a chain 100 nucleotides long being synthesized would have on average one mistake. Most of the chains would thus contain a mistake—not a very efficient process. The chemical reactions need to be on the order of 99.9 percent accurate to give rise to suitable numbers of 100-nucleotide chains. Fortunately, the pieces of DNA necessary to code for insulin fall within this range.

In our example, a piece of DNA encoding one insulin chain was synthesized and cloned using the techniques described above. Another piece of DNA encoding the other chain was synthesized and cloned into a different group of cells. The products of each cloning were then mixed to give rise to active insulin.

Purifying the Product

Once all the other steps of cloning have been completed, it is usually necessary to purify the desired product. This is really a more general chemical question: How can a particular molecule be separated from all the other molecules, similar and dissimilar, present in the cell?

Purification techniques are important in biotechnology. In fact, almost every effort in biotechnology involves some degree of purification. We will take this opportunity to explore one approach that can be used to separate molecules from each other.

In order to separate two molecules, we need to exploit differences between them with respect to their properties. For example, one might be small, while the other is

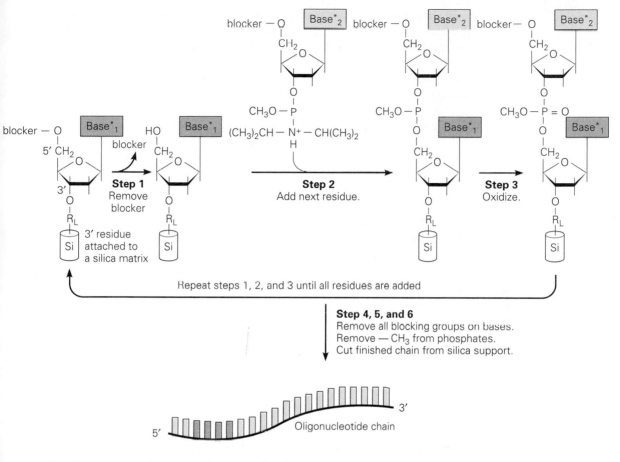

Repeat steps 1, 2, and 3 until all residues are added

Step 4, 5, and 6
Remove all blocking groups on bases.
Remove — CH_3 from phosphates.
Cut finished chain from silica support.

5′   Oligonucleotide chain   3′

*Reactive groups on all bases are blocked by chemical reagents.

$$R_L = -\overset{\parallel}{\underset{O}{C}} - (CH_2)_2 - \overset{\parallel}{\underset{O}{C}} - NH - (CH_2)_3 - O - Si$$

Figure 8.10 The operation of an automated DNA synthesizer. A column containing a solid support (a silica matrix) is set up in such a way that various solutions can be pumped through. Individual nucleotide solutions are prepared such that one end of each nucleotide is blocked by a chemical agent, dimethoxytrityl (DMTr). The blocking agent allows the next nucleotide to be added to bind to nucleotides attached to the support but prevents it from chemically reacting with free nucleotides. As the first nucleotide solution is pumped through the column, nucleotides bind to the support. Following the removal of unbound nucleotides, a deblocking solution removes the DMTr from the nucleotides. The deblocking step allows the addition of the next nucleotides. When the next nucleotides to be added are then passed through the column, they will react with the unblocked nucleotides. The newly added nucleotides are then deblocked to allow the addition of the next nucleotide. These steps are repeated until the DNA molecule is complete. Then the entire DNA chain is removed from the support and exits the column.

large. One might have a positive charge, and one might have a negative charge. One might dissolve readily in water, while another might not. A number of properties can be exploited, but differences in size, charge, and solubility are most often used.

If we want to separate molecules based on their size, we must force the molecules to act differently in some way that results in their separation. Chromatography techniques are designed to do just this. Imagine that we have a mixture of only two types of proteins, one of which is large and the other small. We can employ size-exclusion chromatography (also known as gel-filtration chromatography) to separate them. To begin, we obtain some special chemical beads that have been made with small holes in them. The holes are big enough to accommodate the smaller proteins but not the large. If we place the beads in a column and pass the solution of proteins through the column, the large proteins will pass directly through, since they will not interact with the beads. On the other hand, the smaller proteins will spend some amount of their time in the holes and thus will take longer to come out the bottom of the column. If we collect the material as it comes out the bottom of the column, we can easily separate the two different-size groups of proteins (Figure 8.11).

This is only one example of a technique employed to purify molecules. Often, several different approaches must be used in order to completely remove all traces of undesirable contaminants. We will encounter other methods of separation in later chapters.

ISSUES: SAFETY IN GENETIC ENGINEERING

One of the more apparent public reactions to the development of genetic engineering has been concern about safety issues. What if a genetically engineered organism is released into the environment? Could it cause rampant disease or result in the destruction of environmental habitats? We will encounter many specific examples of safety concerns as we discuss various applications of genetic engineering in the rest of the book, but it seems logical to ask some basic questions at this time.

The primary safety issue concerns the effect genetically engineered organisms might have on the environment or on the population if they were released. Let's look at a hypothetical example. **Cholera** is a serious disease that results in many deaths each year, primarily in tropical, underdeveloped nations. It is caused by a bacterial infection. The bacteria release a particular protein, the cholera toxin, that causes massive diarrhea. If untreated, an infected person might die of dehydration and other complications.

Scientists studying cholera might want to learn more about the cholera toxin. Since it is a protein, it could conceivably be produced in large quantities by cloning its gene into *E. coli* cells. Using the techniques outlined in this chapter, such cloning should be fairly straightforward, resulting in *E. coli* cells that are capable of producing the cholera toxin.

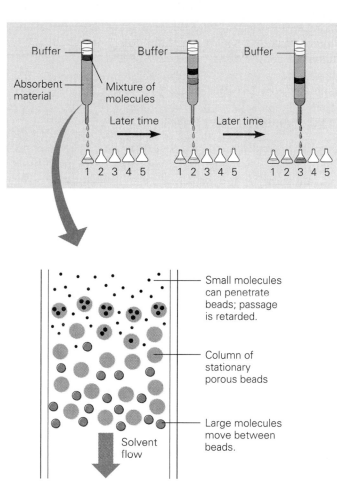

Figure 8.11 Gel-filtration chromatography. Molecules can be separated according to their size by gel-filtration chromatography. Porous beads are prepared that contain holes of a particular size that only small molecules can enter. As a mixture of molecules is passed over the beads in a column, the smaller molecules will spend time inside the holes in the beads, whereas the larger molecules will not. Thus, the larger molecules will exit through the bottom of the column more quickly than the smaller molecules. By collecting the material as it comes out the bottom, molecules of different sizes can be separated from one another.

Since *E. coli* cells are normally found within our intestinal tract, it is conceivable that some of these genetically engineered bacteria could find their way into the intestines of those who work with the bacteria. Someone might forget to wash his or her hands after handling cells, for instance. Could the cells take up residence in the intestines and release the cholera toxin?

This hypothetical example raises fears about the use of genetic engineering. In response to this particular example, handling *E. coli* cells engineered to contain the cholera toxin involves no more inherent danger than handling the cholera-causing bacteria in the first place. However, we can imagine more extreme scenarios. Safety issues need to be considered, and many rules and regulations have been imposed by the federal government pertaining to the use of genetic engineering, with the aim of reducing hazards associated with this new technology. But as informed citizens we need to protect our own health and environment. We should all become involved in efforts to pursue technological advances safely.

....

CLONING CONSIDERATIONS

Thus far we have concentrated on a particular cloning application, the production of insulin in bacterial cells. Along the way we have introduced basic ideas of recombinant DNA and some of the specific tools and techniques involved. However, a wide variety of such tools and techniques is available. We will now examine some of these alternatives.

Choosing a Cloning Vector

We have examined the general use of plasmids as cloning and expression vectors. A number of different plasmids are available, and some of them are optimized for specific purposes. However, plasmids have limitations, the most serious of which is that plasmids cannot carry very large genes. Should it become important to clone large genes, another type of vector—a bacteriophage—could be used. Bacteriophages are viruses that specifically infect bacteria. Two examples are the **lambda (λ) phage** and the **M-13 phage**.

Lambda consists of a DNA molecule contained within a phage head (Figure 8.12a). The head is made of proteins that form a shell or coat surrounding and protecting the DNA. The head is attached to a protein tail, which is responsible for binding to a bacterial cell and delivering the phage DNA into the cell. When lambda infects a cell, it first binds to the outside of the cell. The tail piece then contracts, injecting the DNA into the cell. Under certain conditions, the phage DNA then begins to replicate and directs the synthesis of new phage proteins. Ultimately, new heads and tails are formed, and the replicated lambda DNA is packaged into these heads, forming new phage particles. Finally, the bacterium bursts (lyses), releasing all of the newly formed phages, which go on to infect other cells. This is referred to as the **lytic cycle** of the phage, because it results in cell lysis.

Sometimes, under different environmental circumstances, the phage DNA will not replicate and produce new phage. In this case, the phage DNA is integrated into the bacterial chromosome. In its integrated state, it is referred to as a **prophage**. Under these conditions, the phage genes are turned off, preventing production of new phages. This **lysogenic cycle** does not kill the bacterial cell. Instead, the prophage is replicated along with the bacterial chromosome.

It turns out that a large portion of the lambda DNA is not required for successful infection and can be removed and replaced by other DNA (Figure 8.12b). This DNA will then be carried along with the phage DNA and will be replicated and passed on as if it were a normal part of the phage DNA. The greatest advantage of using lambda phage as a cloning vector is that it can be used to clone larger pieces of DNA than can plasmids.

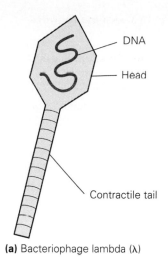

Figure 8.12 Bacteriophage lambda (λ). (**a**) The structure of bacteriophage λ. (**b**) Bacteriophage λ can be used as a cloning vector for DNA fragments up to 15,000 base-pairs (15 kbp) in length.

DNA

Head

Contractile tail

(**a**) Bacteriophage lambda (λ)

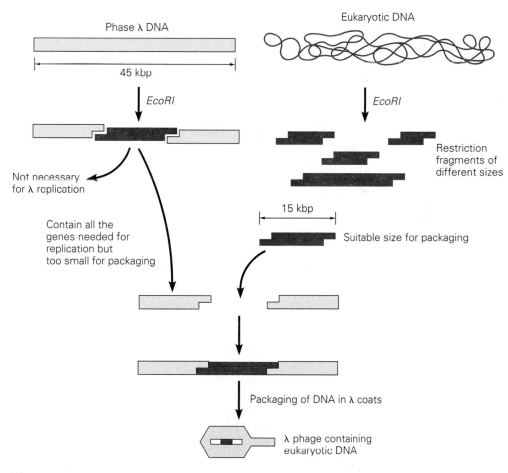

Phase λ DNA

Eukaryotic DNA

45 kbp

EcoRI

EcoRI

Not necessary
for λ replication

Restriction
fragments of
different sizes

Contain all the
genes needed for
replication but
too small for packaging

15 kbp

Suitable size for packaging

Packaging of DNA in λ coats

λ phage containing
eukaryotic DNA

(**b**)

The lysogenic and lytic phage cycles are of great utility in cloning. The lysogenic life cycle provides bacterial host cells that carry foreign DNA. These bacteria can be treated like any other, which means that they can be used in various selection schemes. Alternatively, when it is necessary to produce copies of the cloned DNA, a prophage can be induced to enter the lytic cycle, resulting in the production of many copies of the cloned DNA and phage vector.

However, phage vectors such as lambda are generally not as convenient to work with as plasmids, often requiring more manipulations to accomplish the same task. Thus, for "routine" recombinant DNA experiments, plasmids are usually the vectors of choice. However, for instances in which larger genes must be manipulated, lambda or M-13 may be a logical candidate for use as a vector.

Choosing a Host

We made a case for the utility of bacterial cells as a "universal" host. Their ease of growth, facile manipulation, and ready ability to take up DNA make them ideal hosts for many cloning experiments. Their use is limited, however, particularly when modified eukaryotic proteins are the desired end product. In short, if the goal of a study is to learn more about features of a gene or protein unique to eukaryotes, using a eukaryotic host is probably a better choice.

The closest thing to a universal eukaryotic host organism is a yeast such as *Saccharomyces cerevisiae* (baker's yeast). Transforming yeast cells is relatively easy, and they possess the machinery necessary for the modification of proteins. On the other hand, it is sometimes necessary to clone a gene into a more specific type of eukaryotic cell, for example, into a mammalian cell or a plant cell. Appropriate methods have been developed to enable a wide variety of eukaryotic cells to act as hosts. In later chapters, we will encounter some of the circumstances in which different hosts must be used.

SUMMARY

We spent most of this chapter introducing specific techniques that allow scientists to move genes from one organism to another. No matter what type of genetic alteration might be desired, the same basic ideas apply. The gene of interest must be isolated or synthesized and then moved into a host organism. Cleverly designed vectors are used for shuttling genes from the test tube into an organism and back again, as well as for providing the necessary signals for expression in the particular host organism being used. Selection techniques are applied in order to choose a cell that contains the desired genetic information. Ultimately, if the end result is to be the preparation of a specific product, purification schemes must be designed that will effectively remove contaminating molecules. The same general techniques that we have described are applicable to almost all genetic engineering efforts.

R E V I E W   Q U E S T I O N S

1. Describe what is meant by "genetic engineering."

2. What is diabetes mellitus and how is it related to the hormone insulin? Why do persons with uncontrolled, untreated diabetes have "acetone breath"?

3. The current biotechnological method of producing insulin was given extensive coverage in this chapter. State two factors that led to the development of human insulin production by bacteria.

4. How is *Escherichia coli* used for the cloning and production of human insulin?

5. Explain how each of the following is involved in laboratory cultivation of bacteria: *Escherichia coli*, petri plates, agar and nutrients, colonies.

6. When employing bacteria to clone eukaryotic genes, why is it advantageous to start with mRNA rather than chromosomal DNA?

7. List, in order from the first, the five general steps required to obtain insulin from genetically engineered *E. coli*.

8. List the advantages and disadvantages of using a plasmid and a bacteriophage as a cloning vector.

9. Outline the steps or activities required to insert specific genes into vectors such as bacterial plasmids.

10. How do restriction enzymes make it possible to create recombinant DNA molecules?

11. How is the bacterial enzyme β-galactosidase involved in the selection of bacterial recipients of engineered inserts?

12. Why are antibody molecules that are specific for insulin tagged with radioactive atoms?

13. Describe the problems faced by genetic engineers in getting a eukaryotic gene (such as the one for insulin) expressed in a prokaryotic (bacterial) system.

14. Describe how each of the following is related to obtaining pure active insulin from engineered *E. coli*: preproinsulin, proinsulin, insulin, and "gene machines."

15. Lambda is sometimes used as a cloning vector. How does that usage involve the lysogenic cycle and the lytic cycle?

CHAPTER 9

More Genetic Engineering: Analysis of DNA and Applications

Chapter 8 introduced the basics of genetic engineering, concentrating on putting a human gene into a bacterial cell to produce insulin. The logic for doing this is straightforward—it provides a ready source of human insulin for the treatment of diabetics. We became familiar with the major tools and techniques necessary to accomplish this, including vectors, restriction enzymes, and transformation of bacteria. However, there is more to genetic engineering than just moving genes from one organism to another. In this chapter, we will describe techniques that allow DNA sequencing and also explore how these techniques, as well as the general cloning techniques already described, can be applied in a variety of situations.

Along the way, we will encounter controversy. Some techniques are poised to provide us with an overwhelming amount of information about ourselves, information we may not be prepared to handle. Other techniques allow us to consider changing the genetic makeup of humans. Some of the applications of recombinant DNA technology have met with strong opposition.

Visit your library and browse the recombinant DNA, genetic engineering, and bioethics areas. Many of the books will be strictly scientific, including manuals describing recombinant DNA techniques, reports from scientific meetings, and basic and advanced textbooks. You will also find a collection of books directed at nonscientists, many of which sound a variety of alarms concerning recombinant DNA and genetic engineering. Here are ten titles, along with author and year of publication:

Altered Fates: Gene Therapy and the Retooling of Human Life, Jeff Lyon and Peter Gorner, 1995.

The Biotech Century: Harnessing the Gene and Remaking the World, Jeremy Rifkin, 1998.

The Ethics of Genetic Engineering, Maureen Junker-Kenny (Ed.), 1998.

The Ethics of Human Cloning, Leon R. Kass and James Q. Wilson, 1998.

The Frankenstein Syndrome: Ethical and Social Issues in the Genetic Engineering of Animals, Bernard E. Rollin, 1995.

Genetic Engineering Dream or Nightmare? The Brave New World of Science and Business, Mae-Wan Ho, 1998.

Improving Nature? The Science and Ethics of Genetic Engineering, Roger Straughan and Michael J. Reiss, 1998.

Playing God? Genetic Determinism and Human Freedom, Ted Peters, 1996.

Remaking Eden: Cloning and Beyond in a Brave New World, Lee M. Silver, 1997.

Who's Afraid of Human Cloning? Gregory E. Pence, 1998.

Why is there such a strong reaction against recombinant DNA and genetic engineering? In this chapter, we will seek to understand what is meant by the phrase "the power of genetic engineering" by examining techniques for manipulating and analyzing DNA and then describing some of the myriad applications of the technology. Cloning a gene is just the first step toward understanding what the gene does and how it functions; to move beyond cloning, we need to learn more about DNA.

ANALYSIS OF DNA MOLECULES

We have seen an example of gene cloning. Now we will examine some of the techniques used to analyze cloned DNA. There are many such tools and techniques. However, we will study only those that are most commonly used and are fundamental to understanding applications of recombinant DNA technology. Some tools are useful in most aspects of genetic engineering, while others are more specific in their application, perhaps being useful only during certain phases of gene isolation or for detecting the presence of recombinant molecules. As we lay out these techniques, it is important not to lose sight of why the different techniques are significant. The examples we detail reinforce the importance and utility of the various techniques.

Gel Electrophoresis

From our observations of cloning methods, we recognize the difficulty of identifying a particular DNA molecule from among all the possible combinations that might have formed during the construction of recombinant DNA. Our example took advantage of the ability of bacterial cells to produce insulin and the use of antibodies to detect its presence. The truth is that, in most situations, such straightforward schemes for detection do not exist. Therefore, we must often go through a variety of steps in order to ascertain exactly what DNA fragment is contained within a particular recombinant plasmid. Without already knowing the sequence of the DNA or its function, we can exploit only two properties that make these molecules appear different: DNA fragment size and the location of sites recognized by restriction enzymes (restriction sites). We can examine both of these properties by using **gel electrophoresis**.

Electrophoresis is the process of separating molecules based on their size and/ or electrical charge. Consider the following hypothetical example. Suppose you have two bacterial colonies comprised of cells that you think might carry a particular plasmid containing your gene of interest. For this example, let's assume that the cells in one of the colonies contain a plasmid with your gene and the cells in the other colony contain the same plasmid but with a different gene. You know that the gene you want is approximately 2,000 base-pairs long, but you know nothing else about it. You have no functional assay (test) for the product of the gene, so you must somehow determine the sizes of the inserts cloned in each bacterial cell. The situation is illustrated in Figure 9.1a.

The first step is to isolate plasmid DNA from each bacterial colony. Remember that each cell within a colony is a direct descendant of one original cell, so each will contain the same plasmid with the same insert. Next the DNA is digested (cut) with the same restriction enzyme used to create the insertion site for cloning—in this case, the *EcoRI* restriction enzyme. Cutting each sample plasmid with this enzyme will result in the formation of two DNA fragments from each: One is the original

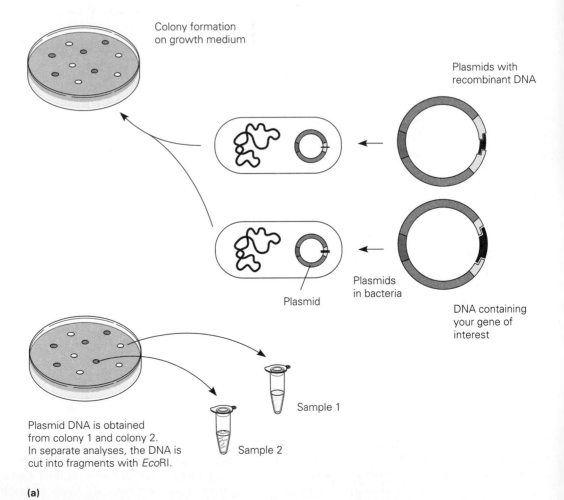

Colony formation on growth medium

Plasmids with recombinant DNA

Plasmids in bacteria

Plasmid

DNA containing your gene of interest

Sample 1

Sample 2

Plasmid DNA is obtained from colony 1 and colony 2. In separate analyses, the DNA is cut into fragments with *EcoRI*.

(a)

Figure 9.1 Determining the location of a DNA insert using agarose gel electrophoresis.
(**a**) Diagrammatic process leading to cut plasmids (DNA samples) with the restriction enzyme *EcoRI*.

Figure 9.1 (**b**) Details of a typical agarose DNA electrophoresis.

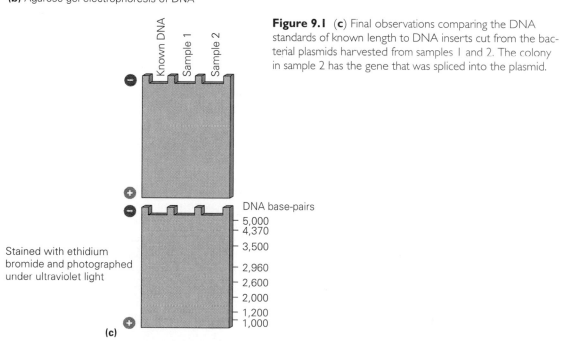

Cathode

Mixture of DNA fragments of different sizes

Power source

Gel

Glass plates

Anode

Longer fragments

Shorter fragments

Completed gel

(**b**) Agarose gel electrophoresis of DNA

Figure 9.1 (**c**) Final observations comparing the DNA standards of known length to DNA inserts cut from the bacterial plasmids harvested from samples 1 and 2. The colony in sample 2 has the gene that was spliced into the plasmid.

Known DNA
Sample 1
Sample 2

Stained with ethidium bromide and photographed under ultraviolet light

DNA base-pairs
5,000
4,370
3,500
2,960
2,600
2,000
1,200
1,000

(**c**)

cloning plasmid, and the other is the insert. Note that the plasmid fragments from the two different plasmid preparations are identical—they should be identical pieces of DNA. On the other hand, the inserts should be different. Finally, we have to separate these pieces on the basis of their size and then visualize them to see which one is the expected size.

To analyze the DNA fragments generated by the restriction digest, we can use a technique called **agarose gel electrophoresis** (Figure 9.1b). Imagine a preparation of a thin layer of a gel-like substance. This substance (agarose) is actually interlocked fibers that leave holes between them to form a matrix (think of a filter or a sieve). At one end of the layer is a series of holes, or wells, into which small amounts of different DNA samples can be loaded. Recall that DNA molecules have sugar-phosphate backbones. The phosphate groups give DNA a very strong negative charge. To take advantage of this, the agarose gel is subjected to an electric field, with the positive electrode (cathode) at the opposite end of the gel from the loaded DNA samples. The DNA molecules will be attracted to the positive pole and will begin to move through the gel matrix. However, because of the relatively small holes in the gel, larger DNA molecules will have a more difficult time passing through and thus will migrate through the gel more slowly than smaller DNA molecules. After a certain period of time, the different DNA fragments will separate—the smaller fragments will migrate farther than the larger fragments. The size of the "holes" in the agarose gel can be adjusted by altering the concentration of the agarose. The higher the concentration, the smaller the holes. This allows for the separation of different-size DNA fragments.

For analysis, the separated DNA fragments must be visualized. What is usually done is to stain the DNA such that it becomes visible. When electrophoresis is completed, the gel can be soaked in a solution of ethidium bromide (EtBr). EtBr is useful because it binds very tightly to DNA molecules and is **fluorescent**, that is, when it is exposed to a particular wavelength of ultraviolet (UV) light, it absorbs the energy of that light and then gives it off as light that we can see. EtBr fluoresces a bright orange color. Thus, when the gel is exposed to UV light, the DNA fragments in the gel show up as orange bands (Figure 9.1c).

If the samples from the restriction digests are electrophoresed (referred to as "running" by molecular biologists and biotechnologists) side by side with DNA molecules of known length (size standards), the sizes of the fragments present in the samples can be estimated. In the example shown in Figure 9.1c, sample 1 has two fragments 4,370 and 1,200 base-pairs long, while sample 2 has two fragments of 4,370 and 2,020 base-pairs. Since we know that both samples should contain the original plasmid, the 4,370-base-pair fragment common to both must correspond to the plasmid, so the other fragments must represent the inserts. Since we are looking for a fragment of about 2,000 base-pairs, it is clear from our experiment that sample 2 has the fragment of desired length and must represent the cloned gene of interest. We can now go back to our original plates containing the bacterial colonies and continue working with colony 2.

We have seen one example of using electrophoresis for size analysis. The technique can be more complex. For example, there can be many more fragments present, some of which may have quite similar sizes. The use of different restriction enzymes and combinations of enzymes would add to the complexity.

In fact, an analysis called **restriction mapping** is often one of the first steps performed on a cloned DNA fragment. The idea is to create a "map" of the cloned DNA fragment on which sites for different restriction enzymes are noted. The information for a restriction map is obtained by digesting DNA samples with different restriction enzymes and combinations of restriction enzymes and then analyzing the sizes of the fragments produced. From the resulting puzzlelike set of information, a map can be created. Figure 9.2 illustrates the construction of a simple restriction map.

Why is this step important? Initial cloning experiments often yield a fragment that contains more DNA than we really want. We usually start by cloning a big piece of DNA that contains what we want and then clone successively smaller pieces until eventually a fragment is cloned that contains only the sequence of DNA in which we are interested. How do we obtain these smaller pieces? The primary tools available to cut DNA are restriction enzymes, so in order to generate smaller pieces of DNA we need to know what restriction sites exist and where they are located in the cloned DNA. Thus, after initially cloning a large piece of DNA that contains what we want, we create a restriction map that identifies suitable restriction sites and their locations

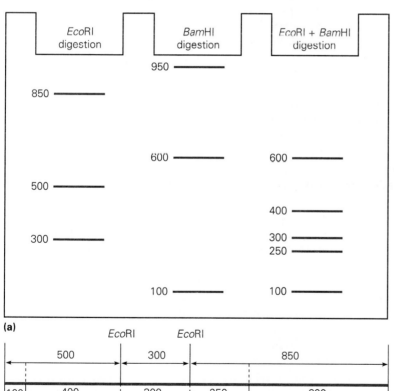

Figure 9.2 Construction of a restriction map.
(**a**) Restriction enzyme digests were done previously and diagrammed after electrophorytic separation of the fragments. The DNA was cut by two different restriction enzymes (*Eco*RI, *Bam*HI) separately and together. The lines under each well represent bands of DNA fragments in the gel following electrophoresis and application of ethidium bromide. The numbers designate the number of base-pairs in each fragment. (**b**) The restriction map produced following analysis of the electrophorytically separated digests of DNA. (Data from Glick and Pastemack, *Molecular Biotechnology*, ASM Press, 1994, Figure 2.7, p. 26.)

relative to one another. Then smaller and smaller pieces can be successfully cloned using the larger fragment as a source of DNA. This process is usually referred to as **subcloning**.

DNA Sequencing

Another basic operation that follows the cloning of a piece of DNA is deciphering the sequence of bases within the fragment. This provides a great deal of information—not only does the sequence allow a determination of the expected protein product, but it also allows construction of a more detailed restriction map that determines the precise locations of all restriction sites. Resolving the exact sequence of a piece of DNA is not a particularly easy task to accomplish. Several different techniques need to be combined in order to sequence DNA.

Figure 9.3 outlines the Sanger method of sequencing DNA. The process of sequencing requires many copies of the single-stranded DNA fragment to be sequenced. These copies will serve as templates for synthesis. Thus, the fragment of DNA to be sequenced must first be isolated as single-stranded DNA. Purified single strands are then distributed into four different tubes, each of which corresponds to one of the four possible DNA bases. To each tube is added a mix of nucleotides (A, C, T, G), DNA polymerase (the DNA-producing enzyme), and starter or primer molecules that have been radioactively labeled and will allow the synthesis of the second (complementary) strand of DNA to take place. Now comes the important part—there is a difference between the tubes. Each tube contains a single modified base called a **dideoxynucleotide** (Figure 9.4). These bases are incorporated into growing DNA chains just like their deoxynucleotide counterparts. However, the absence of an oxygen on a specific carbon means that once a dideoxynucleotide is incorporated into a DNA chain no other nucleotides can be added. Thus, any DNA chain that contains a dideoxynucleotide is terminated at that point. This method of sequencing is called **chain-termination sequencing**, or **dideoxy-sequencing**.

The dideoxynucleotides must be present in the reaction mixes in very small amounts compared to the regular deoxynucleotides so that within each reaction tube only one dideoxynucleotide is incorporated per chain. This means, for example, that an individual chain may incorporate a great many regular As before it incorporates a dideoxy A; conversely, the dideoxy A may be incorporated at the very first opportunity. Completion of these reactions results in a collection of partially double-stranded molecules in which the length of the second strand corresponds to the amount of DNA synthesized before the incorporation of the dideoxynucleotide that caused termination. In other words, each reaction tube contains a mixture of randomly terminated second strands representing all possible lengths when that particular dideoxynucleotide is used as a terminator.

We now need to separate and display the different-size fragments present in each reaction tube. **Acrylamide gel electrophoresis** can be used to accomplish this. First, the partial second strands must be removed from their complementary first strands. The process of separating double-stranded molecules into single strands is called **denaturation**. The denatured contents of each reaction tube are then loaded into a different lane (the path or line along which molecules move through the gel) of a gel

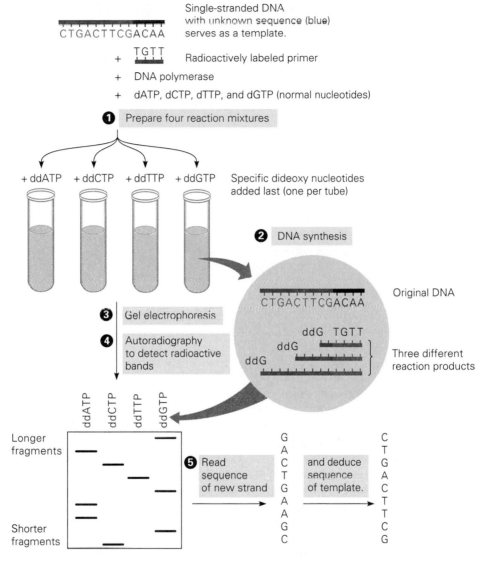

Figure 9.3 Chain-termination sequencing of DNA: Sanger method. Many copies of the single-stranded DNA fragment to be sequenced serve as templates for complementary synthesis. Purified single strands of the DNA are loaded into four different tubes, and to each tube is added a mix of nucleotides (A, C, T, G), DNA polymerase (the DNA-producing enzyme), and starter or primer molecules that have been radioactively labeled. Last, a few single modified bases (dideoxynucleotides) are added. Once a dideoxynucleotide is incorporated into a DNA chain, the chain is terminated. Electrophoresis and autoradiography allow separation and visualization of the prematurely terminated new DNA fragments. The base sequence is determined, and the original template DNA is deduced.

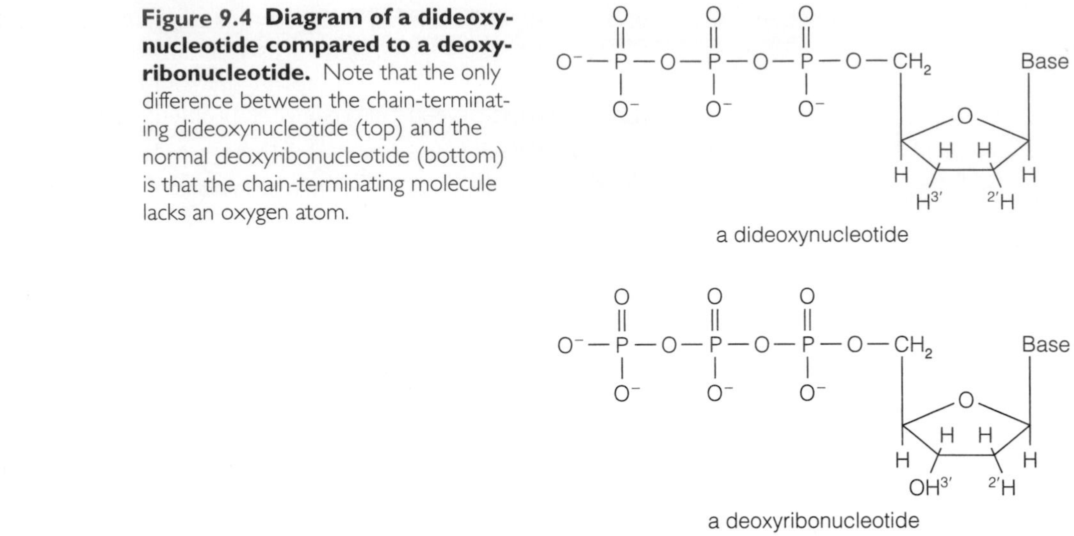

Figure 9.4 Diagram of a dideoxy-nucleotide compared to a deoxy-ribonucleotide. Note that the only difference between the chain-terminating dideoxynucleotide (top) and the normal deoxyribonucleotide (bottom) is that the chain-terminating molecule lacks an oxygen atom.

a dideoxynucleotide

a deoxyribonucleotide

designed to separate these small DNA fragments. After "running the gel," the fragments are separated in order of their size, with the shorter fragments located near the bottom of the gel (Figure 9.3).

We still need a way to visualize the DNA. The most commonly used method is to radioactively label the strands of DNA, usually by utilizing dideoxynucleotides that contain radioactive phosphorus. By this method, every terminated DNA fragment is radioactive and can expose X-ray film. Thus, after the gel is run, X-ray film is laid over it and exposed. Developing the film yields a pattern of bands exactly like the one on the gel itself—a permanent record of what the gel looked like.

Now we can see how the sequence of bases is determined from this picture. Beginning at the bottom of the gel, we begin to scan upward, noting the next larger DNA fragment on the gel (Figure 9.3). If this band is present in the G lane, then we know that the fragment was terminated by the addition of a dideoxy G, and thus G must be the base at that particular position. The next band is in the A lane, meaning that dideoxy A terminated the chain, so A must be the next base in the DNA. We can read the sequence of the entire DNA molecule in a similar manner (Figure 9.3).

The complete DNA sequencing process, from running the reactions to reading the gels, can be automated. A large number of bases of DNA can now be sequenced in a very short period of time using **DNA sequencers**. As we will see, a tremendous amount of DNA sequence information is being generated as part of the Human Genome Project.

Hybridization

One of the more useful techniques employed by genetic engineers is **hybridization**, which is based on the ability of complementary nucleic acid sequences to bind to each other. With hybridization, a particular DNA fragment can be identified and even purified from among many other fragments. In order to illustrate one way that hybridization works, let's go back to the selection steps we used in our cloning example.

Recall from Chapter 8 that one of our greatest challenges was to find the particular sequence that encoded insulin. The way we solved the problem was to use antibodies to look for bacterial cells that actually produced insulin. However, this technique cannot be used in all cases. Often, particularly when we are trying to clone eukaryotic genes into prokaryotic cells, there is little chance of the bacterial cells actually producing the desired product, at least following initial attempts at cloning. If we cannot use antibodies or in some other way look for the product of the cloned gene, then we must look for the presence of the gene itself.

To use hybridization, it is necessary to have some prior information about the gene. For the moment, we'll ignore how we get this information—we'll return to this topic after we go through an example of hybridization. Let's look at our insulin example. We inserted genes into plasmids and then used these plasmids to transform bacterial cells. After using antibiotic and blue-white screening to arrive at a pool of candidates that have inserts, we need to identify those cells (in pure colonies) that contain the inserted insulin gene. Figure 9.5 illustrates the basic steps in the procedure.

We first make a positional replica using the master plate (the plate with separated bacterial colonies on its surface). This replica is made by using a nitrocellulose filter. The bacterial cells that adhere to the replica are then broken open by use of a detergent, and their DNA is released onto the filter. The typically double-stranded DNA is denatured by sodium hydroxide, and the resulting single-stranded molecules remain attached to the filter in the same positions as the original colonies. Thus, the pattern of colonies on the master plate is perfectly reflected by the pattern of single-stranded DNA molecules on the filter.

We next add a radioactive **probe**, a single-stranded DNA that contains part of the insulin sequence. We'll see later how to obtain this probe. The probe is labeled so it can be visualized at the end of the experiment. Labeling probes with radioactivity is common, although nonradioactive methods can accomplish the same goal. When the probe is added, it can bind to complementary single-stranded sequences. Thus, if a gene for insulin is present, the probe will bind to it, forming double-stranded, hybrid DNA. Since the original strand of DNA is still attached, this newly formed double-stranded complex will be attached in the same position. After the hybridization reaction proceeds for a time, it is stopped. Any unbound probe is washed off; thus, any probe that remains on the filter is bound to insulin-encoding DNA. After the filter is washed and dried, X-ray film can be placed in contact with it. The localized, radioactively labeled probe will expose the film, yielding dark spots wherever it is attached. When the film is developed, we need simply to compare the exposed film to the master plate containing the candidate colonies in order to identify those colonies that contain insulin DNA.

Hybridization is an extremely useful technique. It is very sensitive—a small amount of probe bound to a membrane can be detected. It can also be very selective—since the technique is based on binding complementary DNA sequences, only specific sequences will be identified. Hybridization can be applied to any DNA or RNA sequence and can be used on material transferred onto membranes from whole

Figure 9.5 Steps in determining which cells contain the inserted insulin gene using replicas and hybridization.

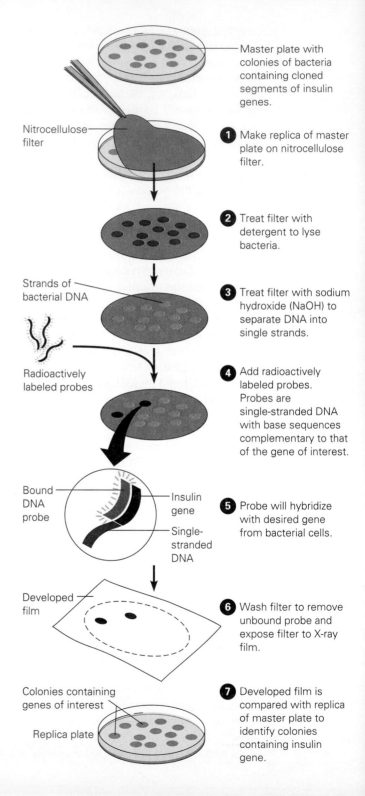

Master plate with colonies of bacteria containing cloned segments of insulin genes.

Nitrocellulose filter

1 Make replica of master plate on nitrocellulose filter.

2 Treat filter with detergent to lyse bacteria.

Strands of bacterial DNA

3 Treat filter with sodium hydroxide (NaOH) to separate DNA into single strands.

Radioactively labeled probes

4 Add radioactively labeled probes. Probes are single-stranded DNA with base sequences complementary to that of the gene of interest.

Bound DNA probe

Insulin gene

Single-stranded DNA

5 Probe will hybridize with desired gene from bacterial cells.

Developed film

6 Wash filter to remove unbound probe and expose filter to X-ray film.

Colonies containing genes of interest

Replica plate

7 Developed film is compared with replica of master plate to identify colonies containing insulin gene.

colonies or from gels. We will encounter several examples of the use of hybridization throughout the rest of this book.

Now let's return to the probe. In order for hybridization to work, we need to have a probe molecule that contains at least part of the sequence complementary to the one being sought. If our purpose in cloning is to get the gene because we don't have it, how can we already have the information necessary to make a probe? One common way is to work backward from the amino acid sequence of the protein.

At the time of the first attempts to clone insulin, much more was known about the protein product than about the gene. The entire amino acid sequence of the protein was known, for example. Since the genetic code was known, the DNA sequence that was used to create the protein could be determined by working backward from this code (Figure 9.6). There is a problem, however—the code is degenerate. Remember that often more than one unique codon can code for a given amino acid. Thus, as we work backward through the code, we come to many instances where multiple codons might have been present in the original DNA. How do we know which codon is the correct one, that is, which codon was originally present? The usual way to circumvent this problem is to construct probes of every possible combination. Very often, probes are obtained by using the DNA synthesis technology described in Chapter 8. If we know that different sequences might potentially code for

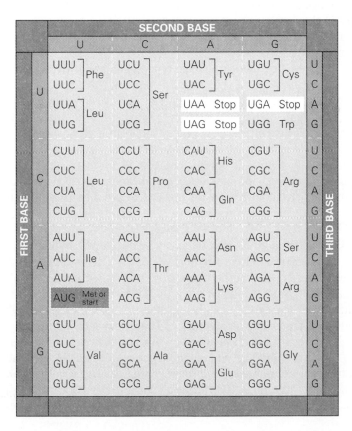

Figure 9.6 The genetic code depicted in mRNA codons. The first, second, and third bases of mRNA codons are designated on the left, top, and right sides, respectively. Note that most amino acids are specified by two or more codons (for example, histidine is encoded by CAU or CAC). Some amino acids, such as serine, leucine, and arginine, are encoded by six different codons. We can consider CAU and CAC synonyms for histidine; UCU, UCC, UCA, UCG, AGU, and AGC are all synonyms for serine.

the same protein, we can synthesize each of them and use a mixture of these probes in our hybridization. The machines used for DNA synthesis automatically create such a mix, given the proper set of instructions. One or more of the probes should be of the proper complementary sequence to bind to the insulin DNA; the others most likely will not bind to any sequences and thus will be extraneous. Therefore, even though we have no direct information about the gene we are trying to clone, we can get information from the gene product via the amino acid sequence.

Other forms of DNA can serve as probes. For instance, the cDNA clone for insulin does not truly represent the gene as it actually exists in the original DNA of the organism, since the introns have been removed and other processing events have modified it. It is often important to be able to isolate the original gene, with introns, in order to see exactly how it is controlled and regulated. In this case, the cloned cDNA copy of the gene, or a portion of it, can be used as a probe. The entire sequence of the cDNA clone of the gene exists in the original gene, which most likely contains introns. Therefore, sequences derived from cDNA clones make perfect probes for finding entire, unmodified genes.

Another major application of hybridization is in gene regulation research. Recall that regulation of gene expression is vital to the livelihood of a particular cell and ultimately is important in establishing the exact role of any cell within a multicellular organism. Intensive research is being conducted to determine how particular genes are regulated to establish different types of cells or to allow cells to respond to changes in their environment. Since regulation of expression is often accomplished at the transcription stage, studying the production of mRNA from a specific gene is pertinent.

For example, let's look at a single type of cell under two very different environmental conditions. Cells of your skin are exposed to all sorts of environmental conditions—cold, heat, wind, water, light, and dark. In response to some of these adverse conditions, the cells synthesize proteins that help protect them against the environment. For example, most cells respond to high dosages of UV light by synthesizing a particular set of enzymes involved in repairing mutations in DNA. Remember that UV light is a potent mutagen. Without repair enzymes, cells would accumulate mutations and perhaps become cancerous. However, under normal circumstances, there is no reason for the cell to produce large amounts of these enzymes, and thus the genes are inactive. The expression of these genes is induced by the UV light.

This induction of gene expression can be traced with hybridization methodology. First, it is necessary to clone one or more of the sets of genes of interest. After this has been done, probes can be made from the cloned copies. Skin cells can be removed and cultured in the laboratory. Both prior to and at different times following exposure to UV light, mRNA can be isolated from samples of the cells. The cells that have not been exposed to UV light should have very few, if any, copies of mRNA coding for DNA repair enzymes, whereas the population from cells exposed to high doses of UV light should have many more copies. The mRNA populations can be separated by gel electrophoresis, and the resulting pattern of mRNAs can be transferred onto a nitrocellulose filter. Hybridization can then be performed using the available probes. Figure 9.7 gives the procedural details and shows a hypothetical result from

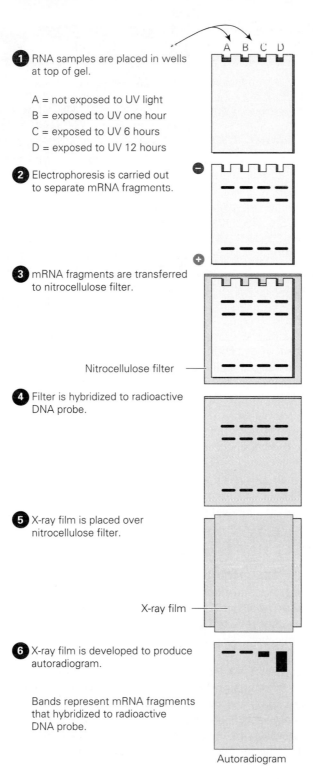

1 RNA samples are placed in wells at top of gel.

    A = not exposed to UV light
    B = exposed to UV one hour
    C = exposed to UV 6 hours
    D = exposed to UV 12 hours

2 Electrophoresis is carried out to separate mRNA fragments.

3 mRNA fragments are transferred to nitrocellulose filter.

Nitrocellulose filter ——

4 Filter is hybridized to radioactive DNA probe.

5 X-ray film is placed over nitrocellulose filter.

X-ray film ——

6 X-ray film is developed to produce autoradiogram.

Bands represent mRNA fragments that hybridized to radioactive DNA probe.

Autoradiogram

Figure 9.7 Example of electrophoresis and hybridization used to study gene expression. In step 1, RNA samples are loaded in wells. In step 2, electrophoresis is carried out; mRNA fragment bands will be seen until step 6. Step 3 involves mRNA transfer to a nitrocellulose filter. In step 4, radioactive DNA probes are added that hybridize with the specific mRNA. In step 5, X-ray film is exposed, and in step 6 the film is developed. Note that the two darkest bands are at the top under the six- and twelve-hour UV exposure. mRNA fragments with the twelve-hour UV exposure have the largest and most intense band.

Figure 9.8 A young boy with xeroderma pigmentosum.

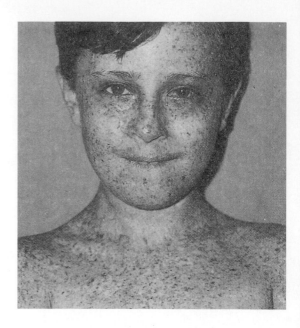

such an experiment by detecting the specific mRNA of interest. As long as conditions are similar for all samples, differences in band intensity reflect differences in the number of mRNAs present. In this hypothetical example, it is clear that much more of the specific message under investigation is being produced following longer exposure of the cells to UV light.

Patients suffering from the genetic disease xeroderma pigmentosum, an autosomal recessive condition, are acutely sensitive to sunlight. Portions of their skin exposed to sunlight usually develop intense freckling and cancerous growths (Figure 9.8). Their high sensitivity to light is due to an impaired ability to repair DNA. Evidence suggests that this high sensitivity to sunlight is due to the improper induction of gene expression following exposure to UV radiation.

Polymerase Chain Reaction

Another technique based on hybridization has become extremely valuable in recent years: the **polymerase chain reaction** (**PCR**). It is a technique used to amplify (create large numbers of copies of) a single piece of DNA. It is particularly useful as a way of producing large quantities of rare DNA fragments, and, as we will see, it has many related applications.

PCR involves the repetition of several steps. Let's assume that the region of DNA identified in Figure 9.9 is to serve as a probe in a series of experiments concerning the regulation of a hormone receptor gene. We need many copies of this region of DNA so that we will have many probes. Having identified the target sequence, we then construct *primers* (small, single-stranded DNA molecules) complementary to regions immediately on either side of this sequence. The primers, the original DNA to be amplified, a specific type of DNA polymerase, and the necessary chemicals

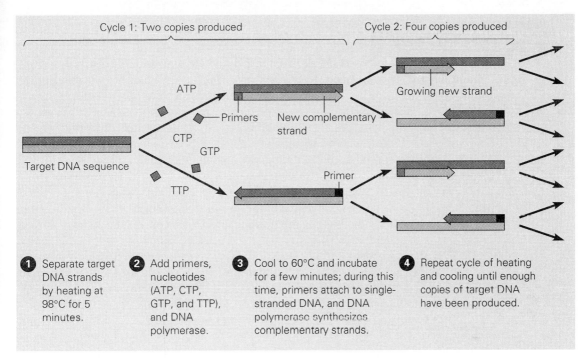

Figure 9.9 Polymerase chain reaction (PCR). PCR is a procedure developed to amplify (increase) the amount of DNA quickly. First, a solution with as little as a single piece of double-stranded DNA (to be amplified) is heated to 98°C to separate the strands, which will act as the initial templates. Second, a supply of the four nucleotides (A, G, C, T), heat-resistant DNA polymerase, and short pieces of nucleic acid known as primers are added, and the solution is allowed to cool to 60°C and incubate. Meanwhile, the primers hybridize with specific sequences of the separated strands and the polymerase makes complementary strands on the templates. The cycle of heating (adding raw materials) and cooling (incubating) is repeated until the DNA has been sufficiently amplified.

for DNA synthesis are mixed. Then the cycle begins. First, the temperature is raised until the double-stranded DNA template denatures into single-stranded molecules. Then the temperature is lowered until the primers can hybridize, or bind to complementary regions on the DNA. The primers are used by DNA polymerase to initiate synthesis, and new complementary strands of DNA are made. Once this synthesis is finished, the steps are repeated: denaturation, renaturation, and synthesis. As shown in Figure 9.9, after each cycle the number of DNA molecules doubles. The result is the production of many copies (potentially millions) of the original sequence. The target sequence has been amplified.

One of the advantages of this technique is that, through the proper selection of primers, specific sequences can be amplified, leaving all other sequences untouched. Since only particular regions of DNA will be amplified, rare sequences can be turned into commonplace ones in a population. This feature has been used on many occasions to increase the availability of rare sequences for cloning, although again we must know something of the sequence before being able to amplify it. We will describe several specific applications of PCR in medical diagnostics later in the chapter.

RFLP Analysis

Another technique based on electrophoresis and hybridization is **restriction fragment length polymorphism** (**RFLP**) analysis. This lengthy name is quite descriptive. Recall that much of the DNA of a eukaryotic organism does not code for proteins. Within the noncoding DNA, there are often many more mutations than within genes, because in many instances such mutations have no effect on the cell, tissue, organ, or organism. While mutations within the noncoding DNA do not result in phenotypes that are apparent, they do reflect differences among individuals and can be exploited in a number of ways.

Imagine two individuals, X and Y, each of whom contains genes *A* and *B*, which are separated by a non-coding region of the chromosome (Figure 9.10). Within genes *A* and *B* in both individuals is a restriction site for a restriction enzyme known as *Hind* III. This is the same for both individuals; there are no mutations within the genes. However, there are also *Hind* III restriction sites between genes *A* and *B*. In these regions, the two individuals differ: X has two sites, while Y has only one. A mutation in Y has changed a single base in one of the *Hind* III restriction sites, preventing the enzyme from cutting at that location. Can the presence of this mutation in Y be detected?

DNA is isolated from both individuals, and specific fragments are separated and then digested with *Hind* III (Figure 9.10, step 1). The resulting fragments are separated by gel electrophoresis (step 2) and subsequently transferred onto a nitrocellulose filter. When the radioactive complement to the left side of the B gene, identified as a probe, is used in a hybridization experiment, the results are shown as at the bottom of Figure 9.10. Note that from individual X a single band is identified (lit up) by the probe; from Y a single band is also identified, but it is closer to the top of the film, indicating that it was larger and heavier and did not show as much electrophoretic mobility. The Y DNA band is present because one of the *Hind* III sites is missing, making a single larger, heavier, and less mobile DNA fragment out of what in X is two fragments (one relatively large and one small). As we can see, this hybridization experiment demonstrates a difference between individuals X and Y even though their phenotypes are identical. The differences in the lengths of the fragments generated by restriction enzyme digestion give us the name restriction fragment length polymorphism.

Why are RFLPs so important? RFLPs behave like any other mutations—they are heritable genetic markers that can be passed from one generation to the next. But they have no obvious phenotype associated with them other than creating differences in the lengths of restriction fragments. As more genes involved in human disease are discovered, the ability to clone these genes and study them becomes more valuable. The more that is known about the location of a gene, the easier it is to clone. Genetic markers provide a means for the mapping of genes, by which the inheritance of markers is studied in order to position the markers relative to each other on a physical map, or model, of the DNA. In cases for which there are few genetic markers with obvious phenotypes, mapping is made much easier by also using RFLPs as markers. Since RFLPs represent heritable changes in DNA sequence, they can be used in mapping studies to more precisely locate other genes of interest. We will see shortly exactly how we can use RFLPs to help identify particular genes as well as particular individuals.

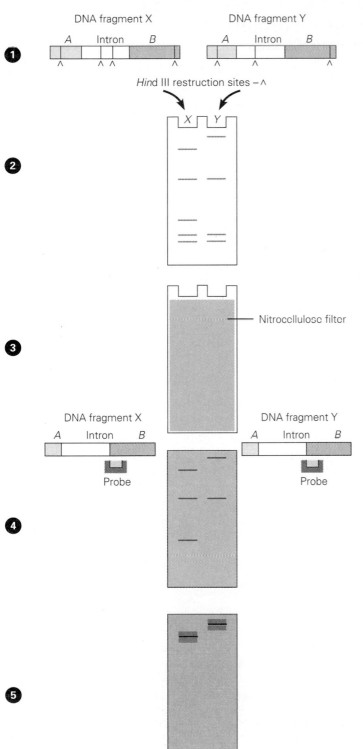

Figure 9.10 Restriction fragment length polymorphism (RFLP). A diagrammatic close-up of the homologous DNA fragments from individuals X and Y shows genes A and B and all Hind III restriction sites. In step 1, restriction fragment preparation, DNA fragments from individuals X and Y are cut at specific places using Hind III restriction enzymes. The resulting DNA segments are then separated by electrophoresis (step 2). The light DNA bands are not actually visible. After the DNA on the gel is denatured (strands separated) by chemical treatment, the single strands are transferred onto nitrocellulose paper through capillary action (step 3). In step 4, a suspension of a single-stranded radioactive probe is added to the paper. The probe is complementary to the left side of gene B and attaches to it (hybridizes) by base-pairing. After the probe hybridizes and the excess is rinsed away, a sheet of photographic film is laid on the paper (step 5). The radioactivity of the bound probe exposes the film, forming an image corresponding to the DNA bands containing the left side of gene B.

APPLICATIONS OF RECOMBINANT DNA: PRODUCTS AND PROCESSES

We have reviewed how genes are cloned and moved between organisms. We have also examined some of the techniques used by genetic engineers to study and manipulate pieces of DNA. We will now survey some specific applications of recombinant DNA technology. Some applications use bacterial cells to produce products, much as scientists use bacterial cells as living factories to produce insulin. Other applications involve changing the genetic makeup of an organism in order to give that organism a capability it did not have, such as degrading certain kinds of environmental waste. Still other applications are only indirectly based on recombinant DNA and involve the use of the techniques described above as well as others to gather specific information regarding genes or their regulation.

Basic Research

The most direct impact of recombinant DNA has been on **basic biological research** activities, that is, research with the clear aim of obtaining knowledge for knowledge's sake. Every field of biological endeavor has benefited from advances in recombinant DNA. As biologists have learned more about how organisms work, it has become apparent that a detailed understanding of an organism requires a detailed understanding of the cells within that organism. Furthermore, understanding cells means, among other things, understanding genes—how they are regulated—and their products.

Rather than trying to summarize all the ways in which recombinant DNA has affected basic biological research, we will outline two scenarios that summarize many aspects of recombinant DNA work. The first scenario involves a study of one or more genes responsible for a particular function within a cell. The second involves trying to discover the function that the product of a particular gene might serve.

For the first scenario, suppose we are studying the genetic disease cystic fibrosis. Evidence suggests that defects in one particular gene are responsible for the disease. In order to understand the disease, it is necessary to understand the normal gene. What protein does it code for? How is its expression regulated? How do mutated genes differ from the normal version? To answer these questions, we must clone the gene. The details of cloning the cystic fibrosis gene are explored in the next chapter. For now it serves simply as an example of how recombinant DNA techniques can advance basic biological research.

The second general application of recombinant DNA to basic research entails essentially the opposite process. Suppose we have been looking at a population of mRNAs present in liver cells that have been exposed to a toxic chemical. We find that a particular mRNA is expressed only following exposure. Logically, this mRNA might encode a protein that is involved in the cell's response to the toxin. What is the specific function of this gene? Is it to encode a protein that detoxifies the chemical? Is there a protein that binds to the toxin and forces it to be secreted from the cell? A number of possibilities exist. How could a genetic engineer determine the function of this particular gene?

The first step is usually to clone the gene in question. Since the particular mRNA involved has been identified, we could clone the gene from the mRNA and then determine the gene sequence. The challenge is to elucidate the function encoded by the cloned gene. This may not be an easy task, but there are ways to approach the problem. The sequence of the gene can be used to predict the amino acid sequence of the protein for which it codes. Knowing the amino acid sequence of a protein helps us to identify it, purify it, and reveal its function. The gene's sequence can also be compared to other cloned sequences in hopes of finding a sequence very similar to something known. Sequence comparisons are commonly used to assign functions to proteins.

The illustrations presented provide two examples from among a myriad of ways in which recombinant DNA technology has enhanced basic biological research. Our knowledge of the inner workings of cells and organisms has grown tremendously in recent years, leading to a greater understanding of human disease, among other things. The technological advances that have allowed recombinant DNA to develop and the advances that have sprung from this development have been nothing short of revolutionary. Nowhere is this more evident than in the study of human genes.

The Human Genome Project

The Human Genome Project is a massive, government-subsidized effort designed to sequence the entire DNA of a human. It is the largest biological research effort ever undertaken and presents many challenging technical problems. Foremost is the problem of the sheer amount of information. It will take thousands of person-hours to complete the sequencing of this much DNA, certainly not a trivial or inexpensive task. Furthermore, since sequencing information is obtained in relatively small pieces, the various small pieces must be arranged into a coherent whole. How is all of this information to be stored? How is the accuracy of the information to be verified?

Several subprojects are part of the greater work. First, since the eventual sequencing of all of the human DNA will require some advances in technology, smaller genomes are being sequenced—not simply for practice and technological improvement, but also because they will be of tremendous value in their own right. The entire DNA sequence of the bacterium *Escherichia coli* is the smallest of these sequences. Yeast (*Saccharomyces cerevisiae*), a member of the mustard family (*Arabidopsis thaliana*), and a roundworm or nematode (*Caenorhabditis elegans*) are also being sequenced. Each of these organisms is a very important experimental system, and knowledge of the entire DNA sequence of each organism will aid future biological studies.

To obtain the sequence of all of the human DNA, a detailed map of the human genome is first needed. Not only will such a map aid in identifying and cloning individual genes, but it will also prove invaluable in arranging sequenced segments. Thus, a second subproject is to obtain genetic markers, suitably spaced throughout the genome, to allow DNA fragments to be easily cloned and arranged.

A third effort involves obtaining detailed restriction maps of the human DNA. As we have seen, restriction maps make it easier to clone specific fragments of DNA of interest. The fourth subproject is to perform the DNA sequencing. As we have noted, the development of automated systems capable of sequencing tremendous amounts of DNA has increased the feasibility of this project. Continued technological improvements will no doubt enhance the rapidity with which the project can be accomplished.

ISSUES: THE VALUE OF THE HUMAN GENOME PROJECT

The decision to embark upon the Human Genome Project generated controversy, much of which arose from general questions that can reasonably be asked. For example, of what possible use is all of the information? Proponents of the project suggest that having the complete sequence of human DNA would allow major advances in medical care by making research into the causes and treatments of diseases easier. Opponents argue that much of the information applicable to health and disease could be obtained without resorting to such an expensive large-scale project. Is the information worth the cost? Completion of the project is conservatively estimated to cost well over $3 billion. Proponents argue that the information gained would be worth almost any cost, since it will make a tremendous difference in medical care. They also point out that more will come from the work than just an accumulation of DNA sequence information. Many technological developments will be driven by this project that presumably will aid all aspects of research. Opponents of the project counter by pointing out that the money being spent on this project had to be diverted from other work, and they wonder aloud whether placing such a huge emphasis on this single project is wise. Opponents also wonder whether we are placing too many high expectations on genes as explanations of our ills.

The Human Genome Project will continue to generate controversy, even after it is finished. It is eye-opening to note that, in an effort to develop policies and study the social implications of this project, 5 percent of the total budget is allotted to studying its philosophical, ethical, and moral implications and social consequences.

Medical Applications

Much of the promotion and early effort in developing recombinant DNA came from the prospect of improvements in medical projects or procedures. For example, pharmaceutical companies spent tremendous amounts of capital either investing in these efforts or developing their own programs. They did so because the market for medicinal products in developed countries is huge, and any way they see of tapping that market is immediately exploited.

Pharmaceutical Products We have seen one example of a pharmaceutical product obtained through recombinant DNA and biotechnology: insulin. Biotechnology provided a method of insulin production that was less expensive, reduced the risk of use

Table 9.1 Pharmaceutical Products of Biotechnology

Product	Use(s)
Adenocorticotropic hormone	Treats rheumatic diseases
Alpha- and gamma-interferon	Possible cancer and virus-disease therapy
Antitrypsin	Treats emphysema
B-cell growth factors	Treat immune disorders
Bone morphogenic proteins	Induce new bone formation; useful in healing fractures and reconstructive surgery
Colony-stimulating factor	Counteracts effects of chemotherapy; improves resistance to infectious disease such as AIDS; treats leukemia
Endorphins	Analgesic agents
Epidermal growth factor	Heals wounds, burns, ulcers
Erythropoietin	Treats anemia
Factors VIII, IX	Treat hemophilia; improves clotting
Hepatitis B vaccine	Prevents hepatitis B
Human growth hormone	Corrects growth deficiencies in children
Human insulin	Therapy for diabetics
Interleukin-2	Possible treatment for cancer; stimulates the immune system
Lymphotoxin	Antitumor agent
Macrophage activating factor	Antitumor agent
Monoclonal antibodies	Possible therapy for cancer and transplant rejection; used in diagnostic test
Nerve growth factor	Promotes nerve damage repair
Platelet-derived growth factor	Treats atherosclerosis
Prourokinase	Anticoagulant; therapy for heart attacks
Relaxin	Facilitates childbirth
Serum albumin	Supplements plasma
Superoxide dismutase	Minimizes damage caused by oxygen free radicals
Taxol	Treats ovarian cancer
Tissue plasminogen activator	Dissolves blood clots; therapy for heart attack
Tumor necrosis factor	Causes disintegration of tumor cells
Urogastrone	Antiulcerative agent

of the final product, and removed the dependence on the availability of animal organs. Other products are being produced through the use of this technology, many of them protein products that are used to treat or cure disease. Table 9.1 lists some of these products, along with descriptions of what they do. Unlike the insulin example, in most cases no other source of the particular material is readily available. Without recombinant DNA technology, such products would be either unavailable or prohibitively expensive.

Diagnostic Tools Accurate and timely diagnosis of disease is one of the most crucial aspects of modern medical treatment. Early identification of an infectious agent can sometimes mean the difference between life and death. It should come as no surprise that recombinant DNA and biotechnology in general contribute greatly to medical diagnosis.

There are two general ways in which diagnosis can be performed using biotechnological tools. The first involves the use of antibodies and is the subject of a more complete analysis in Chapter 12. The second is based on the technique of hybridization and is becoming a major diagnostic tool.

The concept is fairly direct: If an individual is infected with a particular virus, then genetic material from that virus will be present and will be different from the individual's DNA (Figure 9.11). As we have seen, hybridization is ideally suited for the detection of unique DNA sequences. DNA can be isolated from the patient's blood. This DNA sample will include the viral DNA along with the normal components, assuming the individual is infected. If we know what virus we are looking for and if the DNA sequence of at least a portion of that virus is already known, then it should be possible to construct a small probe that will specifically hybridize to the viral DNA. After the DNA sample has been isolated, separated by gel electrophoresis, and transferred onto a nitrocellulose filter, hybridization with the viral probe can determine whether the viral DNA is present in the DNA sample. If it is, then the specific infection can be diagnosed.

One problem with this approach is that the level of infection may be so low that little viral DNA is present—potentially so little that even hybridization is unable to detect its presence. This problem can be solved by using PCR. Recall that PCR is used to amplify a specific DNA sequence. PCR primers can be designed that will specifically amplify the viral DNA present within a DNA sample. After amplification, hybridization should detect the many copies of viral DNA now present. If the patient is not infected, then PCR will amplify nothing, and thus there will be no hybridization signal.

Such diagnostic techniques are very accurate and specific. Often, they are also

Figure 9.11 Radioactive RNA probe.
The RNA probe locates viral DNA, indicating infection.

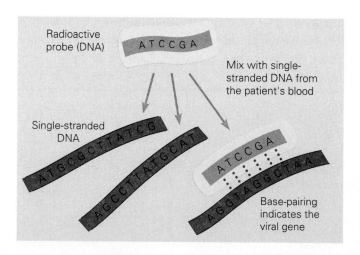

more rapid than more traditional diagnostic techniques such as culturing organisms. In many extremely serious infectious diseases, for example, the wide variety of hemorrhagic fevers found in Southeast Asia and Africa, rapid diagnosis is of the utmost importance. These viral diseases are characterized by such massive hemorrhaging that there is a significant chance that the patient will die if treatment is not begun within a day or even hours of the onset of symptoms. Because these diseases display many of the symptoms of other, unrelated infections, rapid and accurate diagnosis allows immediate and proper treatment.

Diagnosis and Treatment of Genetic Disease The principle of diagnosis through hybridization makes possible the diagnosis of genetic disease. Suppose a specific genetic mutation is found to cause a particular disease, say Alzheimer's disease (this is an oversimplification, but it will illustrate the idea). If we know what specific mutations are involved, then we can construct DNA probes that will allow hybridization to detect the mutations' presence. It is then theoretically possible to prepare DNA samples from patients and determine whether they possess the mutated gene. Thus, genetic diagnosis can be performed using the same techniques used for diagnosing infectious diseases.

Related to the idea that genetic differences can be diagnosed or detected in individuals is the next logical step: If a genetic defect can be detected, perhaps it can be reversed or cured. **Gene therapy** is exactly this process. The concept is fairly straightforward: If a specific gene is identified as defective, why not replace or supplement it with a normal one? Hemophiliacs possess a defective gene for one of the proteins involved in blood clotting. Traditional treatment of hemophiliacs consists of administering the proper quantities of the missing proteins at regular intervals. This is not a cure but only an effective preventative. If it were possible to replace the defective gene with one that functioned normally, this measure would be unnecessary, and the disease could be cured.

Diagnosis of genetic disease and gene therapy are two of the more medically newsworthy biotechnology ventures. They simultaneously offer hope for curing many diseases and raise a number of ethical and moral questions. Because of the importance of these medical advances, we will devote Chapter 10 to them.

Development of Vaccines Another important topic, considered in detail in Chapter 11, is the development and production of new and better vaccines. Enormous effort is being expended to develop vaccines against the human immunodeficiency virus (HIV), the causative agent of AIDS, as well as many other disease agents. In many ways, the biotechnology behind vaccine development is similar to that behind the development of other pharmaceutical products, and the techniques used in those efforts find application here as well.

Forensics

Several highly publicized legal proceedings have featured the use of **DNA fingerprinting**. This technique is a specific application of RFLP analysis and is based on the fact that individuals, even though they possess many common genes, have many differences in their genetic material. Most of these differences occur in noncoding regions of the DNA, which is presumably why they have no effect on the individuals. Since many noncoding regions are quite variable, it is likely that two individuals will differ in the particular noncoding regions they possess unless they are related and therefore share genetic material. DNA fingerprinting seeks to identify these genetic differences in the hope of determining whether two different DNA samples are from the same person or from different persons.

The basic technique employed in DNA fingerprinting is, once again, hybridization. A DNA sample can be prepared from biological material found at a crime scene—blood, semen, or even hair. Often, PCR is used to amplify specific fragments of DNA, since little biological material may be found. Other DNA samples are prepared from suspects. The DNA samples are digested with restriction enzymes, and the DNA is separated and displayed by gel electrophoresis. Following transfer to a membrane, the DNA can then be subjected to hybridization.

The important next step is to select appropriate DNA probes. Ideally, we would like to visualize one or more DNA fragments that are known to be highly variable among individuals. Therefore, the probes will be chosen so that they hybridize to these variable regions. If the DNA in samples from a suspect and from the crime scene show identical bands following hybridization, then within some statistical limit the two DNA samples can be said to be from the same person. If, on the other hand, the DNA samples show different bands, then the DNA is not from the same person. Figure 9.12 illustrates the principle of DNA fingerprinting.

The use of DNA as evidence has been the subject of both legal and scientific debate. Technical difficulties can result in erroneous or improper conclusions, and thus the reliability of the technique can be questioned. Another concern is the statistical validity of the results. What is usually reported as a result of a DNA fingerprint is the likelihood that two DNA samples are from the same individual. In determining this statistical result, several assumptions are used, and comparisons are made with generalized sets of DNA fingerprint data. The statistical analysis is complex and is based on its own set of assumptions—assumptions that are nearly impossible for nonexperts to recognize, let alone judge. How appropriate is it for highly technical, spe-

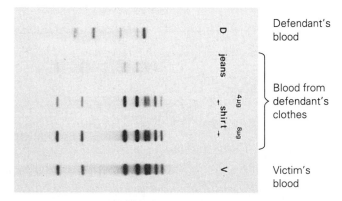

Figure 9.12 DNA fingerprinting. These DNA fingerprints are actually derived from blood samples from a murder defendant, from stains found on the defendant's jeans and shirt, and from the victim. The blood was subjected to RFLP analysis, that is, it was digested by the same restriction enzymes and then the resulting DNA fragments were separated electrophoretically. The comparison between the blood sample taken from the defendant's shirt and the victim's blood shows a perfect match. (Courtesy of Cellmark Diagnostics, Inc., Germantown, Maryland.)

cialized information to be introduced into a courtroom, where presumably no judge or jury is qualified to assess the results of the test? Is the use of "expert testimony" a reasonable way to present such evidence in court? While many questions are raised by the use of DNA fingerprinting, there is no doubt that it has provided valuable information in many court cases.

Environmental Applications

Another area with immense potential for the application of recombinant DNA technology is environmental management. Today it is well known that we have implemented many environmentally unsound policies. We produce tremendous amounts of waste, we try to extract every last bit of precious material from the earth, and we collect and store all sorts of hazardous and toxic chemicals. Radical changes in the way we do business are called for. Although it may be some time before drastic measures are enacted, there is plenty of room for new developments that can help reduce our environmental impact. Recombinant DNA technology can be used in several ways to further this effort.

One promising approach concerns **biomass utilization**. Biomass refers to all the material produced during the production and processing of agricultural and food materials that is usually discarded as waste. This often takes the form of the indigestible portions (sometimes this means most) of plants. Tremendous quantities of these wastes are produced. They are generally not toxic or hazardous; they are wastes simply because they cannot be used in the products. However, there is no reason why these wastes cannot be used in some alternative way. Let's examine how biotechnology might contribute to the reduction of wastes.

The major components of biomass waste are the portions of plants that are not readily processed or incorporated into foodstuffs. Mostly, this consists of the woody portions of plants, which are composed of lignin and cellulose. We will concentrate on cellulose, which we've already described in Chapter 1. Indigestible material left from the processing of foods, waste from the timber industry, and paper products all contain large amounts of cellulose. Recall that cellulose is a biopolymer of connected glucose molecules. If the cellulose could be degraded, it might be a valuable source of glucose, a sugar that can be used readily in a variety of ways.

Many organisms can degrade cellulose through the use of enzymes known as cellulases. The identification, isolation, and purification of cellulases make up an important avenue of research. One example of a potential industrial application is to treat waste paper (office paper, newspapers, cardboard) with cellulases and then use the released glucose in yeast fermentations. One major end product of such a fermentation is alcohol, an important industrial raw material and also a potential fuel source. The process has been tried on a small scale in the laboratory, and the results have been encouraging. On a larger scale, it is estimated that if all the waste paper generated annually were fermented to alcohol and used as fuel, gasoline consumption in the United States could be reduced by 7–20 percent. Not only would the cellulose waste be greatly reduced, but the utilization of fossil fuels would decrease as well. The role of recombinant DNA in this project is similar to its role in other basic research efforts—to help identify genes and gene products that perform certain functions and to transfer these functions to different organisms.

Another avenue of environmental research that may prove to be of great benefit in the future concerns the treatment of hazardous and toxic materials. Treatment of waste material with microbes is not new—sewage treatment plants have utilized microorganisms for as long as there have been municipal treatment plants. However, many materials cannot be degraded by the procedures and organisms used in routine sewage treatment. It is comforting to note that some natural microorganisms can degrade and detoxify many chemicals, such as herbicides and pesticides, organic chemicals, and compounds containing heavy metals. One bacterial genus, *Pseudomonas*, contains many species that are adept at these chemical processes. Generally, compounds are broken into metabolic intermediates and then used as food by the bacteria. One example of such bacteria is the "oil-eating" bacteria that have been genetically engineered to use some components of oil spills as a food source. While these bacteria do not provide a solution for accidents, little doubt remains that continued research will develop new and exciting ways of handling some of our more toxic wastes.

Agricultural Uses of Recombinant DNA

The relatively recent history of modern agriculture has been repeatedly punctuated with a variety of human attempts to improve the breeds of both plants and animals. Until quite recently many of the so-called improvements were unsuccessful or met with limited success. Today we have learned much to increase our gains through recombinant DNA technology.

Animal Applications Two distinct applications of recombinant DNA technology involve the direct modification of farm animals. Both involve the process of inserting "foreign" genes from (1) other animals or (2) humans into farm animals. How is this done? Until now, we have seen how microorganisms—mostly prokaryotes—can be made to take up DNA. The process of transferring genes to mammals is quite different and deserves our attention.

Suppose you have identified a breed of cattle that is genetically resistant to mastitis, a bacterial infection of the mammary gland, and you want to move a mastitis resistance gene into other breeds of cattle. You would first identify the gene and clone it using the standard recombinant DNA techniques. Then you would begin the process of transferring this gene into a cow.

The only feasible way to create offspring in which every cell contains the foreign gene is to transfer the DNA into eggs or sperm. Since the fertilized egg divides to produce every cell in the organism, an egg carrying a foreign gene will result in the formation of offspring in which every cell contains the desired gene. Constructing a **transgenic** animal, that is, an animal into which foreign genes have been inserted (Figure 9.13), requires that embryos be manipulated outside the body and then returned to a mother. Unfertilized eggs are obtained from a cow and maintained in the laboratory. Foreign DNA is inserted into them by the process of **microinjection**, during which DNA is injected directly into the egg through a fine needle. Once inside the

Figure 9.13 A large, transgenic "super-mouse" and a normal-size littermate. The gene for rat growth hormone was microinjected into mouse sperm and *in vitro* fertilization was carried out. The resultant embryos were transferred to the reproductive tracts of foster mothers. Two littermates are shown—on the left is the mouse with the gene for rat growth hormone.

cell, the DNA can become incorporated into the existing chromosomes. Following microinjection, the eggs are fertilized *in vitro* and then implanted into a cow. Since the fertilized egg contains the foreign DNA, every cell of the developing embryo will also contain the DNA. Assuming that the foreign gene is expressed and functions properly, the resulting calf will be resistant to mastitis.

This example illustrates one of the major focuses of transgenic animal research—moving particular genetic traits into livestock to improve resistance or in some other way improve the animals' traits. In the past, such improvements were achieved through traditional selective animal breeding. Individuals that exhibited desirable traits (for example, faster growth or improved yield) were specifically bred with each other. The traits of the offspring were checked, and the process continued until an improved breed was attained.

The successes of traditional breeding of both crops and livestock have been remarkable over the years, resulting in vastly increased yields of crops, milk, meat, and other products. However, traditional breeding processes offer limited returns. Let's assume that we have a particular breed of cattle right now that is very good—high yield of milk, quick growth, and so on. To try to further improve this breed requires that it be bred with another breed with different traits. The offspring may well show improvement in one or more traits, perhaps producing more milk. But the odds are that one of the other desirable traits, such as fast growth, will be adversely affected by the breeding. The more highly bred a particular line, the easier it is to lose desired traits through further breeding. It becomes more and more difficult to improve upon the traits of livestock because when two lines are bred the offspring share all of the DNA of the two parents, whether the genes are desirable or not.

Recombinant DNA technology affords the singularly unique possibility of changing a single trait within an already highly bred line. By insertion of a single gene, a single trait might be affected, while other, desired traits are maintained. As a result, breeding through recombinant DNA can be used to improve livestock and crops in ways that would not have been possible through traditional breeding programs. Results can also be achieved more quickly, since the changes being introduced are much more directed. It is important to note, however, that attempts to alter even a single trait with a single gene are often complicated by **pleiotropic** effects. That is, a change in one gene affects many traits, not just one. Simple processes are often complicated by genetic effects we don't completely understand.

Livestock can be utilized by the biotechnology industry in another way. We saw in Chapter 8 that one of the difficulties of using eukaryotic genes in prokaryotic organisms is that the products of the eukaryotic genes often need to be modified in some fashion after they are synthesized, for example, through the addition of carbohydrate groups to the protein. Prokaryotic cells cannot perform this function, and thus other hosts must be used in order to obtain modified products. In some cases, no organisms besides mammals can perform the modifications required for the synthesis of human proteins. For example, factor IX, one of the proteins found in mammalian blood that is involved in blood clotting and is missing in most hemophiliacs, cannot be produced intact by traditional recombinant DNA hosts. Mammals alone can produce a functional factor IX protein. How could a mammal be used as a host for recombinant DNA?

What if the gene for human factor IX were used to construct a transgenic cow, as described above but with a difference—the recombinant DNA molecule is constructed such that the factor IX gene is controlled not by its own natural promoter but by a promoter for production of milk? Such a promoter will be active only in cells of the mammary glands. In this way, assuming that transgenic animals can be produced using this recombinant DNA construct, the gene will be controlled so that its product is formed in the milk of the host. As the milk is collected, it will contain human factor IX, synthesized and properly modified by a mammal.

The secretion into milk affords at least two advantages besides providing for proper modification of a protein. First, since tremendous amounts of milk are produced by livestock, such secretion should provide an abundant source of the material. The mammary glands of the animal have essentially been turned into a bioreactor to produce a pharmaceutical product! Second, the factor IX is being secreted in a form that will make it easier to purify. Since there are few different proteins in milk, the necessary separation and purification steps are greatly reduced. A number of such "pharm" animals are being used today, and many more are planned.

Plants and Bacteria

The biotechnology involved with plants has been and continues to be productive and exciting. Molecular biologists and biotechnologists who deal with plants recognize what many consider to be a special advantage, since whole plants are often able to be regenerated from individual cells grown in tissue culture (Figure 9.14).

Genes may be introduced into plant cells on a plasmid carried by *Agrobacterium tumefaciens*—a natural plant pathogen that causes a cancerlike condition, crown gall, in a number of plants. Tumorous growth in plants is ultimately caused by **Ti** (tumor-inducing) **plasmids** that have the capability of integrating into the host plant's chromosomes.

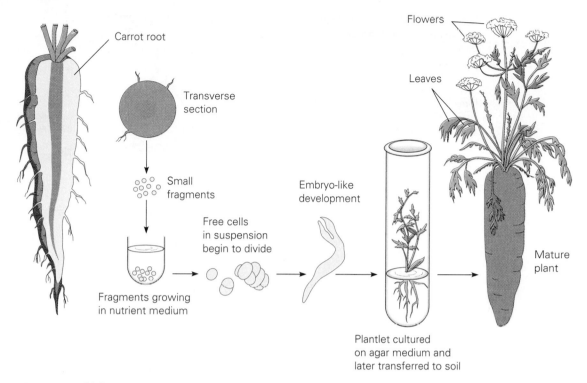

Figure 9.14 Using carrot root cells to produce whole carrot plants in tissue culture. This diagram illustrates the general steps necessary to regenerate a whole carrot plant from a single cell derived from a storage root.

Figure 9.15 illustrates the basic sequence of events that go into the creation of transgenic plants using engineered Ti plasmids and *A. tumefaciens*. Initially, the Ti plasmids are isolated from the bacterial cells, and the specific gene of interest is inserted into a particular plasmid region. The insertion stops the plasmid from inducing a plant tumor. The engineered plasmid is put back into *A. tumefaciens*, and the bacterium is allowed to infect plant cells growing in tissue culture. Once inside a growing plant cell, the inserted gene is integrated into the plant's chromosomes (along with the nonfunctional Ti plasmid). Thereafter, every daughter cell carries the foreign gene. Finally, the cells of the tissue culture are grown separately to produce whole plants, and each cell of every plant has the foreign gene. To add to this already happy state of affairs, when the mature plant produces seeds, each seed will carry the gene. Herbicide resistance, the ability to withstand drought, and increased chemical production are just three traits that can be altered in plants through this process.

Some plants live in a type of **symbiotic** relationship, known as mutualism, with certain bacteria. Symbiosis is the living together of two different organisms. *Mutualism* is a type of symbiosis in which two different organisms associate closely together and each benefits from the association. A notable example of mutualism is the interaction of certain plants with nitrogen-fixing bacteria.

Nitrogen is required by all living things, including plants. Most nitrogen in our

Figure 9.15 Transgenic plant production using *Agrobacterium tumefaciens* and its Ti plasmid. *A. tumefaciens* has a Ti plasmid that is uniquely suited to serve as a gene carrier for inserting foreign genes into plant cells in tissue culture. Details of the first four steps are shown. Step 5 is generally similar to the process shown in Figure 9.14.

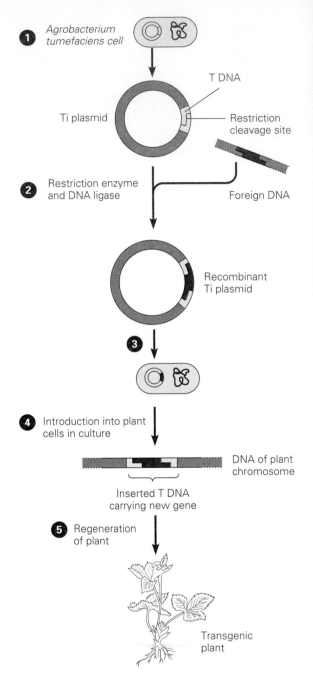

environment is present in the form of nitrogen gas (N_2) in the atmosphere. Plants cannot use nitrogen in this form. Rather, they obtain nitrogen from their surroundings in another chemical form, typically ammonia (NH_3) or nitrate (NO_3^-). Such compounds are usually in very short supply in soil, and thus the growth and development of plants can be limited by the available nitrogen. To overcome this deficiency, as well as others, farmers apply fertilizers to fields. Nitrogen constitutes a major and very expensive element in most commercial fertilizers.

Certain bacterial species present in the soil are capable of "fixing" nitrogen. Fixation is the process of incorporating inorganic molecules, in this case nitrogen gas, into organic compounds such as ammonia. Although sometimes the bacteria will do this while living free in the soil, they generally need to enter into a complex symbiotic relationship with plants such as soybeans, peanuts, or alfalfa (legumes). This relationship begins when the bacteria infect the roots of the plant. The plant responds by enclosing the bacterial cells within a special compartment, or **nodule**. Within this compartment, the bacteria thrive, receiving their nutrition from the plant. The plant, on the other hand, tolerates the existence of these nodules because the bacteria provide it with fixed nitrogen. Thus, both organisms benefit from the arrangement.

Symbiotic nitrogen fixation is limited to certain plants and bacteria. Most plants cannot participate in such arrangements. Imagine how beneficial it would be if the process of nitrogen fixation occurred within plant cells directly, without requiring the symbiotic bacteria. The savings in fertilizer costs alone would be staggering, and the yields of the plants might increase as well. A tremendous amount of research is being done with the goal of incorporating some or all of the genes encoding nitrogen-fixing enzymes into important crops. However, producing transgenic plants capable of fixing nitrogen is a daunting task. First, many genes are involved in the process, not just one or two. Transferring multiple genes into an organism is difficult, and getting them expressed appropriately is even more difficult. Second, the metabolic steps involved in nitrogen fixation are sensitive to the presence of atmospheric oxygen. When fixation is carried out in root nodules by the symbiotic bacteria, resources applied by both the bacteria and the plant root nodule act to exclude oxygen so that it does not inhibit fixation. A separate compartment, the root nodule, is constructed so that its central chamber is oxygen-free. Even if the genes for all of the proper enzymes could be inserted and expressed in a plant cell, too much oxygen would be present to allow fixation to occur.

Another related area of research involves examining why only certain bacteria are able to enter into symbiotic relationships with particular plants. Might it be possible, through recombinant DNA techniques, to modify the range of plants that a particular bacterial species infects? Might it be possible to genetically alter a plant so it becomes susceptible to infection? All of these questions and others are being actively pursued.

Insecticides and Herbicides

In the 1960s, we learned that the predominant insecticides in use were quite harmful to the environment. DDT and other chemical compounds had been used for decades, and initially DDT was found to be an extremely effective insecticide against a wide range of species. However, over the years, ever higher dosages had to be applied to plants to prevent insect infestations. It was found that many insects had developed DDT tolerance or resistance. Worse yet, it was found that compounds such as DDT devastated some sectors of the environment, persisted for years, and accumulated in a wide variety of animals, eventually killing them.

However, the need for some sort of pest control remained. New chemical means of fighting insects and weeds are constantly being developed, as are new natural methods of control. Recombinant DNA technology plays an important role in the development of new tools for agricultural use.

A wide range of more environmentally friendly insecticides and herbicides now exists. They are degraded rapidly and have fewer adverse side effects. Unfortunately, many of these compounds are so effective at killing living things that they attack not only the pests, but the crop plants as well! To use such compounds effectively on pests, resistance to them must be developed in the crop plants. Recombinant DNA techniques can be used to identify and clone suitable resistance genes for the creation of resistant transgenic plants.

A natural insecticidal system shows promise as a general approach to insect control. A number of bacterial species, most notably *Bacillus thuringiensis*, produce proteins that kill particular insect pests. *B. thuringiensis*, commonly called "Bt," produces solid insecticidal proteins and stores them. As the bacteria are ingested by the insects feeding on sprayed crops, the insecticidal proteins are activated by the insects' digestive systems. The activated molecules dissolve the gut epithelium of the insects, killing them. Crops can be protected from these pests by applying the bacteria, usually in the form of a spray, at specific times during the growth of the plants.

Such natural insecticidal systems are now being employed in some instances, but they have strict limitations. First, only specific pests are affected. Second, the bacteria must be ingested by the insect in order for the toxin to become active. Third, *B. thuringiensis* does not normally live in the ground, so insects that attack roots cannot be controlled in this manner. Recombinant DNA techniques might be able to overcome these difficulties.

The study of different toxin genes may make it possible to piece together and clone combinations of toxin genes that might be effective against different insects, or at least against more types of insects. The more general the toxin can be made, the more useful it will be (within limits). Other difficulties might be overcome if the plant itself could produce the insect toxin. If transgenic plants could be constructed that produced and accumulated toxin within their own cells, then any insect that ate the plant would be killed. The toxin could be expressed in all plant tissues, including the roots, affording all of them protection. This has already been done in a particular variety of corn.

ISSUES: GENETIC ENGINEERING AND EVOLUTION

Does genetic engineering change evolution? We might wonder if we are bypassing natural evolutionary mechanisms by creating transgenic animals and bacteria with eukaryotic genes. This is a fairly common concern related to the advent of recombinant DNA. Many think it "unnatural," perhaps even antievolutionary, and believe that we should not pursue the technology or at least should avoid taking some of its paths.

Evolution is a scientific theory that describes how living things change over the course of time. This change is a result of interactions between the biological information carried by organisms in the form of genes and the environmental influences exerted upon individuals or groups. The mechanism through which evolution operates is natural selection, as described in Chapter 3. No outside force operates to control natural selection; it is simply the result of the interaction between individuals possessing different traits and the living and nonliving environments in which the individuals live.

Most organisms have the ability to react to conditions in their environment. For example, many common fungi obtain nutrients and grow very quickly in hydrated organic matter. Growth is almost entirely vegetative until nutrients begin to become scarce and drying begins. Then, instead of continuing the feeding and expansive growth that characterize the vegetative state, the fungus begins to produce aerial spores. These spores can easily be airborne, spread to another location, and reinitiate vegetative growth. Thus, fungi can literally seek out a new home, a new environment. Other living things create shelters in order to survive in particular environments. Many plants give off chemicals that prevent the nearby growth of other plants, thereby decreasing the competition for nutrients. These organisms utilize the raw materials available to them to construct devices that will aid their lives. This is not "unnatural"; on the contrary, it is the most natural thing in the world for an individual organism to try to survive.

People are no different from any other living thing in this respect. We are subject to natural selection. We do not stand apart from evolution and dictate how it will operate; we simply take advantage of the traits we possess to utilize or modify our environment in order to survive. No one would suggest that the use of eyeglasses is so unnatural or antievolutionary that it should be banned. We are simply responding to our environment by using traits we possess in order to modify that environment in order to survive.

Everything we do is part of evolution. It is impossible for us to "alter" evolution, since we are part of the process. All we do is participate in the ever-changing relationship between the traits of organisms and the environment. We cannot bypass natural selection; we can simply alter the specific interactions that are subject to it.

ESSAY: CONCERNS ABOUT THE USE OF RECOMBINANT DNA

To end this chapter, we would like to raise some general questions about how individuals feel about the use and abuse of genetic engineering. Hopefully, the power of recombinant DNA and genetic engineering has become apparent to you in the last few chapters. We have only scratched the surface; there are many applications of the general techniques that we have not mentioned. It is also easy to imagine scenarios, both fantastic and realistic, in which recombinant DNA could be put to use in more questionable ways, such as the development of new weapons for biological warfare.

The power of this technology is what makes it so exciting to some and is also what makes it feared by others.

Why is this technology so exciting? The answer may seem obvious, given that we have just spent two chapters outlining a variety of potential applications. But the applications themselves are only part of the excitement. Other reasons also make for interesting reflection.

The business of biotechnology has become a hotbed of investment activity. Large sums of money are invested in the development of new products or procedures. Lucrative business arrangements are commonplace among the scientists involved, and many pharmaceutical and biotechnology firms are willing to pay enormous sums of money for the rights to exploit particular genes or organisms and to obtain the expertise to do so. No doubt many people in the biology profession are excited about the prospect of financial reward and prestige.

This situation is really no different than in any other business; a lot of scientists and others connected with the world of biotechnology simply happen to be in the right place at the right time to take advantage of this particular scientific revolution. Economic excitement and competitiveness are the basis of our capitalistic system.

Another particularly exciting aspect of recombinant DNA technology is its perceived ability to explain many of the mysteries of life. We are fascinated by who and what we are, and to the degree that studying our genes will help us understand this, we would all probably like to know more. While we may never understand "what it is that makes us individual human beings," as some claim we will as a result of the Human Genome Project and its successors, we will certainly be able to more fully understand human biology and the evolutionary relationship between humans and other living creatures.

The last cause for excitement is one that not many admit to. Recombinant DNA technology affords us the ability to control, at least to some degree, one of the few remaining aspects of life over which we have no control—genes. Having control is important to human beings. Being able to genetically engineer organisms is as close as we can come to creating life, to seemingly overcoming evolution and the natural order of things. This raw ability to control at least some aspects of life is a powerful aphrodisiac, even though we might not like to admit it.

But not everyone is positively excited about the prospects of genetic engineering. Some people are fearful, while others are merely concerned. Why? What is there to be concerned about? Again, this question has some obvious answers related to specific or potential applications of recombinant DNA. Might deadly viruses or other organisms be created, either accidentally or on purpose, and released into the environment? What might be the effect of trying to alter the genes in a human egg? It is proper to have concerns about specific applications, and these should be considered during the development of a product or procedure.

But what drives the more gut-level reactions against recombinant DNA in general? Certainly, many people hold religious or quasi-religious beliefs that some things just shouldn't be altered. The ability to "contradict" nature by mixing genes that might otherwise never be mixed is abhorrent to some. This view can be held even without invoking religious beliefs; it can simply seem unnatural.

A concern that arises with every new discovery and technical revolution is that we just don't know enough about the long-term effects of a new product, process, or organism to justify its use or implementation. Powerful arguments can be made suggesting that we tend to incorporate new technology into our fabric of life without giving careful enough thought to its consequences. Opponents of genetic engineering range from those who simply want to slow the rapid advance so that more discussion and testing can occur to those who would like to see the entire technology banned.

Another concern expressed by many people is that our society is not well equipped to have all the knowledge that recombinant DNA promises to deliver. Are we prepared for the day when the Human Genome Project provides the raw material to allow diagnosis of a multitude of traits? Are we prepared to handle the overwhelming amount of genetic information that could become available about each and every individual? A legitimate concern exists that such information should not be gathered until our society is ready to handle the flood of information and its consequences.

These are some of the more general reasons given for genetic engineering being both welcomed and feared. Each of us has his or her own level of comfort with the introduction of new technology that may deeply affect us and our society. It is our hope that we will engage in an informed debate about the pros and cons of this technology, not simply embrace it just because it is new and exciting or condemn it out of hand because it might alter our perception of the natural order of life.

SUMMARY

Recombinant DNA technology is a set of techniques for effectively manipulating and analyzing DNA molecules. Specific genes can be isolated, moved from one organism to another, modified, and controlled. This technology allows new combinations of functions to be gathered into particular organisms, or particular products to be made, in ways unimaginable a short time ago. Other techniques allow us to determine the sequence of bases in a DNA molecule, yielding information about the product of any gene contained in the DNA. Still other techniques allow for the rapid determination of whether a particular DNA sequence is present in a sample or whether a particular DNA sample is the same as or different from another.

This collection of tools and techniques has significantly changed the face of modern biology. No longer are genes "mysterious"; they can now be effectively studied. No longer are we limited to using living things that possess given traits; we can now engineer traits into many different organisms.

Is this technology in any way unnatural? Some would argue that we are tampering with the natural order of things and are violating or changing the course of evolution by mixing genes from different species. This issue is an important one, and it frequently arises during discussions about recombinant DNA.

REVIEW QUESTIONS

1. For what is gel electrophoresis used? What property of DNA is used during the running of a gel?

2. Explain how chain termination is involved in DNA sequencing.

3. Why is restriction mapping important?

4. How is hybridization used to (a) determine whether a cell has a particular gene, (b) study gene expression, and (c) diagnose viral infections?

5. Describe how genetic engineering of transgenic plants and animals can be beneficial. What are some of the potential dangers involved with the use of this technology?

6. How would knowledge of the sequence of a gene allow us to determine the sequence of amino acids in the protein product?

7. When the amino acid sequence of a protein is used to determine the corresponding DNA sequence of a gene, what problem does the fact that more than one codon encodes most of the individual amino acids present? Show by means of a diagram that you understand the question.

8. Why is PCR an extremely valuable technique? List and describe the rationale behind the major steps involved in PCR.

9. How can PCR be used in conjunction with hybridization in forensics and in the diagnosis of disease?

10. Nitrogen fixation is naturally very efficient when specific bacteria and legumes (clover, alfalfa, peas, beans, soybeans, and peanuts) are involved. What would be the economic value of introducing a nitrogen-fixing gene into a crop such as field corn? (*Hint*: Corn is not a legume.)

11. One area in which biotechnology may be needed is biomass utilization. Environmentally (ecologically) and economically, there are strong concerns: What do we do with the biomass? Give a possible answer to this question.

12. How has biotechnology enhanced medical diagnosis and treatment? Provide some specific examples.

13. Gene therapy as currently practiced is relegated to somatic cell applications. Why is there little or no concentration on human germ line therapy (altering genes in human eggs and sperm)?

14. It is often said that DNA fingerprinting in paternity cases can prove that a given individual is *not* the father, but it cannot prove that he is. Why?

15. The "oil-eating" bacteria that have been genetically engineered to degrade crude petroleum offer an excellent example of bioremediation. In what other situations might bioremediation be useful?

16. List five of the most outstanding examples of how biotechnology can be applied to agriculture.

17. Explain why it might be important to transfer herbicide resistance genes into crops such as corn or soybeans. Note that herbicides are usually employed agriculturally to reduce weed presence and/or growth in and around row crops.

18. What reason would anyone have for putting a gene for Bt poison (a natural bacterial product) into cauliflower and broccoli seeds?

19. How is *Agrobacterium tumefaciens* especially suited for biotechnological applications in agriculture?

CHAPTER 10

Genetic Disease and Gene Therapy

Most of us want our children to be better off than we are. We go to great lengths to ensure that they will be better educated, more competitive in the workplace, and financially more secure. We assume that they will also enjoy better health, primarily through improved health care but also through healthier life-styles. Today we know more about the workings of our bodies than did our parents, and most of us expect to live long, healthy lives. We have gained some measure of control over our bodies and the infectious agents that continually threaten them.

But what of our genes? Between 5 and 10 percent of children inherit an identifiable genetic defect. Some are much more serious than others, of course, but the fact remains that genetic disease is a frequent occurrence in human populations. To some degree, modern medicine has been able to alleviate the symptoms of some genetic diseases, but until recently a cure for genetic disease has been unthinkable. A cure requires that a defective gene be replaced or augmented with an intact, normal gene. Before the development of recombinant DNA, it was impossible to conceive of how this could be done.

Over the last two chapters, we have covered much about recombinant DNA. Most of the description has centered on the technology: how genes are moved around, how we can clone genes, and how we can get cloned genes expressed in other organisms. In this chapter, we will look at the direct application of recombinant DNA to our own genes and see how modern medicine can begin to address impairments in our genes as well as our bodies.

GENES THAT CAUSE DISEASE

Since the ultimate goal of this chapter is to understand how to treat genetic diseases, we first need to understand as much as we can about the causes of genetic diseases. How do we know that a disease is genetic? How do we know what has happened to a gene to alter it? How can we tell if an individual carries genes that will cause disease? All of these questions must be answered before we can think about curing a particular disease.

Inheritance of Disease

To classify a disease as genetic, at least one gene must clearly contribute in a significant way to causing the disease. This does not necessarily mean that only genes are involved; they simply must be involved in some specific way. We have already looked at some of the difficulties of classifying a disease as genetic (Chapters 5 and 6), but here we can explore the issue in more detail.

Many diseases are determined completely by genetics, for example, hemophilia. This disease occurs because of an absence of normal levels of one of the blood-clotting proteins. Without this protein, blood will not clot. It doesn't matter what environmental influences might be present; the patient will still suffer from hemophilia. If a hemophiliac is cut, bleeding will not stop. This disease is determined completely by mutations present within one or more specific genes in the body.

Other diseases have very significant genetic components but also depend on nongenetic influences. Familial hypercholesterolemia (FH) is a good example. This disease arises because of defects within the gene for the low-density-lipoprotein (LDL) receptor, which is responsible for taking up LDL from the bloodstream (Chapter 5). If LDL is not taken up and processed by cells, it accumulates to very high levels in the blood. Since LDL molecules carry cholesterol, cholesterol and fat in the blood can rise to dangerously high levels. FH is a genetic disease in that mutations within the LDL receptor gene are responsible for the decreased ability of cells to take up cholesterol. On the other hand, a significant amount of blood cholesterol can be affected by diet and exercise. Thus, FH cannot be considered a completely genetic disease, since the symptoms can be controlled at least to some extent by changes in the environment.

Given a particular disease, how can we tell whether it is genetic? The primary way to tell is by analyzing the inheritance of the disease. Genetic diseases should be inherited in some manner, either in strict Mendelian fashion or through some other pattern. Pedigree analysis, as well as twin studies, can shed considerable light on whether or not a particular disease is inherited. Note that to perform such studies we employ the usual genetic analysis of traits. A genetic disease is a trait that can be inherited just like any other trait, and it can be studied in the same way. It can be dominant or recessive. It can exist as many different alleles or as a single allele. It can be caused by changes in the interaction of many genes or by a defect in a single gene.

For many reasons, however, studying the inheritance of a disease does not always give a clear indication of whether the disease is genetic in origin. A complete understanding of the genetic contribution to a disease can be gained only by a molecular study of the gene in question. The particular gene (or genes) must be identified, cloned, sequenced, and compared to the "normal" gene(s). Only then can we determine what biochemical function is normally provided by the product of the gene and how the product of the defective gene fails to function.

The thought of this task can be daunting, since there are so many human diseases and genes about which we know absolutely nothing. However, we can use recombinant DNA techniques to help us accomplish these goals.

Finding and Cloning a Human Disease-causing Gene

Assuming that a defect in a single gene results in a disease, can we isolate that gene or even locate it on a particular chromosome (within a particular region)? The more precisely we can locate it, the more easily we can clone it. As we know from our consideration of genetic mapping in Chapter 4, determining the genetic location of a disease trait is a matter of studying the inheritance of the trait in relation to known genetic markers. The mapping that is being performed as part of the Human Genome Project is invaluable to continued success in locating genes involved in human diseases.

Let's begin by looking at the most common lethal autosomal recessive disease among Caucasians—cystic fibrosis. Cystic fibrosis (CF) is characterized by abnormal glandular secretions that cause pancreatic, pulmonary, and digestive dysfunction in children and young adults. The primary symptoms are repeated respiratory tract infections and malnutrition due to an inability to digest food. The infections result from greatly increased secretions of extremely viscous mucus into the lungs, which provide an exceptionally hospitable environment for the growth of microbial agents. Malnutrition results from the absence of pancreatic juices in the digestive tract; the secretion of these juices from the pancreas is blocked, and thus the digestive enzymes responsible for the breakdown of foods are missing. Patients usually die from respiratory failure; as a result of chronic infection, fibrous tissue develops in the lungs, gradually preventing normal gaseous exchange. The disease is treated by supplementing the diet with digestive aids, taking antibiotics to prevent infections, and using physical therapy to remove excess mucus from the lungs. The disease affects about one in every 1,800 Caucasian children, and patients rarely survive beyond the age of forty. The disease is rare in other races. Pedigree analyses of affected families (Figure 10.1) have firmly established CF as an autosomal recessive genetic disease that becomes manifest in early life, with more than 50 percent of patients dying before they reach age twenty-one.

A combination of two recombinant DNA techniques was used to identify and clone the CF gene. Recall that the closer two particular genes are to each other on a chromosome, the more likely they are to remain together through the reproductive process—a situation known as genetic linkage. Genes that are tightly linked will almost always be found together in individuals. Genes that are not linked will be randomly distributed.

The first phase involved in identifying the CF gene was to establish a linkage between cystic fibrosis and some genetic marker (Figure 10.2). Since so few genetic markers in the human genome were known, RFLP markers were used in these early mapping studies. Many families are affected by CF, so researchers could obtain DNA samples from many individuals. Each individual was checked for various RFLP

Figure 10.1 Pedigree analysis. This pedigree analysis depicts the inheritance pattern of the autosomal recessive disease cystic fibrosis. Shaded individuals suffer from CF.

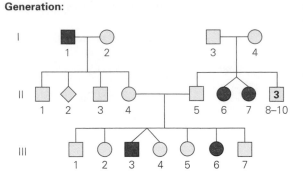

Figure 10.2 RFLP markers close to a gene. When a gene causing a disease has not been cloned and its location is unknown, sometimes its presence can be detected by testing for the presence of RFLP markers close to the responsible gene. This diagram shows homologous DNA from a family in which some members have CF and others do not. In this family, different versions of an RFLP marker are associated with different genes. If a person has inherited the RFLP marker with one restriction site rather than two, there is a high probability that the individual has also inherited the disease-causing gene.

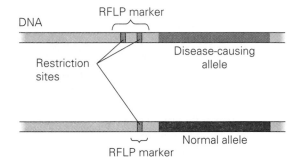

markers whose presence or absence was correlated with the presence or absence of the disease. The intent was to identify one particular RFLP that was tightly linked to the disease; in other words, scientists were searching for a unique DNA sequence that was located very close to the CF gene. After much work, scientists identified one RFLP that was tightly linked and therefore located close to the unknown CF gene.

At that time, it was known only that the CF gene was located near this particular RFLP marker, and all that was known about the gene was that it caused CF. How was it to be identified? The process began with the cloning of successive pieces of DNA leading from the known sequence (Figure 10.3), through a process called **chromosome walking**. A large piece of DNA was first cloned using the RFLP sequence as a probe. Once this fragment was sequenced, a small region at the very end was used as a probe to help clone another fragment. After this fragment was cloned and sequenced, a probe derived from this sequence was used to clone yet further pieces of DNA. In this way, an overlapping set of DNA fragments was cloned, and the fragments could be placed in order, reconstructing the sequence of the chromosome in the region near the original marker sequence. If chromosome walking continued, eventually the nearby CF gene would be cloned in one of these DNA fragments.

Of course, in chromosome walking, you never know when you are done—when you have finally cloned the desired gene. How can we tell if we have the CF gene?

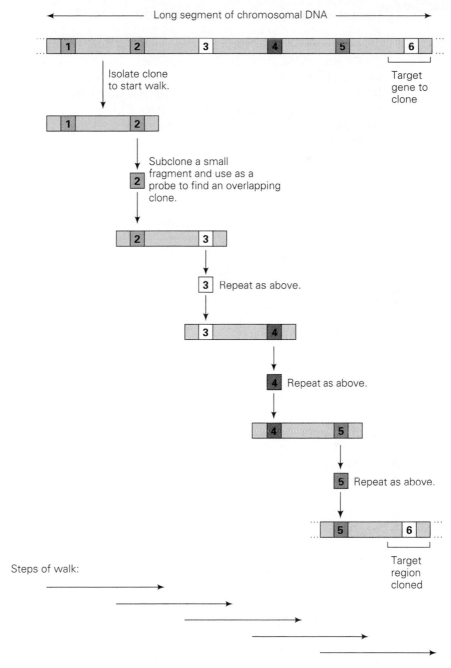

Figure 10.3 Chromosome walking. This technique is used to clone successive fragments of DNA in order, beginning at a specific location. The purpose is to be able to "walk" from an RFLP marker to an unknown gene (known to be near this RFLP).

Open reading frames, or sequences that can potentially code for proteins as determined by the genetic code, are identified within the cloned sequences. All open reading frames are potential CF genes. If any of these open reading frames correspond to the CF gene, these genes should be expressed differently in CF patients than in other individuals; mRNA corresponding to the defective CF gene should be present in at least some tissues of CF patients. Sweat glands were chosen for study, because it was known that these glands were always affected by CF. Probes corresponding to each potential open reading frame were constructed, and these probes were used in hybridization analysis of the mRNA present in the sweat glands of patients. One gene was identified that was expressed abnormally in CF patients but not in unaffected individuals. This became the prime CF gene candidate.

The best way to ensure that this was the CF gene was to study it in a variety of individuals, some suffering from the disease and some not. In all CF patients this gene should be mutated, whereas in normal individuals the normal gene will be intact. Such was the case with the potential CF gene; in every CF individual examined, a mutation was found within that gene.

Further evidence that this was the CF gene came from continuing biochemical and recombinant DNA studies. Researchers found that the transport of chloride ions was deficient in secretory cells of CF patients. Some of these cells were cultured in the laboratory, and a normal copy of the CF gene was introduced into them by recombinant DNA techniques. Chloride transport was restored in these cells, providing convincing evidence that the identified gene really was the CF gene.

The entire process of cloning the CF gene required about four years of intense work in several laboratories, not a trivial undertaking. However, the general strategy used, **positional cloning**, has proved to be of great value in cloning genes that cause human diseases. Many other genes have been cloned in this manner.

The information obtained from the cloned gene can be used in many ways. For example, when the DNA sequence of the CF gene was obtained, the amino acid sequence encoded by the gene could be determined. A comparative study of the protein product suggested that it was a membrane protein and was most likely involved in transport. This supported the hypothesis that a defect in chloride transport was at least one biochemical problem responsible for CF.

After the actual protein responsible for causing CF had been identified, further studies aimed to determine why it was defective, in hopes of being able to treat the disease by reversing the situation. Also, the DNA sequence itself can be used in several ways, which we'll outline next.

GENETIC DIAGNOSIS

Genetic diagnosis means determining whether or not a particular genetic defect is present in an individual. It often takes the form of a test designed to detect the presence or absence of a particular enzyme. Phenylketonuria and Tay-Sachs disease are two genetic conditions that can be diagnosed through detection of missing enzyme activity. Alternatively, a diagnostic test can be performed that determines directly whether or not a mutated form of a gene is present. In such a directed search, DNA from an individual is isolated and subjected to either RFLP analysis or hybridization analysis, using a probe constructed from the known sequence of the defective gene. If the probe is constructed so that it will hybridize only to defective copies of the gene, then the probe can be used for genetic diagnosis. The intensive effort under way to clone genes responsible for human genetic diseases has resulted in an ever-increasing number of such probes becoming available. In theory, any individual at any stage of life can be tested for the presence of any known gene. However, a number of technical difficulties are associated with designing diagnostic tests, particularly those involving hybridization.

Specificity of Diagnosis

As with any hybridization process, the final result is only as good as the probe specificity. If a probe will bind to many different sequences, then the result will not be specific. Being able to construct good probes is an integral part of designing good hybridization procedures. How do we determine the specificity of a probe?

Recall that hybridization is the binding of a small, single-stranded DNA probe to its complementary sequence, which is usually contained within a much larger piece of DNA. The binding is governed by the base-pairing interactions between the two sequences. A number of factors ultimately determine how well the two molecules bind to each other. The first is the length of the probe. The longer the probe, the more bases there are available to bind to the target molecule. The more base-pairs that can form, the stronger the interaction between the two molecules (Chapter 2). This is what results in increased specificity—stronger interactions will persist where weaker ones will not.

Another factor governing specificity is the experimental conditions under which molecules are allowed to hybridize. In particular, two factors are often varied: the temperature and the salt concentration. Higher temperature tends to denature, or split apart, double-stranded nucleic acids. The stronger a particular interaction, the higher the temperature must be in order to disrupt the interaction. Similar effects are caused by increases in salt concentration. The higher the concentration of ions in the solution surrounding DNA molecules, the harder it will be for a probe to bind to the target molecule.

The third factor (and the one most important for genetic diagnostic procedures) is the degree to which the probe is complementary to the target molecule. Consider the situation illustrated in Figure 10.4. If a probe that has a base mismatch is used in a hybridization procedure, the attraction will be lower. Depending on the size of the probe and the nature of the conditions, this mismatch may be enough to prevent the probe from binding tightly.

The ability to vary the conditions under which hybridization takes place has made it possible to determine single base changes in DNA sequences. This is useful because a defective gene may differ from the normal gene by only a single base. But such is not always the case; in many instances, mutations leading to disease are a result of more substantial changes, sometimes even large deletions or insertions. Clearly, one problem that arises in designing diagnostic procedures is choosing

Figure 10.4 The use of hybridization to distinguish between target sequences containing only a single base-pair difference. The probe is designed to be perfectly complementary to the defective gene. This same probe, when used to hybridize to the normal gene (**a**), contains a base mismatch. The conditions of hybridization can be adjusted such that this single difference is sufficient to prevent the binding of the probe to the normal gene but allow binding to the defective gene (**b**). Thus, in a hybridization analysis, the defective gene will be detected and the normal gene will not.

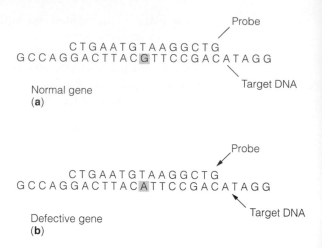

appropriate probes and establishing conditions under which the proper specificity can be obtained.

Another difficulty, related to specificity, arises in the design of genetic diagnostic tests. Are all cases of a particular disease caused by the same mutation? In other words, does patient X, who suffers from a genetic disease, carry the same mutation as patient Y, who also suffers from the same disease? Maybe not; after all, many different types of mutation can result in the loss of a protein's function. Cystic fibrosis provides an excellent example of this problem. After the gene for cystic fibrosis was identified, CF genes from many different individuals were sequenced. It was found that about 70 percent of those suffering from the disease carried a single particular mutation but that the other 30 percent of patients carried numerous other mutations. (Over a hundred different CF alleles have been identified.)

To develop an effective diagnostic test, we should be able to determine whether *any* of these mutant sequences are present. This is particularly difficult to determine through hybridization, considering the vast number of potential mutations. Since each mutation requires its own unique probe, screening for all forms of CF would require using a great many probes mixed together. Even then, some rare form of the gene might be missing from the collection. While it is certainly possible to use a mixture of probes in a hybridization procedure, it is difficult and costly to include all those necessary in this instance. Increasing the number of probes also increases the chances for some nonspecific interaction between one of the probes and some similar DNA sequence outside the CF gene. How reliable would such a test be?

Alternatively, we could develop a very accurate test using a probe for the predominant mutation, the one responsible for 70 percent of CF cases. This particular test, though very accurate, would provide an answer that was only 70 percent reliable. In other words, if a particular individual were tested and showed negative, all that we could say is that there is a 70 percent chance that the individual does not have a gene that will result in cystic fibrosis. How useful would this type of test be?

Two issues need to be addressed during the development of any hybridization-based genetic diagnostic procedure. First, the nature and number of the mutations responsible for the disease must be established. Can a reasonable test be devised using a small number of probes? Second, the necessary probes must be selected, and experimental conditions must be established in which they will provide the specificity necessary to draw accurate conclusions. An unreliable test is not worth the effort involved in developing it.

Prenatal Diagnosis

Assuming that a reliable diagnostic test can be developed, whether through hybridization or enzyme assay, how might such testing be applied? Again, in theory, it is possible to test any individual at any stage of life. One application of genetic diagnosis is to examine the genetic makeup of a fetus in order to determine whether the individual will suffer from a debilitating disease.

Consider a married couple who are both carriers of Tay-Sachs disease. Recall that this is an autosomal recessive disorder that affects the nervous system and causes death by the age of three; no treatment is available for the patient, and there is no hope of survival. A carrier has one normal copy and one defective copy of the gene; since the defective copy is recessive, the carrier does not suffer from the disease. There is a 25 percent chance that any baby born to these parents will inherit both copies of the defective gene and have the disease. If the fetus could be accurately diagnosed as carrying both defective copies, therapeutic abortion could be considered if the evidence was discovered early in the pregnancy.

Prenatal diagnosis through hybridization requires that DNA be obtained from the fetus. There are two ways of doing this: **amniocentesis** and **chorionic villus sampling** (**CVS**). Amniocentesis involves inserting a needle through the abdominal cavity of the mother into the sac that holds the fetus (Figure 10.5) and withdrawing a small volume of amniotic fluid containing fetal cells. Some of these cells are cultured in the laboratory, resulting in enough cells to be able to perform a variety of tests, including karyotyping (Chapter 3). However, culturing fetal cells is time-consuming. In addition, amniocentesis is recommended between the sixteenth and twentieth weeks of pregnancy. Any delay, which is normal in culturing fetal cells, might preclude the use of therapeutic abortion, which is usually not performed after the twenty-fourth week.

A relatively new way of performing amniocentesis involves guiding the needle by ultrasound into the amniotic fluid. This allows the procedure to be performed four to six weeks earlier, allowing more time before a decision must be made regarding abortion. Diagnostic tests can be performed on the cells directly harvested from the amniotic fluid through a combination of PCR (to amplify target sequences) and hybridization. Therefore, it is possible to apply hybridization-based diagnostic tests to fetuses, resulting in prenatal diagnosis.

Chorionic villus sampling involves removing pieces of the chorionic villi, embryonic tissue that anchors the fetus to the uterine wall and serves a nutritional function until the placenta develops (Figure 10.6). These pieces can be obtained either by puncturing the mother's abdominal wall with a needle, as in amniocentesis, or by inserting a catheter through the cervix. CVS has the distinct advantage of allowing earlier diagnosis—tissue can be obtained from the eighth week through the twelfth week

Figure 10.5 Amniocentesis—a method for prenatal diagnosis of some genetic defects. A needle is carefully inserted through the abdominal wall of the mother into the amnion surrounding the fetus. About 30 ml of fluid is withdrawn. Biochemical and hybridization tests can be performed on the fluid and on cells taken from it. Fetal cells are separated from the rest of the amniotic fluid and grown in culture. After several weeks, a sufficient number of cells have grown to allow karyotyping.

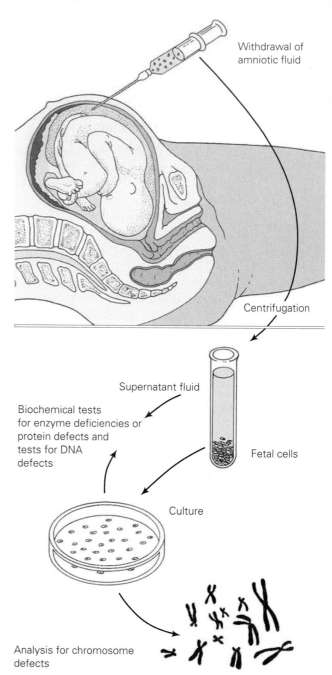

Withdrawal of amniotic fluid

Centrifugation

Supernatant fluid

Biochemical tests for enzyme deficiencies or protein defects and tests for DNA defects

Fetal cells

Culture

Analysis for chromosome defects

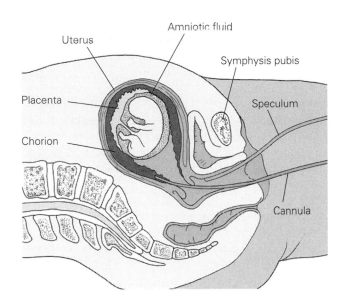

Uterus
Amniotic fluid
Symphysis pubis
Placenta
Speculum
Chorion
Cannula

Figure 10.6 Chorionic villus sampling. A tube is inserted through the cervix to contact the chorionic villi. Portions of the villi are then removed by suction. Fetal cells are separated from fluid and from any maternal tissue. There are usually enough cells in this sample to directly perform karyotyping, as well as other biochemical and hybridization tests.

of pregnancy, allowing therapeutic abortion during the first trimester. This sampling method is not without its problems, however. CVS has a slightly increased risk of miscarriage compared to amniocentesis (1 percent versus 0.5 percent), although some argue that this is an acceptable risk since such early diagnosis is possible.

Even our current level of technology allows the fairly routine examination of fetal tissues, and our capabilities are likely to increase. Fetal tissues can be examined for gross chromosomal aberrations, such as Down syndrome; biochemical defects (missing enzymes); or the presence of specific mutations in genes. Only cost and the difficulty of developing a reliable test for a specific disease prevent widespread use of this technology.

Embryonic Diagnosis

It has become possible in recent years to perform genetic diagnosis using material obtained from a single cell. The DNA of the cell is isolated, and PCR is used to amplify the desired target sequence or sequences. Following amplification, the target DNA can easily be tested by using hybridization procedures.

This capability leads to another, even more controversial way to diagnose the genes of an individual before birth. To some extent, the moral dilemma associated with therapeutic abortions is avoided; however, concerns of a different nature can be raised. This technique is built on our ability to manipulate eggs and embryos outside the womb.

In vitro fertilization is a fairly common and routine procedure. Unfertilized eggs are removed from the mother. They are then mixed with sperm outside the body. Following fertilization, the embryo begins to divide. After a certain number of cell divisions, the dividing embryo is returned to the mother's body, where it can implant in the womb. The mother must be treated with hormones in order to simulate the effects of early fertilization. The process is not perfect; multiple attempts are often required before successful implantation occurs. *In vitro* reproductive techniques have allowed many couples who, for whatever reason, were experiencing difficulties reproducing to have children.

The cells of dividing embryos, while still outside the mother's body, can actually be separated from one another and cultured. From an eight-cell embryo, for example, a single cell can be removed and transferred to a separate culture dish. The remaining seven-cell embryo will continue normal development. The removed cell, while it does not continue normal development, can be maintained in culture and used as a source of material for genetic diagnosis.

The following scenario is then possible. Prospective parents (let's assume both are heterozygous for cystic fibrosis, so there is a 25 percent chance that an individual offspring will suffer from the disease) donate eggs and sperm for *in vitro* fertilization. Many eggs are fertilized, but only some survive to begin dividing. The survivors are cultured until they reach the eight-cell stage. At this time, one cell is removed from each embryo. The seven-cell embryos are further cultured, while the removed single cells are subjected to genetic testing for cystic fibrosis. About 25 percent of the embryos should be found to have two copies of the cystic fibrosis gene. These embryos can be destroyed without ever being implanted—as are many embryos during the normal course of *in vitro* fertilization. The other embryos can be implanted into the mother. What this amounts to is a way to diagnose the embryos before they are returned to the mother. Using this procedure, parents heterozygous for CF could not only assure themselves of never having children who suffer from the disease, but also avoid having to abort to achieve this goal. In fact, they could, if desired, choose to implant only those embryos (the 25 percent) that have no copies of the CF gene. Then not only the individuals arising from these embryos but also their offspring, the next generation, would be free of CF.

Our ability to manipulate dividing embryos has other implications. For example, an embryo can be split into two, with each remaining fully capable of developing normally, thus generating identical twins from a single *in vitro* fertilization. One of the separated embryos could be stored frozen in a state of suspended animation. That embryo could be removed later and returned to culture to continue development. In essence, twins of different ages could be generated.

The implications of such technological advances are enormous. It is easy to imagine a future time when many genes can be routinely disgnosed, ensuring that children do not suffer from a number of genetic disorders. It is equally easy to imagine such capabilities being used to diagnose other genetic traits—for example, to choose embryos that will produce children with a particular hair color.

 ## Diagnosis of Adults

Presently, diagnosing genetic disease in adults is much more commonplace than diagnosing it in fetuses, for numerous reasons. First, it is relatively simple since a blood sample can provide the material for some enzyme diagnostic tests, as well as the cells necessary for hybridization-based tests. Second, screening can be applied to individuals with clear informed consent. Third, the results are usually not used to make decisions about abortion. As with other forms of diagnosis, tests can be performed to detect biochemical differences or differences in genetic makeup.

Why would individuals choose to be diagnosed? An adult is usually diagnosed in order to determine whether he or she is heterozygous for a particular trait. To prospective parents, this is potentially useful information that allows them to make reproductive choices. For instance, parents who are both heterozygous for a particular recessive gene might elect not to have children, knowing that a child would have a 25 percent chance of inheriting both copies of the defective gene. Alternatively, the same prospective parents might choose to conceive and then pursue prenatal diagnosis in order to determine whether the fetus will suffer from the disease.

One example of adult genetic diagnosis deserves specific attention. Recall that Huntington's disease is an autosomal dominant genetic disease of the nervous system. There is no treatment, and following the onset of the disease it progresses for ten to twenty years until the patient can no longer breathe. This disease is unusual in that it does not generally manifest itself until after age forty.

The gene responsible for Huntington's disease has been identified, so it is now possible to perform a nearly foolproof diagnostic test based on hybridization. Since the disease symptoms are much delayed, there is a long period of time during which an individual can be diagnosed before significant problems arise. For some people, deciding whether or not to be tested is a difficult choice. Instead of simply finding out whether they are heterozygous for a particular trait or whether a fetus will have a disease, people being tested will be finding out directly whether they will suffer from the disease.

ISSUES: TO KNOW OR NOT TO KNOW?

Having diagnostic testing done for a disease that will definitely affect you later in life would most likely place you in an unprecedented situation. In the case of Huntington's disease, you would stand to learn whether you will suffer from a long, debilitating, untreatable, costly, and ultimately fatal condition. If the test is positive, the knowledge gained is this: You will have and die from Huntington's disease.

Some people might choose to be tested, especially if there is a history of Huntington's disease in the family. They might feel that knowing their status will prepare them or others for the inevitable. Others might avoid diagnosis, deliberately choosing not to know—perhaps so they can remain hopeful. The decision about whether to be diagnosed is personal. Should anyone else have the right to know your genetic makeup? Who might want to know, besides family members and loved ones? Surely, most insurance companies would like to know. Most health insurance providers will not cover "preexisting" conditions. Does carrying the gene for Huntington's disease constitute a preexisting condition? What constitutes the preexisting condition, the actual onset of the symptoms or simply having the gene that will lead to the disease?

It might be argued that a health insurance company, in the long-standing business of predicting medical outcomes, has the right to all information that might allow more accurate predictions. The more accurate the predictions, the more accurately insurance companies can assign the costs of insurance to different risk categories. Individuals with high blood pressure or high cholesterol are sometimes forced to pay higher premiums, on the justification that statistically these individuals will incur much higher medical costs as a result of these conditions. Similar arguments have also been made regarding some behaviors; for example, smokers usually pay higher health insurance premiums. Our society has accepted these conditions of insurability. In fact, we have often demanded differentiation into risk categories. Shouldn't genetic information also be provided to allow insurance companies to more accurately categorize individuals?

Should an employer have the right to know about genetic conditions such as Huntington's disease? Given two otherwise completely equal job candidates, doesn't it make more economic sense to invest resources in the person who will potentially serve the company for the longer period of time—the person not carrying a Huntington's disease gene? On the other hand, might such a choice be discriminatory?

There is a growing concern that our ability to diagnose genetic defects will create a new category of disabled persons—the genetically disabled. Individuals carrying the Huntington's gene would fall into this category. While they might not be suffering from the disease now, they will suffer from it in the future. People carrying genes for predispositions to cancer and heart disease, or to Alzheimer's disease and schizophrenia, might also fall into this category.

At the present time, the Americans with Disabilities Act of 1990 is the primary legislation that protects individuals against genetic discrimination. However, this legislation was not written with genetic disease in mind and can be interpreted in many different ways. Many feel that stronger antidiscriminatory legislation is needed. Some states have enacted their own legislation, often in conjunction with workers' compensation programs. Where these laws have been enacted, they are usually very narrow—most mention only discrimination based on the diagnosis of the heterozygous sickle-cell condition. This is just one example of how the implementation of a new technology often lags behind its development.

• •

GENETIC SCREENING

Once a diagnostic test has been developed and shown to be accurate and reliable, **genetic screening** programs can be considered. The goal of a screening program is to test for a genetic defect in as many individuals as possible within some specified target group. The target group is usually individuals at increased risk for the disease. For example, cystic fibrosis occurs much more frequently within the Caucasian population, and Down syndrome occurs more frequently in children born to mothers over the age of thirty-five. Sickle-cell anemia occurs much more frequently within the African American population. Tay-Sachs disease occurs almost exclusively in Ashkenazi Jews. Because subpopulations tend to reproduce within their own groups, particular diseases are often associated with these groups.

Since genetic screening programs are costly and must be highly organized, they arise only through the development of a specific social policy. Exactly because of this, any genetic screening program can be called into question. We readily accept several screens performed on newborns, including one for phenylketonuria. Because both our knowledge of genetic diseases and the potential for curing such diseases are increasing dramatically, increasing pressure may be exerted to enact more genetic screening programs.

Enacting a genetic screening program involves many decisions. We need to understand the scientific basis for diagnosing genetic diseases. How accurate and reliable are the tests? How expensive will they be if they are applied to a large number of people? Scientifically, genetic screening is no different from performing individual genetic diagnoses. Thus, as improved diagnostic tests are developed, they will also be available for use in screening programs.

It is theoretically possible, even given our present technology, to establish a screening program that would diagnose almost all fetuses having the gene for Huntington's disease. Assuming that a positive diagnosis would be followed by therapeutic abortion, it follows that it is theoretically possible to establish a program that would, in effect, drastically reduce the number of individuals in the next generation who will suffer from Huntington's disease. Granted, embarking on a genetic screening program on such a grand scale is not practical or necessarily even desirable, for many reasons. But the technology to do it already exists.

What specific issues will determine whether or not genetic screening programs are enacted? We will consider genetic screening for Tay-Sachs disease and sickle-cell anemia in order to examine this question.

Tay-Sachs Screening

A voluntary screening program designed to detect persons heterozygous for Tay-Sachs disease is in place. The program has proved to be quite successful. The screening test consists of a straightforward enzyme assay that can be used to determine the level of hexosaminidase A, the enzyme missing in Tay-Sachs patients. Individuals with two normal genes produce high levels of the enzyme, while heterozygous persons produce significantly lower levels. Thus, it is usually possible to clearly identify heterozygous individuals.

Since Tay-Sachs disease affects primarily individuals of Ashkenazi Jewish descent, the voluntary adult screening has been organized through Jewish religious centers. Males about to become or considering becoming fathers are the initial target group; if the father is not heterozygous, the mother does not even need to be tested, since it will be impossible for the offspring to inherit two copies of the Tay-Sachs gene. If the father is heterozygous, then there is good reason to test the mother. If the mother also proves to be heterozygous, then their fetus should be tested. Since there is no hope of curing the debilitating and ultimately fatal disease, affected fetuses are usually aborted. This testing regimen minimizes the stress and emotional trauma for all involved and also provides reliable information of practical use to prospective parents.

The success of this particular voluntary screening program has been remarkable, resulting in a substantial (greater then 80 percent) decrease in the number of births of Tay-Sachs children in the past two decades. Since the voluntary program is accompanied by intensive, effective educational efforts instituted at the request of the group at risk, it has been widely accepted and its effects have been well understood.

Sickle-cell Anemia Screening

A remarkably unsuccessful screening program was enacted to diagnose individuals heterozygous for sickle-cell anemia. This autosomal recessive disease is caused by a mutation within one of the globin genes, the genes that direct the synthesis of hemoglobin. HbS, the sickling form of hemoglobin, causes drastic shape changes in red blood cells, often causing them to burst, as well as blocking the flow of blood in many small capillaries. Sickle-cell anemia is a painful disease with very few or no effective treatments. Like many genetic diseases, it is particularly associated with specific groups, in this case African Americans and a particular subpopulation of Mediterranean people.

Sickle-cell anemia is characterized by one specific change in the sequence of the gene that encodes the β-chains of hemoglobin. The morphological change expressed as "sickle-shape" stems from a single amino acid change in each β-chain. Only a single sickle-cell allele exists, and a blood test can reveal the presence of the HbS variant. Heterozygous persons will display both normal and defective Hb, and screening is easy to perform.

In the 1960s, a genetic screening policy was initiated. Many states passed laws requiring African American males about to be married to be screened for HbS. The intent was to identify heterozygous people in order to prevent the birth of children with sickle-cell disease. Some programs were even extended to include schoolchildren, in the mistaken belief that the heterozygous condition would be harmful. This program seemed bound to fail for at least four reasons.

First, parents who were both identified as heterozygous could use this information only by choosing not to have children. In the 1960s, there was no simple way to perform a prenatal diagnosis to determine whether a fetus would suffer from the disease. Restriction enzymes had not been discovered, and the DNA-based test that is available today was not available then. The only test that could be used to screen fetuses was to look for the presence of different hemoglobin proteins in the blood. This required drawing fetal blood; amniocentesis could not be used, because the amniotic cells do not express the globin genes and therefore do not contain any hemoglobins. While it was possible to obtain a fetal blood sample, the procedure was both risky and expensive, not one that could be applied routinely.

Second, no effective education programs accompanied the screenings. There was little to combat the stigmatization of heterozygotes, an inherent risk in all forms of genetic screening. Questions were raised. Would the results become public? Would governmental agencies keep records of the condition? In general, who had the right to the information?

Third, the programs were mandated by the government and targeted at the African American population during a time of great racial tension. Even though many such programs were initiated by African American politicians, many people reached the conclusion that the screening program was racist and discriminatory.

Fourth, the program was mandatory. Individuals were not free to choose to participate.

In hindsight, we can conclude the following: In order for a screening program to be effective, the target population must accept the program, be willing to learn the consequences, and have some choices as a result. Forced participation in a misunderstood program that provided few benefits only invited disaster. The differences between the Tay-Sachs and sickle-cell screening programs provide models for both effective and ineffective ways of pursuing genetic screening policies.

ESSAY: WHEN SHOULD A SCREENING PROGRAM BE INSTITUTED?

How much control should society have over individual genetic screenings? Are there diseases for which a screening program is warranted? On one hand, it seems logical to try to eliminate genetic disease from a humanitarian standpoint. It is also argued that screening programs that result in a significantly lower incidence of genetic diseases would have great economic benefit, since society would not have to bear the medical care burden (Chapter 6). On the other hand, genetic screening programs are probably eugenic at some level.

Almost every baby born in the United States is screened within the first few days of birth for the genetic disease phenylketonuria. It is hard to argue against such a program: It is easily implemented, requires only a small amount of blood, and poses little risk or emotional trauma. Of greatest importance, if the disease is diagnosed early, the infant's diet can be modified to prevent the mental retardation caused by the disease. In other words, early diagnosis is necessary for safe, effective treatment. Finally, even a positive diagnosis does not require a decision by the parents regarding abortion.

Any successful screening program must offer immediate, real, and tangible benefits to the diagnosed individual. This is the case with phenylketonuria and in screening for congenital hypothyroidism. If useful therapies or treatments for Huntington's disease were available, then it might be worth considering a screening program. Even then, there might be insufficient public sentiment for establishing a screening program, because such decisions involve economic considerations. A program's cost is weighed against the number of individuals that would benefit. A monetary valuation of life is sometimes difficult to accept, in spite of the economic realities that drive it.

Genetic Counseling

Our increased knowledge of genetic diseases and the improved capacity to diagnose and screen for them have produced a new medical specialty, **genetic counseling**. Genetic counselors have the responsibility of educating people about genetic diseases, available treatments, and possible reproductive choices. Genetic counselors face enormous challenges. Not only must they be extremely well informed regarding the rapid changes in our knowledge of genetic disease, but they must also be able to deal effectively with parents or individuals who often have preconceived notions about genetic disease. This truly unique blend of characteristics includes being able to deal effectively with both extreme medical problems and the severe emotional strain that results from receiving adverse genetic information.

TREATMENT OF GENETIC DISEASE

Until recently, no thought was given to actually curing a genetic disease, since no one could alter an individual's genetic makeup with any precision. Thus, treatment of specific genetic diseases relied on traditional methods of reducing the intensity of symptoms and alleviating suffering. Developments in gene therapy have led to the possibility of curing genetic disease.

Traditional Treatments

In cases in which the biochemical deficiency involved in a genetic disease is known, it is sometimes possible to effectively treat the disease. We've seen that phenylketonuria can be effectively treated through dietary modification.

Hemophiliacs are missing one of a series of clotting factors. Without this factor, blood does not clot. The treatment of hemophilia involves injecting some of the missing clotting factor whenever bleeding occurs. In this instance, it is possible to provide the missing protein (now prepared through biotechnological means) to the patient. Patient's life-styles remain relatively normal, but this particular treatment is not perfect. Wounds are still to be avoided at all costs, and the administration of the missing factor sometimes leads to complications, potentially including an immune reaction against the injected protein.

While there are often ways of treating the disease symptoms, no traditional treatment can provide an effective cure for a genetic disease. The only way to accomplish a cure would be to replace the defective gene.

Gene Therapy

We saw in Chapter 9 that creating transgenic plants and animals is becoming fairly routine. The process is useful for producing particular materials or transforming the properties of an organism. Is it possible to introduce new genes, or at least intact copies of normal genes, into humans? The only way to completely cure hemophilia, for example, is to provide the individual with the gene responsible for producing the missing clotting factor.

No theoretical boundaries prohibit the transfer of genetic material into humans through the same mechanisms used with other mammals. However, many technical and ethical boundaries do exist. One of the primary technical reasons why transgenic humans are not produced is the fact that, when transgenic animals are created, many individual animals receive the foreign DNA and only the few that survive and display the appropriate properties are kept. In other words, many future animals are sacrificed to obtain one transgenic animal. It is difficult to conceive of approaching human genetic manipulation in this manner. Therefore, we must apply different techniques in order to accomplish more effective transfer of genes into humans.

Also keep in mind that there is a basic technical difference between gene therapy aimed at curing recessive diseases and gene therapy aimed at curing dominant diseases. Recessive disorders result from the inheritance of two alleles of a gene that fail to function or fail to produce enough functional product. Recessive diseases thus are caused by the lack of a particular product. Gene therapy in this case must provide a missing function, so adding an intact gene to an organism might suffice. Dominant diseases, on the other hand, result not from a lack of function but rather from the addition of an altered function. Products of genes causing dominant diseases generally act in some harmful manner to disrupt metabolic processes. Gene therapy in this case must either return the defective gene to a normal state or inactivate it, in addition to providing the normal gene. Our technology is still far from being able to replace a particular gene, or even to alter the expression or function of a specific gene, as we will see. Thus, all the ideas about gene therapy we discuss here will pertain to the treatment of recessive disorders.

W. French Anderson, one of the leading developers of human gene therapy, has established five requirements that should be satisfied before human gene therapy is attempted:

1. The gene to be transferred must be available. This criterion has been satisfied through recombinant DNA, which has resulted in the availability of hundreds of cloned human genes.

2. An effective method of introducing the gene into human cells is necessary.

3. The target tissue or target cells must be accessible to the gene transfer technique. This is one of the major technical difficulties inherent in current gene therapy procedures.

4. The gene therapy procedure must not harm the patient. As we will see, there is a realistic possibility of negative effects resulting from attempts to perform gene therapy. We must do everything possible to avert these potential effects before we can mount an effective gene therapy treatment.

5. The treatment must result in a significant improvement in the health of the patient. Only continued study will tell if improvement can be achieved.

Experiments in human gene therapy are proceeding so quickly that it is difficult to keep abreast of the most current developments. Rather than describe all the possible approaches that are being pursued, we will concentrate on some of the more promising ones and describe the general techniques.

Transfer of Genes into Human Cells The most common way to transfer genes into human cells is through the use of **retroviruses**. Retroviruses have a unique replicative cycle (Figure 10.7). The genetic material of the virus is RNA, not DNA. When a retrovirus infects a cell, the coat of the virus is removed, releasing the RNA and some enzymes into the host cell's cytoplasm. One released enzyme is reverse transcriptase, which we encountered when we constructed cDNA from mRNA. Reverse transcriptase uses the viral RNA as a template for a single strand of DNA. The RNA is then removed, and the single strand of DNA is used as a template for its complement. This double-stranded DNA is then inserted into the chromosome of the host cell. The virus replicates by producing copies of the RNA and required proteins. These assemble into new viral particles and are released from the infected cell. We will examine a particular retrovirus in more detail when we study AIDS in Chapter 13.

Retroviruses can be adapted for use as cloning vectors in much the same way that bacteriophage lambda serves as a cloning vector. However, several modifications are necessary. First, the retrovirus must be able to accept a nonviral gene and incorporate it into the viral RNA. Second, any deleterious genes in the retrovirus must be removed or inactivated. (For instance, as we will see in Chapter 14, some retroviruses contain genes associated with cancer.) Third, to maintain the transferred gene inside a host cell, the ability of the retrovirus to replicate should be destroyed so that no new viral particles are created, and the host cell remains intact.

Engineered retroviral vectors provide the most efficient method of introducing genes into mammalian cells and thus are used for the majority of gene therapy experiments. But there are problems associated with their use. Chief among these problems is that retroviral vectors are able to carry only relatively small pieces of RNA; technical improvements will help address this. Another, more weighty difficulty stems from the fact that the host cell must be actively dividing in order for viral DNA to be incorporated into its chromosome. Most cells in an adult's body have fully differentiated and have stopped dividing.

Target Cells and Organs Do particular cells need to be genetically altered to provide the missing functions? The answer to this question depends on the disease. Our current technology greatly limits the kinds of cells that can reasonably serve as recipients of gene transfers.

Gene transfer is most successful when many potential cells are available to receive the retroviral vector. In this way, even after inefficient transfer of DNA, many cells will have received the vector. For this reason, among others, a particular kind of gene therapy approach called **ex vivo gene therapy** is more common. *Ex vivo* means "outside the body." *Ex vivo* gene therapy involves first removing cells from the patient (Figure 10.8). These cells are cultured in the laboratory and are then used as recipients for retroviral vectors containing the desired gene. Within the vector is an additional gene that can be used to select cells that have incorporated the vector, much as we used antibiotic resistance genes during the selection of transformed bacterial cells. Cells that have incorporated the vector are then cultured and returned

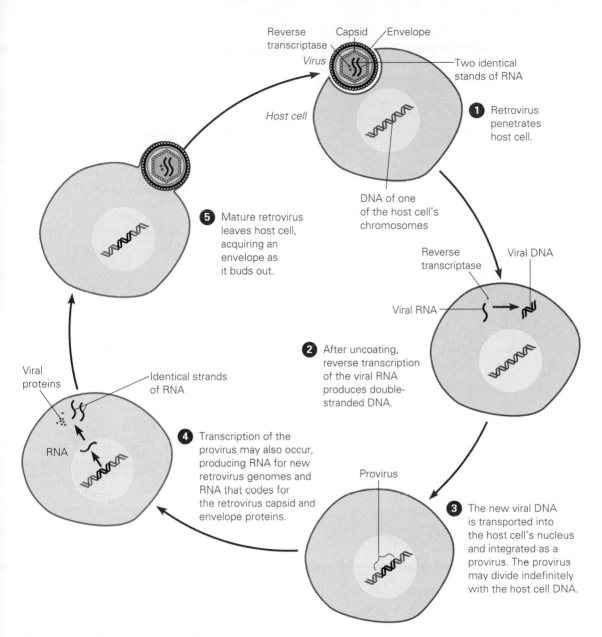

Figure 10.7 The replicative cycle of a retrovirus.

to the patient. In this way, at least some of the patient's cells will now contain the missing gene. Often, cells from the bone marrow are used as recipients, because some can be cultured outside the body and will continue to divide and develop into a number of different cells when they are returned to the body (as in a bone marrow

Figure 10.8 *Ex vivo* gene therapy. Cells, usually from bone marrow, are removed from a patient and cultured. The retroviral vector containing the desired gene is then added to the cells. Some of the cells take up the retrovirus and incorporate it into their genome. These cells are then returned to the patient, either through the blood or into the bone marrow.

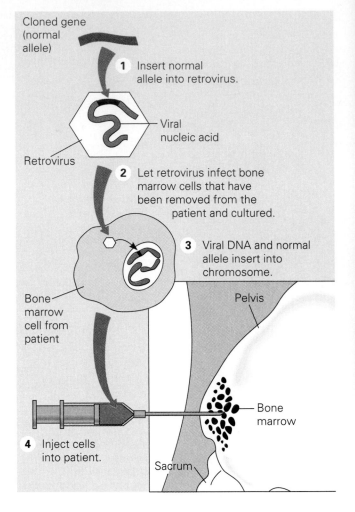

Cloned gene (normal allele)

1 Insert normal allele into retrovirus.

Viral nucleic acid

Retrovirus

2 Let retrovirus infect bone marrow cells that have been removed from the patient and cultured.

3 Viral DNA and normal allele insert into chromosome.

Pelvis

Bone marrow cell from patient

Bone marrow

4 Inject cells into patient.

Sacrum

transplant). Alternatively, certain preexisting cells in the blood can also be fairly easily modified and returned to the patient. Thus, genetic diseases affecting the blood or cells in the blood are the most common targets of gene therapy.

Many potential difficulties are inherent in this approach. If few modified cells survive in the patient, there might not be enough of the missing protein made. Or the modified cells might not survive for a very long time, requiring frequent treatments. Or there may be insufficient expression of the gene product, resulting in not enough protein being made.

Another gene therapy approach, called *in vivo* ("in the body") gene therapy, involves the direct application of retroviral vectors to human tissue. The most notable attempts in this regard have been to package the vectors as a nasal mist. When the mist is sprayed through the nose, some of the retroviruses can infect cells lining the respiratory tract, including the lungs. This approach has met with limited success in attempts to develop a gene therapy for cystic fibrosis.

Potential Risks of Gene Therapy A number of significant risks are involved in the approaches outlined above, all of which must be studied further to determine whether they will affect potential gene therapy treatments. One possibility is that cells modified *ex vivo* and returned to the patient will cause an immune reaction. Such a reaction is highly unlikely, given that the cells were originally taken from the patient; in fact, this is one of the chief advantages of the *ex vivo* approach.

Potential problems are introduced by the use of retroviral vectors. Unfortunately, it is not yet possible to control the site on the chromosome at which retroviral DNA incorporates; in fact, the process appears to be random. Since the vast majority of DNA in a chromosome is noncoding, the odds are that a random insertion will have no effect. However, it is possible that the retroviral vector will insert directly into some unrelated yet necessary gene, destroying its function. It is also possible that insertion of engineered retroviral vectors could stimulate the expression of a nearby gene, as is known to happen with naturally occurring retroviruses, often resulting in cancer (Chapter 14). Until more directed methods of incorporating DNA into the chromosome become available, these risks will always be part of any gene therapy experiment.

There is also the risk that the production of the missing protein will in itself create problems. Consider gene therapy for hemophilia. In principle, providing the missing protein factor seems relatively straightforward. However, having too much of a clotting factor in the bloodstream might prove quite serious, causing unwanted clotting. A major technical challenge for the future will be to develop ways in which the expression of transferred genes can be properly regulated.

Early Gene Therapy Experiments The first human gene therapy trial was approved (after a lengthy review process) in 1990. The patient suffered from adenosine deaminase (ADA) deficiency, a disease that results in a nearly complete loss of the entire functional immune system (Chapter 11). ADA patients used to be confined to sterile environments (in the form of plastic "bubbles") that protected them from infectious agents. Even though ADA deficiency is rare, early experiments provided tests for the first gene therapy.

The first trial, an *ex vivo* approach, involved removing certain cells from the patient's blood, transferring an intact ADA gene into them, and then returning those cells that expressed the gene to the patient. The idea was that these altered cells would produce enough adenosine deaminase to relieve the disease symptoms. The first test was a remarkable success, and the patient now leads a virtually normal life. Unfortunately, the blood cells used for this trial last only a few months, requiring regular transfusions of ADA-producing cells about every three months for the patient to remain healthy.

Table 10.1 Gene Therapy: Clinical Trials Approved and Probable

Some Diseases for Which Gene Therapy Trials Have Been Approved and/or Conducted Since 1990

ADA deficiency	Gaucher disease	malignant melanoma
advanced cancers	hemophilia B	neuroblastoma
AIDS	hypercholesterolemia	ovarian cancer
brain tumor	kidney cancer	
cystic fibrosis	lung cancer	

Other Diseases Being Considered for Treatment Using Somatic Gene Therapy

acute myeloid leukemia	emphysema	macular degeneration
atherosclerosis	hemophilia A	osteoporosis
breast cancer	leukemia	Parkinson's disease
colon cancer	liver cancer	sickle-cell anemia

Hundreds of gene therapy trials have received approval. Some of these are described in Table 10.1. Many of the trials are aimed at developing ways of curing specific cancers. We will describe some of these attempts in Chapter 14.

SOCIAL IMPLICATIONS OF TREATING GENETIC DISEASE

The prospects of developing viable gene therapy approaches to curing disease are exciting, although it may be years before full-scale testing can begin. The success of some of the early trials suggests that gene therapy may advance incredibly quickly. As we have seen, however, the prospect of widespread gene therapy raises many social and political issues.

Regulation and Approval of Gene Therapy Trials

The rather stormy beginning of recombinant DNA technology resulted in the establishment of a number of groups and procedures intended to oversee and review the growing technology. Prominent among them is the Recombinant DNA Advisory Committee (RAC), established by the National Institutes of Health and consisting of scientists, government officials, ethicists, philosophers, and physicians, among others. Its primary role has changed somewhat over the years, but it remains the committee that reviews applications to use novel or newly developed recombinant DNA procedures. Gene therapy comes under its purview.

Anyone intending to pursue gene therapy experimentation must submit a detailed proposal to local ethics and biosafety committees set up by the institution involved or by local communities. If the proposal is approved, it then moves to the Food and Drug Administration (FDA), which has responsibility for approving all new medicinal drugs and therapeutic procedures. The FDA makes a decision as to whether the proposal needs to be reviewed by RAC; if so, RAC considers it and makes a recommendation to the director of the National Institutes of Health. The FDA then considers the proposal, along with any recommendations that have been made, and makes a decision.

Until recently, RAC was required to review all proposals involving certain experimental recombinant DNA procedures. However, this requirement was changed because NIH and FDA officials felt that most research proposals involved procedures that had become quite common, and the mandatory review by RAC was considered unnecessary. RAC would now be used to evaluate proposals involving new techniques or the use of new recombinant DNA vectors, for example.

However, RAC is the only oversight body required to conduct public hearings. Local ethics and biosafety committees are not required to operate publicly (although some do), nor is the FDA required to hold public hearings. Thus, making the RAC evaluation optional means that the only required public hearings on new recombinant DNA experimentation (not just gene therapy) have been bypassed.

Germ Line Gene Therapy

The entire thread of our overview of gene therapy has pointed to somatic cell therapy, the process of adding to or altering the genes of some of the nonreproductive cells of the body. The techniques we have described would not result in any of the genetic changes being passed on to offspring, because the reproductive cells of the individual are not being altered. However, we could consider altering the germ line of humans, just as we create transgenic animals by altering the DNA of eggs. Theoretically, nothing should prevent the same technology from being applied to humans.

At the present time, technical difficulties prevent even the consideration of germ cell gene therapy in humans. Numerous eggs are required for the necessary *in vitro* manipulations, and only a small percentage of these would survive each required step: genetic alteration, fertilization, selection, and implantation. But the most difficult aspect is that we cannot cause DNA to insert at specific locations in a chromosome.

Genetically engineering eukaryotic cells presently involves the random insertion of DNA into the chromosome. As we have seen, this entails a certain risk, since the insertion might inactivate or activate a gene and cause serious difficulties for the developing embryo. Since we know so little about the insertion process and even less about the potential ramifications of random insertion, it is unrealistic at this time to consider using such an approach on humans. While no laws in the United States prevent germ cell gene therapy, none of the regulatory agencies involved will even consider a proposal in this area.

Our ability to deliver DNA to specific locations in eukaryotic cells should improve quickly. If it becomes possible to replace defective genes with intact copies, with no other genetic changes—a possibility probably much closer to reality than we think—should germ cell gene therapy then be pursued?

Clearly, the implications of germ cell gene therapy are tremendous, far greater than those of somatic cell therapy. The technology to perform such therapy will surely become available; the only question is when. Society faces some difficult issues in dealing with this prospect. Imagine the time when it becomes possible to use germ cell gene therapy to correct for a whole host of genetic diseases, including cystic fibrosis, phenylketonuria, and Tay-Sachs disease. When it becomes commonplace to correct these defects, how long will it be before other kinds of changes are attempted? Daniel Koshland, past editor in chief of the journal *Science*, wrote:

> If a child destined to have a permanently low IQ could be cured by replacing a gene, would anyone really argue against that? It is a short step from that decision to improving a normal IQ. Is there an argument against making superior individuals? Not superior morally, and not superior philosophically, just superior in certain skills: better at computers, better as musicians, better physically. As society gets more complex, perhaps it must select for individuals more capable of coping with its complex problems.

ESSAY: A RESPONSE TO KOSHLAND

Our concern with the thoughts expressed by Koshland (and shared by others) can be illustrated by considering some of the language Koshland used in his statements. The word *destined* carries with it an attitude of determinism. We've already seen the tremendous difficulties associated with studying the inheritance of IQ (Chapter 6). We're not prepared to say that anyone is "destined" to have a particular IQ level—but in Koshland's scenario someone clearly would be deciding this. The word *cured* implies that a low IQ is a disease, a defective trait. How low does the IQ have to be in order to classify the person as "diseased"? Should we revive the classification of individuals as idiots, imbeciles, and feebleminded? These questions bring to mind the arguments against eugenics. What Koshland proposes is a eugenic desire to improve the human race.

Koshland says that it is "a short step . . . to improving a normal IQ." He then goes on to say that such improvements would produce "superior individuals." Doesn't this involve arbitrary judgment? It seems rather naive to assume that changes could be limited to raising IQ or improving computer skills.

This argument could have occurred in the 1920s, during the eugenics debate. In fact, it probably did. Our new technology introduces nothing new into the argument. The question is not whether the technical expertise is available to consider such genetic manipulations, but who will decide what changes are appropriate. Advances in the application of recombinant DNA technology to medicine have been rapid, with sometimes impressive results. Such success encourages the expansion of the technology to broader medical concerns and influences the formation or reformation of social policies. It is then only a short step to a situation in which the technology is used in ways that conflict with decisions that were once individual and private.

SUMMARY

Some human diseases are purely genetic, and others have gene contributors to the diseased condition. Just a few years ago, the diagnosis of a genetic disease was considered by many to be a life and death sentence. Nothing could be done except make the patient as comfortable as possible. Today we have begun the direct application of recombinant DNA human genes by first seeking to understand the causes of genetic diseases. We have conducted studies to answer several questions. Is the disease genetic? What change(s) occurred that resulted in the altered gene? How can the carrier state be established? How can biotechnology be used to treat and cure a particular genetic disease?

Today the determination of genetic causation of a particular disease is accomplished by detailed study of its inheritance, including pedigree analysis and twin studies. If these results prove inconclusive, we turn to molecular identification, cloning, sequencing, and comparisons with normal genes. If the research provides accurate and reliable testing methods and if the disease is frequent and severe, genetic screening may be attempted. When or whether to apply genetic screens depends on many complicating factors—scientific, medical, moral, ethical, economic, and religious. And as scientists, medical practitioners, and patients work through these complexities, we shouldn't miss the major point: Today we have diagnostic tools and treatment methods for some genetic diseases; just a short time ago we didn't.

REVIEW QUESTIONS

1. FH cannot be considered a completely genetic disease even though there are known mutations within the LDL receptor gene. Why not?

2. Explain how completion of the Human Genome Project may benefit the search for cures for some genetic diseases.

3. Cystic fibrosis (CF) is known as a genetic disease. What type of inheritance is involved? What are the major symptoms? List three different types of treatment presently employed to aid patients.

4. How were RFLP markers used in the identification of the CF gene?

5. Outline the process known as chromosome walking. For what is it used?

6. Define or describe an open reading frame.

7. Why were sweat glands selected for the study of CF?

8. In older, traditional amniocentesis, fetal cells must be cultured in a laboratory. What is the difficulty, given the range of weeks of pregnancy for which the procedure is recommended and the no-later-than date generally prescribed for a therapeutic abortion?

9. Describe the specific moral/ethical problem that arises if a couple heterozygous for CF decides to undergo *in vitro* fertilization.

10. How is it possible to produce twin babies that actually have birthdays three years apart?

11. List three major reasons why diagnosis of genetic diseases in adults is much more common than it is for fetuses.

12. Put yourself in the position of a young adult whose mother has just been diagnosed with Huntington's disease. Provide arguments both for and against having genetic testing done on you. Further, let's say that your test is positive—you are heterozygous for the trait. Do you tell your health and life insurance companies? Your family members? Your prospective husband or wife? Why or why not in each case?

13. Describe the moral barrier against creating transgenic humans.

14. Clearly distinguish between potential genetic treatments for dominant genetic diseases and those for recessive genetic diseases. Which type of treatment should be easier?

15. Describe the traditional treatment of hemophilia and contrast it with the potential gene therapy for the same condition.

16. What is (are) the basic argument(s) against producing transgenic humans using the current level of technology?

17. There seems to be a major difference in attempting to cure recessive and dominant genetic diseases. What is this general difference?

18. List the five requirements, according to W. F. Anderson, that must be satisfied prior to attempting gene therapy.

19. How are retroviruses used to transfer genes into human cells? List two main problems with using retroviruses as cloning vectors.

20. What do bone marrow transplants have to do with *ex vivo* gene therapy?

21. List three general risks associated with gene therapy.

22. Technically, what interferes with using germ line gene therapy in humans?

CHAPTER 11

Immunology

We live in a world inhabited by many organisms besides humans. Environmental awareness has risen to new heights in recent years, driven, in part, by increased recognition of the interdependence of all organisms. Many people have become much more interested in protecting endangered species and the diversity of life.

Most microorganisms are not pathogenic, or disease-causing. In fact, many organisms, known as **commensals**, live and reproduce in or on us in relationships in which the human is neither benefited nor harmed but the commensal benefits. Examples of commensals include most of the millions of bacteria that inhabit our mouth and lower intestines, as well as the thirty or more species that live on the surface of our eyes and do us no apparent harm. Other microorganisms are more intimately involved with us and may even provide nutrients or vitamins. One example of a close, mutually beneficial association is *Escherichia coli*, which thrives in our lower intestine living on the food that we consume while in turn providing us with certain vitamins (for example, K) and possibly some nutrients. Recall from Chapter 9 that this kind of symbiotic arrangement, in which two different kinds of organisms coexist and both organisms benefit, is called mutualism. Conversely, when a microorganism takes up residence in or on a human body, reproduces in large numbers, and causes harm, an infectious disease results.

In order for a disease to result from infection, two things must happen. First, an organism must make contact with, invade, and colonize a part of the body. Second, the continued growth and reproduction of this organism must adversely alter the normal function of the body, causing disease. Our bodies wage war against potentially infectious agents by using a multilevel system comprised of three major lines of defense.

The first line of defense is made up of the skin and mucous membranes. These two components act as barriers that prevent the penetration of invading microbes. If an invader does gain entrance into the body, it encounters the second line of defense—circulating cells that patrol the body and destroy foreign invaders using a battery of *nonspecific defenses*. A third line of defense employs two types of cells.

One type marks invaders to identify them as *foreign*, while the other attacks and kills any microbe so identified; together, these two types of cells make up the immune system.

FIRST LINE OF DEFENSE

The skin is the largest organ in the human body, accounting for approximately 15 percent of an adult human's weight. The skin defends the body against invading microbes in several ways. First, intact skin provides a nearly impenetrable barrier that prevents the entry of viruses and microorganisms into our blood and deeper tissues. Second, skin contains sweat and oil glands, the secretions of which give the skin's surface a pH of 3–5, acidic enough to inhibit the growth of many microorganisms. Furthermore, sweat contains lysozyme, an enzyme that destroys bacterial cell walls.

In addition to the skin, invading microbes have two other potential routes of entry: the digestive tract and the respiratory tract. Many microorganisms are present in food. Many of them are killed by saliva, which, like sweat, contains lysozyme. Many more are killed by the very acidic environment of the stomach and by digestive enzymes present in the stomach and intestines. Microbes are also present in the air we breathe, and our warm, moist lungs provide ideal breeding grounds for them. However, the respiratory tract is lined by two types of defensive cells. One type secretes a sticky mucus that traps most microorganisms before they can reach the lungs. The second type has cilia that continually sweep the trapped microbes upward toward the glottis, where they can be swallowed and subsequently destroyed by the digestive tract.

NONSPECIFIC DEFENSES

Despite the protection afforded by the first-line defenses, microorganisms occasionally manage to enter the body. The body uses a battery of nonspecific defenses to defend itself against microbes that successfully gain entry. These defenses are said to be nonspecific because they respond to any infection, without first determining the identity of the infecting agent. We will look closely at three of the most significant nonspecific defenses: cells that ingest invading microorganisms, antimicrobial proteins, and the **inflammatory response**.

Cells That Kill Invading Microorganisms

The cells that play a role in nonspecific defense are blood cells. The principal types of blood cells are shown in Figure 11.1. Only white blood cells, or **leukocytes**, play a role in defense. They are perhaps the most important of the body's nonspecific defenses. Three major types of leukocytes circulate through the body and attack invading microorganisms in the circulatory system and within tissues, and each type kills invading microbes in a different manner.

Two of the three types of leukocytes are **phagocytes**, cells that engulf microor-

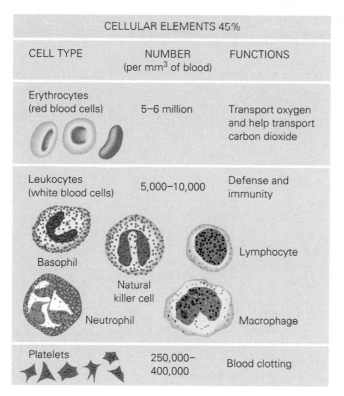

CELLULAR ELEMENTS 45%		
CELL TYPE	NUMBER (per mm³ of blood)	FUNCTIONS
Erythrocytes (red blood cells)	5–6 million	Transport oxygen and help transport carbon dioxide
Leukocytes (white blood cells)	5,000–10,000	Defense and immunity
Platelets	250,000–400,000	Blood clotting

Basophil

Natural killer cell

Neutrophil

Lymphocyte

Macrophage

Figure 11.1 Types of blood cells. Blood is composed of a fluid (plasma) and several different types of cells that circulate within that fluid. Erythrocytes, or red blood cells, transport oxygen throughout the body, while platelets play an important role in blood clotting. Leukocytes, or white blood cells, defend the body against invading microorganisms and other foreign substances.

ganisms or other particles. **Macrophages**, or "big eaters," are large cells that ingest and kill microbes. Macrophages send out cytoplasmic extensions that adhere to the microbes and pull them inside the macrophage by a process called **phagocytosis** (Figure 11.2). Inside the macrophage, the microbe is enclosed in a membrane-bound vesicle that fuses with an organelle, called a lysosome, that contains digestive enzymes. Fusion activates the lysosomal enzymes, which proceed to destroy the microbe. **Neutrophils**, like macrophages, ingest and kill bacteria. However, while macrophages kill microbes one at a time, neutrophils release chemicals that kill any microorganisms in the vicinity—as well as the neutrophils themselves.

The third type of leukocyte that participates in nonspecific defense is the **natural killer cells**. These cells do not attack invading microbes directly but instead kill the body's own cells if they become infected, especially if they are infected by viruses. Unlike neutrophils and macrophages, natural killer cells do not kill by phagocytosis. Rather, they create pores in the membrane of an infected cell (Figure 11.3). These pores allow water to rush into the infected cell, which then swells and lyses. Natural killer cells can also attack and destroy cancer cells, often before the cancer cells can develop into a tumor. Natural killer cells constitute one of the body's most effective defenses against cancer.

Figure 11.2 Phagocytosis by a macrophage. This electron micrograph shows a macrophage probing its surroundings with cytoplasmic extensions. Bacteria that come in contact with the extensions are drawn toward the macrophage and engulfed by phagocytosis.

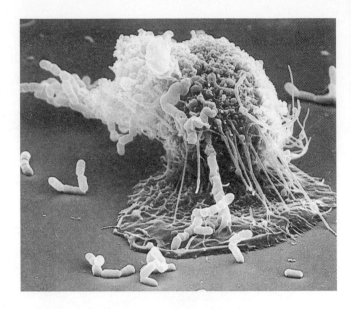

Figure 11.3 How natural killer cells kill target cells. Natural killer cells bind tightly to infected or abnormal cells. Binding causes vesicles loaded with perforin molecules to move to the plasma membrane of the natural killer cell and release their contents into the intracellular space over the target cell. The perforin molecules insert into the plasma membrane of the target cell and form a pore that allows water to rush in and rupture (lyse) the cell.

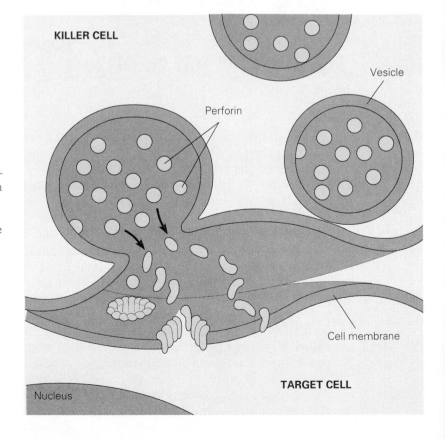

Antimicrobial Proteins

In addition to cellular defenses, the nonspecific defenses include chemical defenses. One type of chemical defense involves the **complement system**, which consists of approximately twenty different blood serum proteins. The complement proteins circulate freely in the blood. When they encounter an invading microorganism, they aggregate to form a **membrane attack complex** that inserts itself into the membrane of the foreign microorganism, forming a pore (Figure 11.4). Water rushes into the foreign cell through these pores, causing the cell to swell and burst.

The complement proteins also play a role in bolstering other nonspecific defenses. Some complement proteins can amplify the inflammatory response (described in the next section). Some act as **chemoattractants** that attract macrophages to the site of infection. Others coat invading microbes, making them easier for macrophages to engulf.

Another class of proteins that plays a role in defending the body is **interferons**, which are released by virus-infected cells. They are called interferons because, after release, they diffuse to neighboring cells and interfere with the ability of the virus to infect those cells.

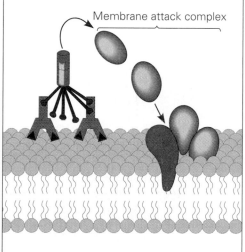
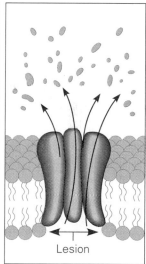

① Complement proteins attach to pair of antibodies.

② Activated complement proteins attach to pathogen's membrane in step-by-step sequence, forming a membrane attack complex.

③ Complement proteins lyse target membrane, resulting in lesion and death of pathogen.

Figure 11.4 How complement proteins kill infected cells. As shown in the diagram, complement proteins form a membrane attack complex, which then forms a lesion or pore in the plasma membrane of an infected cell, causing cell lysis.

The Inflammatory Response

When cells become infected or are injured, they release chemical signals that stimulate the body's defenses. These chemicals, most often histamine and prostaglandins, induce a localized response known as the **inflammatory response** (Figure 11.5). Histamine and prostaglandins cause local blood vessels to dilate, resulting in greater blood flow to the site of infection or injury. They also increase the permeability of local capillaries, causing the recognizable signs of inflammation: localized redness, swelling, and pain. The increased permeability of capillaries enables macrophages and neutrophils to leave the bloodstream so that they can attack the infecting microbes.

The Fever Response

Chemical signals are also involved in the fourth nonspecific defense, the fever response. Fever is triggered by substances, such as bacterial endotoxins and **interleukin-1**, that act as **pyrogens**. Anything that causes an increase in body temperature (fever) is a pyrogen. Interleukin-1 is released by macrophages after they encounter an invading microbe. Interleukin-1 is transported by the blood into the brain, where it stimulates the hypothalamus to raise the body's temperature. Fever plays a role in defense in two major ways. First, the elevated temperature stimulates phagocytosis

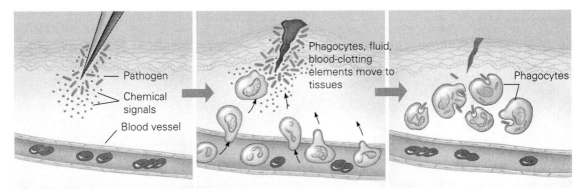

(a) Tissue injury; release of chemical signals (histamine, prostaglandins).

(b) Vasodilation (increased blood flow); increased vessel permeability; phagocyte migration.

(c) Phagocytes (macrophages and neutrophils) consume pathogens and cell debris; tissue heals.

Figure 11.5 The inflammatory response. (a) The inflammatory response is triggered when cells of injured tissue release chemical signals such as histamine and prostaglandins. **(b)** These chemical signals induce increased permeability of the capillaries and increased blood flow to the affected area. Chemoattractants that attract phagocytes are also released. **(c)** At the site of injury, the phagocytes (macrophages and neutrophils) engulf pathogens and cell debris, and the tissue heals.

by macrophages and neutrophils. Secondly, fever causes a reduction in the levels of iron in the blood. This helps fight infection because bacteria need large amounts of iron for growth.

THE IMMUNE SYSTEM

The nonspecific defenses are very effective against a wide range of microbial infections. Invading microbes rarely evade the nonspecific defenses, but when they do they encounter the third, most potent line of defense—the immune system. This system is an intricate mix of many different components. Four questions will frame our discussion of the immune system:

1. What is the immune response, and how does it protect us from infection?
2. How does the immune system recognize so many different infectious agents?
3. Why doesn't the immune system attack and kill our own cells? That is, how can it recognize what is foreign and what is not?
4. How does a vaccine work?

The Immune Response

The *immune response* is the set of reactions carried out by the immune system following the detection of a foreign invader in a healthy human. Immune responses are triggered by substances (usually proteins or polysaccharides) called **antigens** that exist on the surface of invading microbes, where cells of the immune system can interact with them. Such interactions are necessary for recognition of invading microbes and the initiation of the appropriate specific defensive reactions. Some of these are chemical reactions, while others involve the activation or production of specialized cells known as lymphocytes that are found in the circulatory system, primarily in blood or **lymph**. The lymph system is like a drainage system in our bodies. Its role in and relationship to the rest of the circulatory system are described in Figure 11.6.

The important components of the immune system are cells and chemicals released by cells. We will first consider the major types of cells involved in immune responses. As we will see, leukocytes, which play a role in nonspecific defenses, also play a role in immune responses. Leukocytes, like all blood cells, are formed in the bone marrow. In addition to the leukocytes already mentioned (macrophages and neutrophils), two types of lymphocytes, **T cells** and **B cells**, are critical to immune responses.

After we have become familiar with the major components of the immune system, we will examine the events that occur during immune responses by considering what happens when a virus infects a cell. The different responses of the immune system affect different targets. The cell-mediated response kills cells of the body that are

Figure 11.6 The human lymphatic system. The lymphatic system returns fluid from the spaces between tissues to the circulatory system. The lymphatic system includes lymphatic vessels and various satellite organs that play important roles in the body's defense system, including the adenoids, tonsils, numerous lymph nodes, spleen, and appendix. The bone marrow and thymus, sites of white blood cell development, are also shown.

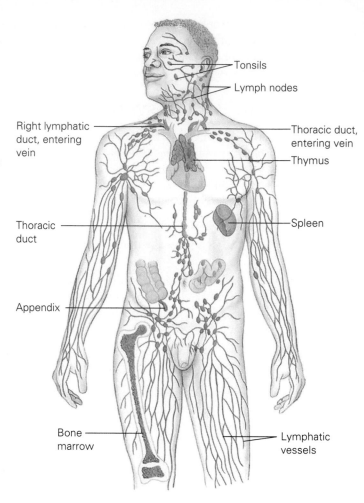

Tonsils

Lymph nodes

Right lymphatic duct, entering vein

Thoracic duct, entering vein

Thymus

Thoracic duct

Spleen

Appendix

Bone marrow

Lymphatic vessels

somehow altered, primarily through infection or mutation. The humoral response inactivates freely circulating viruses, toxins, and microbes. Together with various phagocytic activities, these active defense systems are able to provide protection against most infectious agents.

T Cells T cells develop initially in the bone marrow but then move to the thymus gland, where they finish their development (maturation). These lymphocytes are called T cells because they mature in the *thymus* rather than in the bone marrow, where B cells mature. During the maturation process, T cells acquire the ability to identify viruses and microorganisms from the antigens displayed on their surfaces. Following maturation, the T cells are released into the blood and lymph, where they circulate.

There are at least four types of T cells. **Inducer T cells** mediate the development of T cells in the thymus. **Cytotoxic T cells**, or T_C cells, kill infected or abnormal cells. **Helper T cells**, or T_H cells, initiate immune responses. **Suppressor T cells** mediate suppression of immune responses. For our purposes, we will focus on the helper T cells and the cytotoxic T cells.

B Cells B cells develop and mature in bone marrow. Even before birth, a population of B cells is already formed. They have not yet been exposed to antigens, so they are referred to as *naive B cells*. As naive B cells develop, they migrate into the lymph and take up residence in various lymph nodes, where they will encounter various antigens that have gained entrance to the body. During immune responses, B cells mature into plasma cells that secrete antibodies, proteins that specifically bind to antigens and mark or label them for destruction. Each naive B cell has the capability of producing antibodies that will recognize only a single, specific antigen. Different B cells, however, can produce antibodies that will recognize and bind to different antigens. Thus, in a normal adult, millions of different B cells have the capability of producing millions of different antibodies.

Naive B cells produce membrane-bound antibodies. The stems of the antibodies are anchored in the B cell membrane, while the antigen-binding regions are exposed to the extracellular environment. As antigens circulate through the lymph, they will be exposed to the population of B cells that exists in the lymph nodes. If a particular antigen is recognized by the antibodies on one of the B cells, it will bind to that cell.

Antigen-presenting Cells We will see shortly that activation of immune responses requires regulatory molecules, known as **cytokines**, that are released by activated helper T cells. It is of the utmost importance that the activation of T_H cells be carefully regulated, because inappropriate T_H cell responses to the body's own components can have fatal consequences. To ensure appropriate regulation, T_H cells are activated by an antigen only if the antigen is displayed together with specific molecules called *major histocompatibility complex* (MHC) molecules on the surface of antigen-presenting cells (APCs).

The major antigen-presenting cells are macrophages. A macrophage takes in an antigen by phagocytosis and then displays a part of that antigen, together with the MHC molecule, on its surface. A T_H cell then recognizes the antigen associated with the MHC molecule on the membrane of the antigen-presenting cell, binds to it, becomes activated, and initiates an immune response. Activated macrophages also release **interleukin-1**, which serves to activate T_H cells.

Immediate Responses When a virus infects a cell, the cell responds by secreting interferons, which circulate through the body and stimulate natural killer cells and macrophages to attack virus-infected cells. Macrophages engulf the infected cells, degrade the infecting virus, and display viral antigens on their surface for presentation to circulating helper T cells. Activated macrophages also secrete cytokines, such as **γ-interferon**, which stimulate the production of more macrophages.

Helper T cells that recognize the antigen presented by macrophages will bind to the antigen. After binding, T_H cells become activated and respond to the interleukin-1 secreted by antigen-presenting macrophages by simultaneously activating

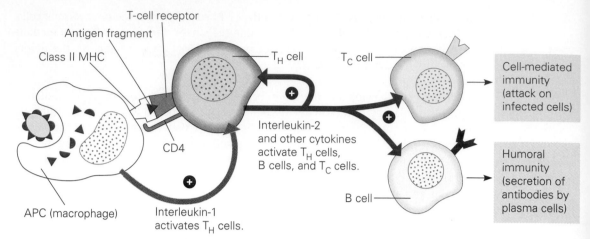

Figure 11.7 The central role of T$_H$ cells in the immune response. Helper T cells activate both cell-mediated and antibody-mediated (humoral) responses. Receptors on the T$_H$ cell bind the class II MHC–antigen complex displayed on the surface of an antigen-presenting cell (APC), usually a macrophage. The macrophage secretes interleukin-1, which activates the T$_H$ cell. An activated T$_H$ cell grows and divides, producing clones of T$_H$ cells, all of which possess receptors specific for the MHC-antigen complex that triggered the response. The T$_H$ cells then secrete interleukin-2, which amplifies the cell-mediated response by stimulating the proliferation and activity of other T$_H$ cells, all of which are specific for the same antigen. Interleukin-2 secreted by T$_H$ cells also activates B cells, which function in the antibody-mediated response, and natural killer cells, which function in the cell-mediated response.

two types of responses (Figure 11.7), one involving T cells (cell-mediated response) and the other involving B cells (humoral response). Both responses act in concert to clear the body of infection: T cells by recognizing and killing virus-infected cells, and B cells by producing antibodies against the virus.

Cell-mediated Immune Response An overview of the cell-mediated response is shown in Figure 11.8. Activated T$_H$ cells initiate the cell-mediated immune response by releasing regulatory molecules known as **lymphokines**. The most important lymphokine for cell-mediated responses is **interleukin-2**. Interleukin-2 stimulates helper T cells and cytotoxic T cells that have bound a viral antigen to divide, producing a large number of clones of T cells (both T$_H$ and T$_C$ cells) capable of recognizing that specific viral antigen. This proliferation greatly enlarges the population of T cells that can recognize and combat the infecting virus.

Activated T$_H$ cells also release a second lymphokine known as macrophage **migration inhibition factor**, which attracts macrophages to the site of infection and inhibits their migration away from it. The recruited macrophages contribute to defense by attacking the circulating virus as well as virus-infected cells. Furthermore, activated T$_H$ cells activate inducer T cells, which stimulate immature lymphocytes to become mature T cells, thus adding to the pool of T cells already combating the infection.

After the body has been cleared of the infecting virus, the cell-mediated responses are suppressed by suppressor T cells, which block the response of cytotoxic

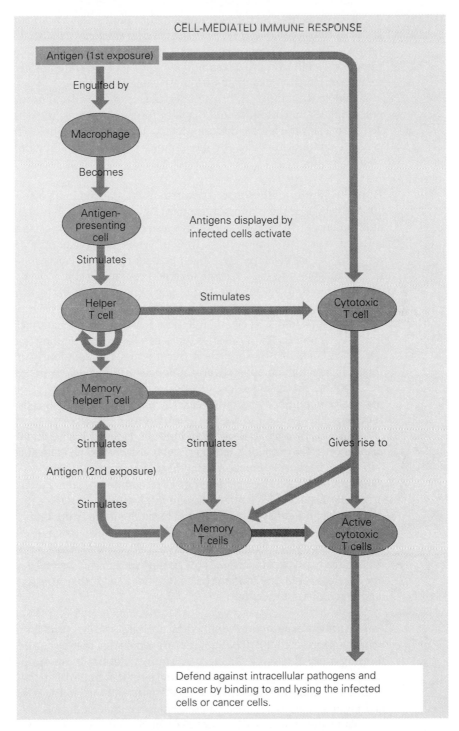

Figure 11.8 An overview of the cell-mediated immune response. In the cell-mediated immune response, T cells defend the body against invading pathogens.

T cells to antigens. After the cell-mediated responses have been suppressed, a population of T cells capable of recognizing specific antigens associated with the infection persists. This population, which includes both T_H cells and T_C cells, is known as **memory T cells** and provides the body with an accelerated cell-mediated response to any subsequent encounter with these same antigens.

Note that tumor cells, or precancerous cells, are also abnormal in many ways. For example, they produce and display "foreign" antigens on their surfaces. Tumor cells are also targeted by the cell-mediated immune response. Many precancerous cells are undoubtedly killed by the immune system before they have a chance to become harmful.

Humoral Immune Response Activated T_H cells also initiate a second defensive response known as the humoral response (Figure 11.9). Recall that B cells express specific antibodies on their surfaces. They also display a specific class of MHC molecules referred to as MHC-II molecules. Unlike T cells, which can only recognize and bind antigens that are displayed by antigen-presenting cells, B cells can recognize and bind to free antigens and then become antigen-presenting cells.

After binding an antigen, the B cell engulfs it, processes it, and displays portions of it on its surface, together with an MHC-II molecule. An antigen-specific T_H cell then binds to the B cell and the Class II MHC–Antigen complex and initiates the humoral response. The T_H cell becomes activated after binding and secretes a number of lymphokines that induce activation and proliferation of the B cell. Several things then begin to happen nearly simultaneously. The activated B cell begins to divide rapidly, producing many identical cells (clones), most of which mature into **plasma cells** (antibody-secreting cells), while the remaining B cells become **memory B cells**.

There are two major differences between memory B cells and plasma cells. First, plasma cells live only for a few days, whereas memory B cells may last for decades. Second, plasma cells function as antibody-secreting factories; each produces up to 2,000 antibody molecules per second. These antibodies do not attach to the cell membrane but instead are secreted into the fluid surrounding the cell. This allows huge numbers of antibodies to be circulated throughout the entire body. As they circulate, the antibodies bind to specific antigens wherever they encounter them, rendering the antigens harmless. In contrast, memory B cells do not secrete antibodies but instead remain ready to quickly form a large number of clones of new plasma cells and memory B cells should one of them come in contact with the same antigen that initially stimulated their production.

How Antibodies Function Antibodies are large serum proteins made up of four subunits (Figure 11.10). The subunits are joined together by specific covalent bonds, forming a broad Y-shaped molecule. The two "arms" of an antibody each have an identical terminal region (antigen-binding site) that binds to an antigen. The "stems" of antibody molecules are the same in all antibodies of a particular type, but the antigen-binding regions vary even among the same type of antibodies.

Antibodies can inactivate antigens in several ways. They may simply prevent the antigens from interacting with a host cell; for example, binding of antibody to a virus might prevent the virus from infecting a host cell. Alternatively, a network of

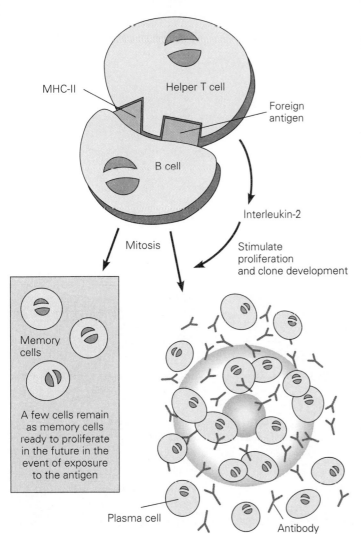

Figure 11.9 The humoral immune response. An activated T$_H$ cell interacts with a B cell that has engulfed antigens and displays a fragment of the antigen along with class II MHC proteins. The T$_H$ cell then secretes interleukin-2, which stimulates the B cell to divide repeatedly and differentiate into plasma (antibody-secreting) cells and memory cells. The secreted antibodies will specifically bind to and inactivate the antigen that stimulated the immune response, as described in the text.

antibody-antigen complexes might form, assembling a larger particle that can be destroyed following phagocytosis. Third, binding of antibodies also serves to "flag" antigens, making them more likely to be recognized and phagocytosed. Fourth, antibodies label antigens for destruction by the complement system, described earlier.

The Diversity of the Immune System

We have seen how our immune system functions to protect us from infectious microbial agents and other foreign molecules. But how can one person produce antibodies and T cells that will recognize millions of different antigens?

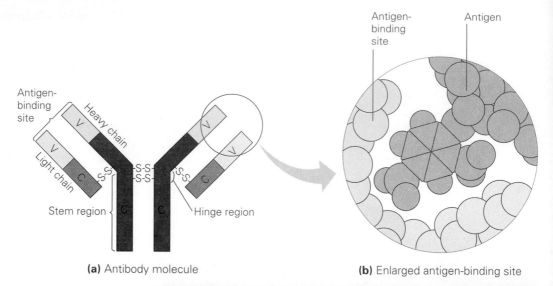

(a) Antibody molecule

(b) Enlarged antigen-binding site

Figure 11.10 The general structure of an antibody molecule. (**a**) A schematic drawing of an antibody molecule. Each antibody is composed of two identical light chains and two identical heavy chains, with two identical antigen-binding sites at the arms of the Y. (**b**) A close-up view of the antigen-binding site in which each amino acid is represented by a small sphere. (**c**) A computer model of an antibody molecule depicting how the four chains wind around one another to form a Y shape.

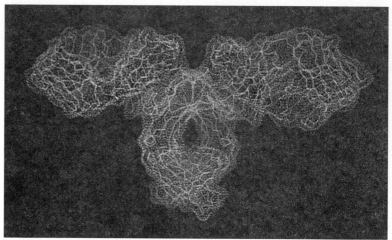

The problem is really one of information capacity. It is conservatively estimated that every human can generate 100 million different antibodies or T cell receptors. It may seem as if there should be 100 million different genes for these proteins. If our DNA did carry this many genes, it would have to be much larger than it is. The problem of how to store such a tremendous amount of information must have a different solution.

The once-popular Cabbage Patch dolls provide an example of a manufacturing scheme somewhat similar to that used by the immune system. Uniqueness was an important feature of the Cabbage Patch dolls; each was packaged with a certificate that assured the purchaser that the doll was truly unique. How could millions of very similar looking dolls be produced in a way that ensured that each was unique?

To produce a doll, different parts need to be put together according to certain rules. For example, the first step might be to put arms onto a torso. Legs, hands, and feet might follow. Then a head would be added, followed by eyes, ears, nose, and mouth. Each component goes in a particular place, but there might be a set of, say, ten different hands to choose from. Or perhaps there are ten different-shape noses or ten different-color eyes. How many unique dolls could be assembled? Assuming that only ten different parts (arms, legs, hands, and so on) are to be assembled and that each part has ten different possible shapes, there are 10^{10} different possible combinations of these parts, or 10 billion different dolls!

Note that a similar result occurs wherever a variety of building blocks is involved in the construction of larger molecules. Recall our analysis of how many different proteins could be made from the twenty available amino acids (Chapter 1). The information for the immune system is contained not in millions of different antibody-encoding genes but in a number of different pieces of genes that are assembled to produce different combinations. The mixing and matching occur during the development of the cell; analogous processes occur in both B and T cells, resulting in the specific antibodies and receptors that each produces. Figure 11.11 illustrates how the combination of several gene pieces can generate such a diverse array of final products.

This arrangement helps explain some other features of the immune system, particularly the fact that the immunity of each individual is unique. Although you inherit the genes encoding antibodies and T-cell receptors from your parents, your cells will not rearrange these genes in the same way your parents' cells did, and therefore you will have a different population of immune cells.

Generating an extremely large number of antibodies is more complex than simply rearranging pieces of genes. The content of those pieces is altered in a number of ways during the rearrangement process, resulting in "mutations" within the final genes. However, these changes are not detrimental. Rather, they serve to generate even more diversity, allowing the immune system to react against a wider variety of antigens.

We need to emphasize that this is one of the few programmed rearrangements of genes that takes place in human chromosomes. For the most part, we believe that our DNA is static, or unchanging, except during cell division, when recombination can occur. But the rearrangements that occur as part of the development of immune cells are an exception. Specific enzymes and regulatory mechanisms serve to keep the rearrangements tightly controlled, allowing only certain rearrangements to occur during the development of the cell.

Self-recognition by the Immune System

How do the cells of the immune system manage to coexist with the other cells of the body? In other words, how do B cells and T cells develop in such a manner that they do not contain antibodies or receptors that will recognize molecules present on the normal cells of the body? Unfortunately, we do not have a complete answer to this question, although we can propose what must happen.

Imagine a B cell developing in the bone marrow of a fetus just before it is born.

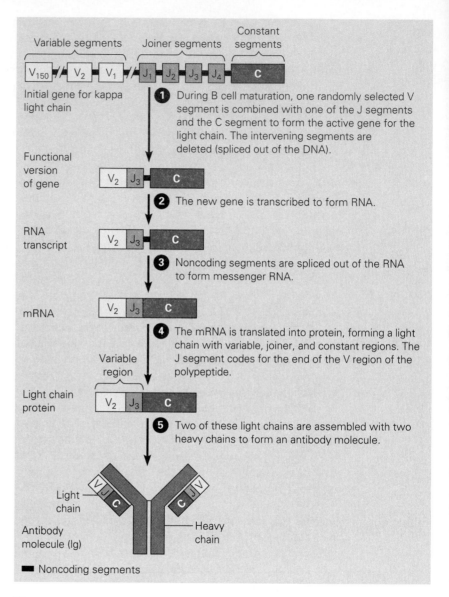

Figure 11.11 The genetic basis of antibody diversity. An overview of the mechanism by which many different antibodies can be made using a limited number of building blocks. Variety stems from the ability to combine any of several hundred unique variable (V) segments with any of four joiner (J) segments to form unique light chains.

It is at this time that the immune system must become active, since the fetus will soon be exposed to infectious agents outside the womb. As part of this activation, millions of B cells have been developing, each one undergoing gene rearrangements responsible for producing a single, unique antibody. Some of these randomly generated antibodies undoubtedly bind to self molecules and would wreak havoc if they became a permanent part of the active immune system, since the body would literally be at war with itself. Therefore, these particular B cells must somehow be eliminated from the developing population.

During the development of a B cell, it is exposed to the circulatory system of the host. Since this happens in the developing fetus, nothing foreign should be present. Any B cell that binds to any molecules or cells normally present in the body is destroyed. In this way, only B cells that do not bind to anything normally present in the body are kept as part of the virgin B cell population.

A similar situation must exist for T cells, although it differs in one important respect. Any T cell that recognizes self molecules will be destroyed, just like B cells. However, proper T-cell function requires that the cell bind not only to antigen but to an MHC molecule. So not only must there be a negative selection, during which all future T cells that recognize self antigens are destroyed, but there must be a stage during which only those T cells that can bind to the individual's MHC are retained. T cells that cannot bind to MHC are of no value to the organism, so positive selection removes them from the population. As a result of the combination of positive and negative selection, the final T-cell population will function properly.

Thus, by the time an infant is born or very shortly thereafter, the baby has a full set of virgin B cells that do not bind to self molecules, as well as a full set of virgin T cells that can bind to MHC but not to self molecules. These cells will provide much of the immune system activity throughout the individual's life.

Autoimmune Diseases An individual's immune system, for one reason or another, sometimes develops the ability to react to molecules or cells normally found within the body. When this happens, the immune system begins to attack its own body cells, a situation referred to as **autoimmunity**. Many diseases are caused by autoimmune activities and are referred to as **autoimmune diseases**. Some examples of autoimmune diseases follow.

Myasthenia gravis is a neuromuscular disease characterized by a gradual loss of muscle strength. It is caused by an autoimmune reaction that develops against particular receptor molecules involved in the transfer of signals from the nervous system to the muscles (Figure 11.12). As the immune response against these molecules grows, the patient weakens. Eventually, the diaphragm and other muscles that control breathing cannot function properly, resulting in death.

Rheumatoid arthritis is caused by an immune reaction against molecules present in joints. Antibodies begin to accumulate in the joints, producing inflammation that eventually results in the progressive destruction of joint tissue.

Insulin-dependent diabetes is an autoimmune disease. Recall that patients suffering from this type of diabetes cannot produce insulin. The reason is that the cells responsible for the synthesis of insulin, the β-cells of the pancreas, are destroyed by a cell-mediated autoimmune response that develops during childhood.

Rheumatic fever, which causes inflammation of the heart, was once a killer of school-age children. At one time, in the United States and other countries, it killed more children than all other diseases combined. Rheumatic fever develops following strep throat, a bacterial infection caused by *Streptococcus pyogenes*. By itself, strep

Figure 11.12 Myasthenia gravis disrupts signaling between neurons. Muscle activation is controlled by signals that are carried from the brain to the muscles by neurons. (**a**) Signaling between neurons is mediated by neurotransmitters, acetylcholine receptors, and ion channels, as shown. (**b**) Myasthenia gravis is a neuromuscular disease in which the body produces autoantibodies against the acetylcholine receptors, thus inhibiting muscle activation. (**c**) Treatment for myasthenia gravis involves the production of antibodies against the acetylcholine receptor antibodies. These antiidiotype antibodies will bind to the acetylcholine receptor antibodies, preventing them from binding the receptor.

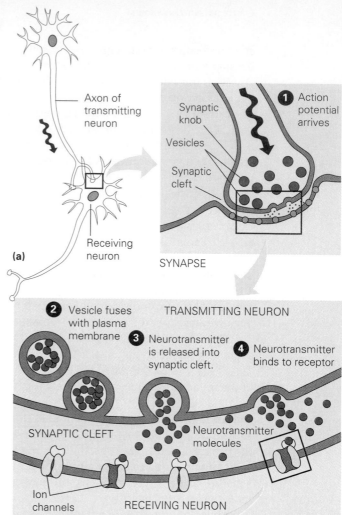

Axon of transmitting neuron

Receiving neuron

(a)

Synaptic knob

Vesicles

Synaptic cleft

1 Action potential arrives

SYNAPSE

2 Vesicle fuses with plasma membrane

TRANSMITTING NEURON

3 Neurotransmitter is released into synaptic cleft.

4 Neurotransmitter binds to receptor

SYNAPTIC CLEFT

Neurotransmitter molecules

Ion channels

RECEIVING NEURON

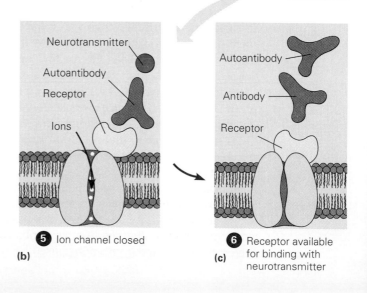

Neurotransmitter

Autoantibody

Receptor

Ions

5 Ion channel closed

(b)

Autoantibody

Antibody

Receptor

6 Receptor available for binding with neurotransmitter

(c)

throat is seldom serious—within a week, the body mounts an effective immune response and the disease is terminated. Treatment with antibiotics can bring the disease under control even more quickly, sometimes in less than a day, by killing the growing *S. pyogenes* population.

Despite the fact that a normal immune response will almost always destroy the *S. pyogenes* population, it is necessary to treat cases of strep throat with antibiotics. *S. pyogenes*, like all microorganisms, is composed of many antigens. One of these antigens closely resembles a protein present in human heart muscle. In a small percentage of people who have strep throat, the antibodies produced in response to the bacterial antigens also attack the heart valves. The result is inflammation that can result in permanent damage to the heart valves. If patients suffering from strep throat are left to fight the disease by developing their own immune response, rheumatic fever may result. On the other hand, if the infection is quickly controlled by antibiotics, any immune response will be limited, and thus the risk of rheumatic fever will be lessened. Thus, while tremendously powerful and effective, the immune system is not perfect.

Autoimmune diseases are very difficult to treat and even more difficult to cure, because the causal agent is a normal part of the body—usually a set of immune system cells that become active when they ordinarily would not be. How can we prevent such cells from becoming active while still allowing all other cells of the immune system to function normally? Agents that suppress the immune system in general will relieve some symptoms of an autoimmune disease, but they will also put the patient at greater risk of incurring other infections. The trade-off may or may not be desirable. There are a few ways to more directly fight some autoimmune diseases. The tools to do this are the subject of the next chapter, but we will lay out the general idea here.

Myasthenia gravis results from the action of immune system cells that produce antibodies against acetylcholine receptors. These receptors are important components of nerve transmission to the muscles. **Autoantibodies** bind to the receptor, preventing it from binding the neurotransmitter acetylcholine, and thus inhibit muscle activation (Figure 11.12b). The autoantibodies also induce complement-mediated destruction of the receptor, which leads to progressive weakening of skeletal muscle. Death usually results from an inability to breathe due to weakened diaphragm muscles.

The agents that cause the disease are the autoantibodies produced by a particular subset of B cells. Trying to eliminate the B cells that have gone astray is impossible. However, it is possible to combat the offending antibodies, allowing for at least some effective treatment.

The next chapter describes ways to construct specific antibodies in the laboratory rather than in the body. Imagine that we have been able to identify and purify some of the antibodies that cause myasthenia gravis. Now think of these antibodies as antigens. What would happen if these antibodies were injected into a laboratory animal? They would be recognized as foreign agents, since they are not normally found within the animal. The animal's immune system would begin to produce antibodies to inactivate the offending antibody (now acting as an antigen) by binding to it and starting all the processes that normally destroy foreign invaders. Some of the antibodies produced will recognize a single **antigenic determinant**, or **idiotype**, of the variable region of the antigenic antibody that is responsible for binding to the acetylcholine receptor (Figure 11.12c). These antibodies are referred to as **anti-idiotype antibodies**. Thus, within the population of antibodies produced by the animal will be some that could bind to the region that ordinarily binds to the receptor.

If it were possible to identify, purify, and mass-produce this particular antibody, it could be used to treat myasthenia gravis. Following injection into the patient, the anti-idiotype antibodies would traverse the body, binding to the acetylcholine receptor antibodies. Once bound, they would prevent the myasthenia gravis antibodies from binding to the receptor, effectively neutralizing it. Such anti-idiotype antibodies are now used to treat a number of autoimmune diseases.

Organ Transplantation Organ transplants have become common in recent years. However, since MHC proteins are unique to each individual, the immune system of the transplant recipient usually recognizes the "foreign" MHCs and mounts an immune response against them. In this process, called *tissue rejection*, the body responds to the transplanted tissue as if it were an invading foreign organism. Can this response be avoided? Physicians attempt to match tissues as closely as possible to prevent major reactions. Tissues from closely related family members have the most closely related MHCs and are used whenever possible.

Two other ways to help prevent tissue rejection are aimed at reducing the immune response in the transplant recipient. First, a number of drugs (cyclosporin is the most common) effectively reduce the entire immune response. When these drugs are delivered before and after transplantation, the body accepts the new tissue more readily. Patients are usually weaned gradually from the drugs to allow the immune system to develop tolerance for the transplanted tissue.

A second approach takes advantage of the capability of producing great quantities of antibodies in the laboratory. Imagine producing an antibody that will react against T cells. Such an antibody would not normally be found in the body—it would have been eliminated during fetal development. But we can artificially produce it in mass quantities. Following organ transplantation, the antibody can be administered to the patient. This antibody effectively attacks T cells, the very cells that are part of the tissue-rejection process. Again, weaning the patient slowly off the antibodies usually results in greater tolerance for the transplanted tissue.

Of course, with both approaches, care must be taken to prevent undue exposure of the patient to infectious agents. The patient's compromised immune system would be unable to respond strongly to any infection during treatment.

Vaccines

We have described the workings of the immune system in some detail. We know how an immune response is generated and why this response is specifically targeted against particular infectious agents. We have also seen how immune responses lead to the formation of memory cells. These partially activated, long-lived cells enable the immune system to "remember" what foreign molecules it has been exposed to in the past. If the same antigen is encountered again, the response will be greater and more rapid, perhaps even strong enough to prevent infection.

What if it were possible to somehow expose the immune system to an antigen associated with a particular disease without causing an actual infection? The result would be to activate the immune system, in essence generating a **primary immune response**. The individual would never have been exposed to the actual **pathogen**, or disease-causing organism. Memory cells would be created without a primary infection. Following exposure at some later time, the immune system would react as if it had already seen the pathogen, generating the more powerful **secondary immune response**.

What we have just described is the principle of **vaccination**. A vaccine is some form of antigen from a particular disease-causing microbe or toxin that can be introduced into the body without significant risk of disease. While the first vaccines developed were produced from related pathogens or by simply killing pathogens, biotechnology provides ways to produce safer versions.

Vaccines Based on Inactive Organisms The first intentional vaccination was performed in 1798 when Edward Jenner deliberately exposed people to a virus that causes cowpox in order to prevent infection by the related smallpox virus. Smallpox was a deadly human disease, whereas cowpox resulted in only minor infections in humans. This program of vaccination was based on the observation that individuals who had contracted cowpox never contracted smallpox. They seemed to be immune to smallpox by virtue of having had cowpox.

We now know that the reason this vaccine worked is that the viruses that cause both cowpox and smallpox are antigenically similar. Thus, exposure to the cowpox virus generated a primary response against similar antigens and subsequent infection by the smallpox virus was met by a secondary response even though the immune system had never before encountered smallpox.

Jenner thus launched the development of vaccines based on whole entities. Exposing the body to small amounts of a particular infectious agent, usually in some inactive form, induces the body to mount a primary response. This results in the production of circulating antibodies and memory cells, effectively preventing future infection.

The inactivation of organisms or viruses is usually accomplished in one of two ways. The organism can be killed or the virus deactivated, usually through chemical treatment, and then used as a vaccine. Even though the organisms or viruses are dead or deactivated, they are composed of the same antigens as were the living organisms or the active viruses. Thus, the immune system would treat the dead or deactivated materials as living foreign pathogens. But since they are dead or deactivated, there is little chance that the vaccinated individual will become infected or diseased.

Another way to inactivate an organism is to alter it so that it is still alive or active but is incapable of causing a disease. Reducing an organism's ability to reproduce or reducing the ability of a virus to enter cells results in greatly reduced infections and disease-causing capability. Exposure to such an *attenuated* organism or virus (weakened either infectiously or pathogenically) generates the primary response, but there is little chance of infection because the microbe is significantly weakened with respect to infection and disease causation.

Both methods of inactivating infectious, pathogenic viruses are in use today. For example, rabies vaccine is prepared by chemical treatment of the causal viruses to render them inactive, while measles and mumps vaccines are composed of attenuated viruses.

Subunit Vaccines Another approach to developing vaccines is becoming more common. It is based on the idea that the whole infectious organism or virus is not required in order to trigger an immune response. Rather, individual antigens that comprise the organism specifically trigger the response. A virus's coat contains a number of different proteins that act as antigens. What the immune system recognizes and responds to are these antigens, not the entire virus.

What would happen if only the viral coat protein, rather than the entire virus, were injected into a person? The antigen should still trigger the primary response, even though it is not associated with a whole organism. This might result in an **immunization**, or rendering of immunity, as effective as that obtained by using the whole virus. Since only the antigen is being used, there is absolutely no chance that the disease will be contracted by the patient. Although the risk inherent in using whole-organism vaccines is low, subunit vaccines are even safer.

The development of recombinant DNA techniques has allowed mass production of individual proteins that can be used as subunit vaccines. In some cases, it is much easier and cheaper to produce a subunit vaccine than to produce one using a whole organism. The hepatitis B vaccine was developed by expressing a portion of the hepatitis B viral coat protein in yeast cells. Besides the safety factor, there are two additional advantages to using recombinant DNA methods to produce subunit vaccines. Once a recombinant DNA production system has been established, it is easy to mass-produce the antigen. Also, the ability to manipulate the gene for a given antigen allows changes to be made in that gene, perhaps increasing the stimulatory effect the antigen will have on the immune system. This is a great advantage, since one of the primary difficulties with subunit vaccines is that there is generally a weaker immune response than with a whole-organism vaccine.

Creating two different subunits and mixing them can provide much better protection than using either subunit alone. Such **multivalent vaccines** are also necessary in order to generate immunity against organisms that exist in different forms or that change their outward appearance (that is, change their surface antigens) and evade the immune system.

THE STATE OF SOME INFECTIOUS DISEASES

We will now examine some infectious diseases, looking at what causes them, whether vaccines are available, what kind of vaccines are available, and why vaccines are not available in some cases. Vaccines have existed for some time for a number of diseases. For the most part, these diseases are no longer significant health threats in industrially developed nations, although they may be serious health concerns elsewhere. There are also many diseases for which no effective vaccine or treatment exists (for example, AIDS; Chapter 13). Five specific diseases will serve as examples of our progress (or lack of progress) in the battle against infectious diseases.

ESSAY: VACCINES FOR THE THIRD WORLD

Preventing infectious disease is a challenging task. Many diseases that arose through unhealthy living conditions have been all but eliminated in countries and societies that can afford to implement sufficient sanitation standards and provide suitable sources of drinking water.

This is not true everywhere in the world. This simple difference is but one aspect of the stark contrast between developed, developing, and underdeveloped countries. In regions of the world encompassing underdeveloped countries, about 15 million children under the age of five die each year due to acute respiratory infections, diarrheal diseases, measles, malaria, tetanus, and other diseases that are relatively rare in the United States. How can we reduce such high levels of disease, and what are the obstacles?

It is presumptuous for those of us in industrialized, developed countries to implement doctrines elsewhere simply because we see great value in them. Cultural differences and variations in basic beliefs may be strong enough to prevent the free exchange of knowledge and its utilization in different parts of the world. Cost is always an important consideration—initiating the changes that we have found to be effective in preventing many infectious diseases costs a great deal. These changes include not only improvements in sanitation and water supplies, but also the development and distribution of vaccines.

The United Nations has sponsored a variety of global vaccination programs over the years. In general, they have been remarkably successful. The eradication of smallpox twenty years ago is a testament to the global ability to unite against a particular disease. Unfortunately, such complete success is rare.

There are two major problems. First, for many significant diseases, no vaccine exists. A recent survey of worldwide vaccine researchers showed that vaccines against HIV and malaria were listed as most important for the Third World, followed by an improved tuberculosis vaccine (the one currently available is not entirely adequate). Millions die each year as a result of these diseases.

Second, even diseases for which good vaccines exist have not been eradicated. Delivering the vaccines to all regions of the world is not a simple task. It costs a great deal, requires a tremendous effort on the part of trained medical personnel, and necessitates the use of vaccines that can withstand the rigors of travel to remote regions where, for example, refrigeration may not be readily available. The World Health Organization's Expanded Program on Immunization aims to vaccinate every child in the world against six infectious diseases: measles, diphtheria, pertussis (whooping cough), tetanus, polio, and tuberculosis. With the exception of the tuberculosis vaccine, each of the other diseases is effectively prevented by vaccination—the same vaccines that have been delivered readily to children in the United States for many years. Unfortunately, the success of such a program cannot be ensured, even with a large influx of money. Delivering improved health care to politically unstable countries is a serious challenge, as is educating a population that does not have a history of Western medical care.

What about developing new vaccines? The bottom line is always economic. Few manufacturers and pharmaceutical companies are in the position to be able to develop new or improved vaccines. Those that are have precious little incentive to do so, because research and development costs are enormous. To develop a new vaccine that might greatly reduce the incidence of cholera, for example, would be of little value to a pharmaceutical company because those who could afford to pay for the vaccine are seldom victims of the disease.

We might frame the question in global terms. What responsibilities do developed nations have to provide improved health care to developing or underdeveloped nations? How much of our tax money should be appropriated to pay for research? These difficult questions will continue to plague all efforts to improve global human health.

Smallpox

The smallpox virus is transmitted through air to the respiratory system. The virus eventually infects the blood and then the skin, causing the formation of lumps or lesions (Figure 11.13). In the Middle Ages, most of the human population suffered from smallpox at one time or another, and millions died as a result. Smallpox ran

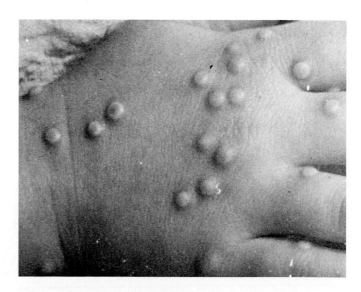

Figure 11.13 Smallpox lesions. Smallpox is caused by a poxvirus known as the smallpox (variola) virus and is distinguished by characteristic lesions.

rampant through the Native American population as well after Europeans arrived and may have been responsible for the death of more Native Americans than any other cause.

Smallpox was also the first disease for which a vaccine was developed, as we described earlier. Through a concerted effort by the World Health Organization, smallpox was certified as having been eliminated as a naturally occurring infectious disease in 1980. Since the smallpox virus does not inhabit organisms other than humans, it proved possible to eradicate the disease by global distribution and administration of vaccine. Smallpox viruses now exist only as laboratory specimens.

Measles

Measles is a potentially deadly disease caused by the *Rubeola* virus. The virus is highly contagious and is spread through the air before any symptoms are apparent in the infected person. Infection begins in the respiratory tract, and after about two weeks coldlike symptoms begin to appear. Then rashes break out, usually starting on the face and the spreading elsewhere (Figure 11.14). In the United States, the fatality

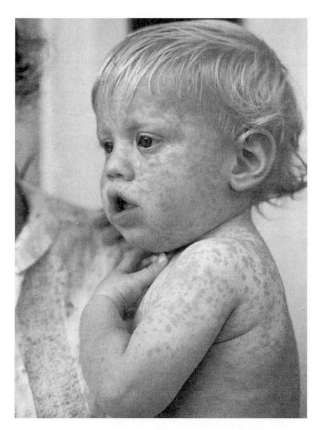

Figure 11.14 The rash of red spots characteristic of measles. Measles (rubeola) is a highly contagious viral disease that is characterized by a rash that typically begins on the face and spreads to the trunk and extremities.

rate is about two or three in every 1,000 cases. However, in some developing nations, the fatality rate is 30 percent and is highest in infants and the elderly.

The first measles vaccine was developed in 1958. It was based on a mutated form of the virus that produces only a low-level infection in most individuals. Within five years of the introduction of mass vaccination programs, the number of measles cases in the United States dropped drastically, from almost 500,000 per year to fewer than 20,000. Since that time, continued mandatory vaccination programs have made measles a rare disease in the United States.

Despite the overall effectiveness of the vaccine (about 95 percent), two interrelated problems remain. First, vaccination results in some mild symptoms of the disease, and some people react more strongly than others. Second, there have been rare cases of people reacting so strongly that they develop full-blown measles. There is room for vaccine improvement but little economic impetus to develop a better version.

We have learned much from the results of the nationwide measles vaccination campaign. The vaccine is not effective until fifteen months of age. If it is delivered before then, little long-lasting immunity results. Infants younger than this must rely on the immunity imparted by their mothers. Mothers share antibodies with developing fetuses, providing them with some level of resistance until their own immune systems kick in. The mother's milk provides additional antibodies to the infant—one of the advantages of breast-feeding. But since mothers have now all been vaccinated against measles, the only antibodies against measles that infants receive are those resulting from the mother's vaccination. These antibodies are presumably much less numerous and perhaps less effective than those resulting from actual infection by the virus. For whatever reason, in recent years the number of measles cases of children under the age of one year has increased.

Polio

Poliomyelitis is caused by a virus that is usually spread through drinking water contaminated with human feces. Infection begins in the throat and intestines and spreads to the lymph system. In most cases, the infection does not spread further, and the patient displays few if any symptoms. In other cases, the infection spreads to the central nervous system, sometimes causing paralysis. In the early 1950s, about 50,000 cases of polio were reported annually in the United States (Figure 11.15).

In 1954, the *Salk polio vaccine* was introduced. It was based on the inactivation of the virus through chemical means, specifically, treatment with formaldehyde. This particular vaccination requires a series of treatments, and boosters are required every few years. In 1963, the *Sabin vaccine*, based on attenuated viruses, was developed. This vaccine is administered once orally. Through the use of these two vaccines, polio has been virtually eliminated in developed countries. A problem with the Sabin vaccine is that very rarely (about once in every million vaccinations) some of the attenuated viruses mutate back to a virulent form and cause the disease. The effort to completely eradicate polio will presumably require the use of both of these vaccines to ensure that no new cases arise.

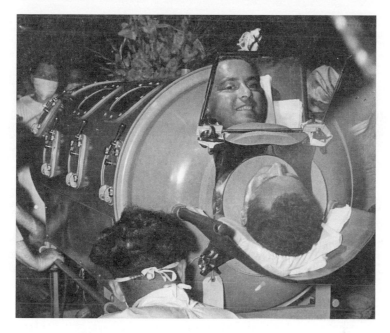

Figure 11.15 Iron lungs at a March of Dimes respiratory center in Los Angeles in the 1950s. For many polio victims, breathing was possible only with these mechanical aids.

Tetanus

Clostridium tetani, the bacterial species responsible for causing tetanus, is found in soil and dust everywhere. It enters the body through even the most trivial wounds. Once it is established under conditions where oxygen is absent, it produces a potent neurotoxin that effectively blocks the signals that cause muscles to relax following contraction. The result is that opposing muscle groups contract continuously, causing tetanus (Figure 11.16).

The tetanus vaccine was actually developed not against the bacterium but against the purified, deactivated neurotoxin. The detoxified toxin (*toxoid*) serves as the vaccine, eliciting the production of antibodies. Should an actual infection occur, the circulating antibodies present as a result of vaccination will bind tightly to any toxin produced, preventing it from inhibiting muscle relaxation. Tetanus booster vaccinations are necessary in order to ensure an adequate supply of circulating antibodies against the toxin.

Malaria

Malaria is a complex disease caused by four different species of protozoans. The complexity stems from the life cycle of this parasitic protozoan, *Plasmodium*, illustrated in Figure 11.17. Three different forms of the parasite reside in two different

Figure 11.16 An advanced case of tetanus. This drawing shows a British soldier afflicted with tetanus during the Napoleonic wars. The tetanus neurotoxin blocks muscle relaxation pathways, causing muscle contractions that result in the characteristic muscle spasms depicted in the drawing.

host organisms. The **sporozoite** form of the protozoan is found in the salivary glands of infected female mosquitoes. The parasite has no effect on these mosquitoes—they simply serve as carriers. When an infected mosquito bites a human, some of the saliva containing the sporozoites is transferred into the blood of the human. Once in the human blood, the sporozoites infect liver cells within minutes.

After some time, during which the protozoans reproduce and change form, **merozoites** are released from the liver cells. Merozoites infect red blood cells, where they reproduce. Eventually, they cause the parasitized red blood cells to burst and release the new merozoites. These infect other red blood cells, which eventually burst and are killed. This scenario explains the periodic bouts of pain and fever that malaria patients suffer: Periodically, some of the red blood cells burst and release merozoites.

Some of the merozoites, as they reproduce within the red blood cells, change into yet another form, *gametes*. The gametes are involved in sexual reproduction, but they require a mosquito host. An uninfected mosquito picks up gametes by biting a human suffering from malaria; some of the gametes are transferred to the mosquito in the human blood. Within the mosquito, these gametes reproduce and infect the intestinal wall of the mosquito, eventually taking up residence in the salivary glands and changing into the sporozoite form.

This is a complex life cycle involving two different host organisms, humans and mosquitoes, and three different forms of the parasite itself, sporozoite, merozoite, and gamete. The three forms are necessary, as are both hosts. Multiple forms and hosts make it very difficult to eradicate the disease.

In the most serious form of malaria, untreated patients die. Even the less severe forms of the disease are worrisome and extremely unpleasant. The traditional medicinal treatment for malaria consists of antimalarial drugs, all of which are variants

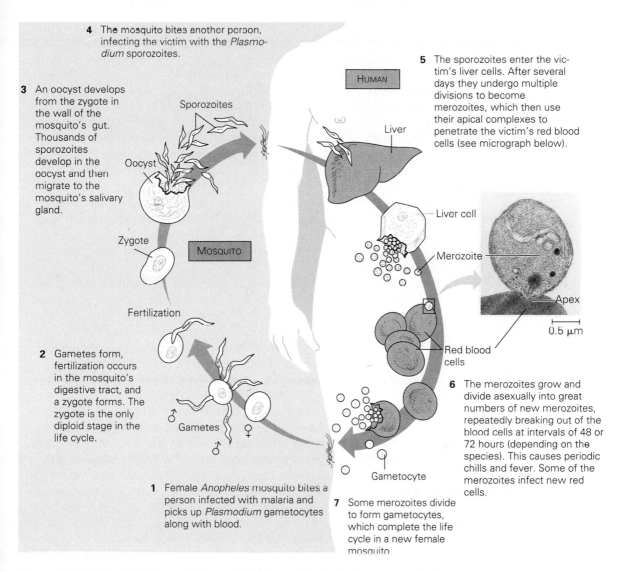

4 The mosquito bites another person, infecting the victim with the *Plasmodium* sporozoites.

3 An oocyst develops from the zygote in the wall of the mosquito's gut. Thousands of sporozoites develop in the oocyst and then migrate to the mosquito's salivary gland.

Sporozoites

Oocyst

Zygote

MOSQUITO

Fertilization

2 Gametes form, fertilization occurs in the mosquito's digestive tract, and a zygote forms. The zygote is the only diploid stage in the life cycle.

Gametes

♂

♀

♂

1 Female *Anopheles* mosquito bites a person infected with malaria and picks up *Plasmodium* gametocytes along with blood.

HUMAN

Liver

5 The sporozoites enter the victim's liver cells. After several days they undergo multiple divisions to become merozoites, which then use their apical complexes to penetrate the victim's red blood cells (see micrograph below).

Liver cell

Merozoite

Apex

0.5 μm

Red blood cells

Gametocyte

6 The merozoites grow and divide asexually into great numbers of new merozoites, repeatedly breaking out of the blood cells at intervals of 48 or 72 hours (depending on the species). This causes periodic chills and fever. Some of the merozoites infect new red cells.

7 Some merozoites divide to form gametocytes, which complete the life cycle in a new female mosquito

Figure 11.17 The life history of *Plasmodium*, the parasite that causes malaria.

of quinine. The two most common forms of the drug used today, chloroquine and mefloquine, slow the growth and reproduction of the merozoite within red blood cells. While these drugs are usually effective, some mosquitoes carry drug-resistant protozoans. Individuals infected with resistant protozoans are usually at risk, since all of the quinine-based drugs have reduced effectiveness against them.

In the 1960s, attempts were made to eradicate malaria by eliminating the mosquito vectors. The reasoning was that if female mosquitoes are required as hosts for the protozoan, the eradication of the mosquito population would prevent the reproduction and spread of the parasite. Massive insecticidal programs were launched in the tropical countries where malaria flourished. The insecticide DDT was the primary weapon in the fight to eradicate the mosquito. However, the programs were dismal failures, for four reasons. First, it proved to be virtually impossible to kill a substantial portion of the mosquito population. There were just too many places for mosquitoes to breed. Second, DDT was found to have severe detrimental effects on birds and nontargeted insects. Third, and most important, the mosquitoes quickly developed resistance to DDT, making them almost impossible to kill. Fourth, after a number of years of trying to reduce mosquito population with little success, poor countries participating in the venture ran out of patience and money. Substantial investments of time and money had not resulted in a significant decrease in the number of mosquitoes, and some countries had essentially bankrupted themselves in the massive effort.

Eradication of the host mosquito is not a realistic goal, and little progress has been made in the development of new antimicrobial drugs to prevent the spread of the protozoans within humans. The best bet for developing treatments or cures for malaria is the production of vaccines. However, developing a vaccine against malaria presents a multitude of challenges.

Consider again the life cycle of the parasite. Following the transfer of the sporozoite form to human blood, it very rapidly infects liver cells. This leaves virtually no time for the immune system to mount a response—infection happens so quickly that the immune system might not even be activated. Once the sporozoites enter the liver cells, they are "hidden" from the immune system in many ways. Certainly, circulating antibodies, the early mainstay of any immune response, have virtually no chance to even recognize the invading sporozoites, let alone develop a response sufficient to prevent infection.

Sporozoites in liver cells then change into the merozoite form, which is released into the blood again. Even if the invading sporozoites did trigger an immune response, the merozoite form that next encounters the immune system does not carry the same set of antigens as the sporozoites. The immune system needs to mount a response against the merozoite form. As the immune system does this, the merozoite form rapidly infects the red blood cells, again "hiding" from the immune system. Finally, the production of yet another form, the gamete, elicits yet another immune reaction.

Each of the three forms of the parasite is antigenically distinct from the others, making it very difficult for the immune system to mount an effective response. Another difficulty encountered by the immune system is the speed with which the sporozoite and merozoite forms infect their respective host cells. Also, significant mutations occur within the parasite during the development of the merozoite, resulting in numerous antigenic forms of the parasite. This allows at least some individual parasites to escape the immune system.

Thus, an antimalaria vaccine has yet to be developed. Furthermore, growing and studying the protozoan in the laboratory have proved to be difficult. Growth of the parasite requires both mosquitoes and humans, as well as systems that can yield the large numbers of parasites necessary for the development of treatments and vaccines.

Molecular biology and genetic engineering methods provide real hope for the development of a vaccine against malaria. It will first be necessary to study the three forms of the parasite that exist in human blood in order to determine exactly which antigens are present in all forms and are not subject to wide variation through mutation. Once these antigenic structures have been identified, trial vaccines can be developed. Presently, trial vaccines consist of a mixture of different antigenic agents, often involving several antigens from each form of the parasite. None of the trial vaccines has been effective enough to be considered a success. An additional complication stems from the fact that four different parasitic species cause malaria. Trial vaccines that have been found effective against one species have shown essentially no activity against the others.

We are clearly a long way from developing the means to effectively treat and cure malaria. Only the continued search for both new pharmaceutical treatments and better recombinant vaccines can provide any hope of bringing this disease under control.

SUMMARY

This introduction to the immune system and the nature of infectious disease should impress you with how well the system protects us from most of the infectious agents around us. The immune system is a wonderful example of complexity that has arisen through evolution. You have also discovered more examples of the applicability of recombinant DNA technology. Much of what we know about our immune system is based on research involving recombinant DNA. The development of new and better vaccines and treatments could not move forward without this type of research. This chapter also noted several uses for mass-produced antibodies. In the next chapter, we will see how antibodies can be mass produced and what applications antibodies produced in this way might have.

REVIEW QUESTIONS

1. List the major components of the first, second, and third lines of human defense against infectious microbes.

2. List and outline the functions of three components of human nonspecific defense.

3. Distinguish between the killing methods of macrophages and neutrophils.

4. Describe the components of the inflammatory response.

5. How does the assembly of Cabbage Patch dolls serve to explain the human ability to recognize and respond to millions of different antigens?

6. List four ways that antibodies act against foreign invaders such as viruses and bacteria.

7. Explain the ultimate value in having two components (cell-mediated and humoral responses) in the specific immune system in humans.

8. Explain how T_H cells act to regulate (turn on) both the cell-mediated and humoral (antibody-mediated) immune responses.

9. What are the advantages and disadvantages of (a) killed or inactive vaccine material and (b) attenuated vaccines?

10. How does MHC ensure appropriate regulation of T_H activation?

11. Outline the cell-mediated immune response.

12. Outline the humoral immune response.

13. How does the existence of memory T and memory B cells result in our having most infectious diseases only once?

14. Describe the general structure of antibodies.

15. What is the relationship among strep throat, *Streptococcus pyogenes*, rheumatic fever, and damaged heart valves?

16. How would anti-idiotype antibodies be used to treat myasthenia gravis?

17. Why are subunit vaccines considered very safe for human administration?

18. Describe the present status of vaccine development against each of the following diseases or intoxications: tetanus, polio, smallpox, measles, and malaria.

CHAPTER 12

Monoclonal Antibody Technology

Biotechnologists constantly search for new and useful ways to use the resources that nature provides. In particular, biotechnologists often will learn about a particular system that operates in living organisms and then modify and expand that system for greater use. As we delved into recombinant DNA technology, we gained some understanding of the tremendous possibilities offered by the exploitation of natural systems. Chapter 11 dealt with some of the details of our immune system. The ability to synthesize on demand molecules that bind to a particular antigen is an amazing natural capability and provides us with a potent specific defense system.

The immune system has invited the curiosity of biotechnologists. We have alluded to several situations in which it would be desirable to have immediate access to large quantities of specific antibodies. Could we tap the immune system and make it work according to our directions? Could we mass-produce antibodies that will bind to a specific antigenic molecule? If so, how? This chapter addresses these questions.

PRODUCING ANTIBODIES

One approach to obtaining antibodies against a particular antigen might be to collect lots of blood from animals that have been exposed to that antigen. The immune systems of the host animals will have reacted against the antigen and made antibodies to it as part of their normal immune response. We should be able to purify these antibodies from the animals' blood.

After we obtain blood, we can separate its components by **centrifugation** (Figure 12.1). Centrifugation acts to artificially increase the gravitational force on the sample so that cells or other particles that eventually would have settled to the bottom of the tube will do so much more rapidly. A variety of cells are present, primarily red blood cells but also many white blood cells. The cells will be forced to the bottom of the tube by centrifugation. Plasma is the fluid portion of the blood remaining after the cells are removed. Within the plasma are found circulating antibodies.

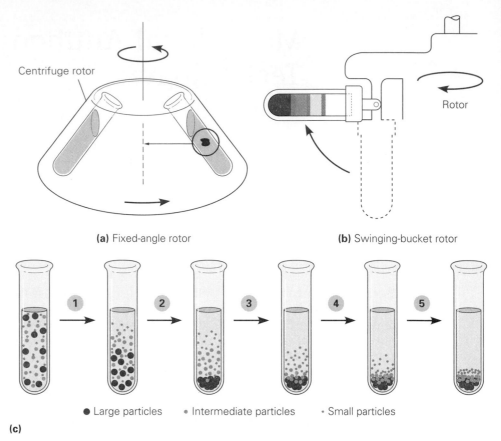

(a) Fixed-angle rotor   **(b)** Swinging-bucket rotor

● Large particles   • Intermediate particles   · Small particles

(c)

Figure 12.1 Centrifugation. A centrifuge consists of a rotor powered by an electric motor. The rotor either holds tubes at a fixed angle (**a**) or is equipped with swinging buckets that allow the tubes to swing outward during acceleration (**b**). Centrifugation provides a means of separating particles based on size, shape, and/or density in response to centrifugal force. (**c**) Here particles of three different sizes are subjected to a centrifugal force for five successive periods of time (circled numbers). Larger particles accumulate first at the base of the tube, followed by those of intermediate size and finally by the smallest particles.

Polyclonal Antibodies

To separate the desired antibodies from the other molecules, including other proteins present in plasma, we use a technique known as **affinity chromatography** (Figure 12.2). Affinity chromatography takes advantage of the ability of different molecules (say, antibodies) to bind to a specific agent (say, specific antigens). Specific target molecules (specific antigens) are chemically attached to a solid support, usually beads. The target-bead complexes are put into a column. When a solution containing the mixture of proteins to be separated (the plasma in this case) is allowed to pass through the column, only those molecules that bind to the target will be retained, while all others will flow out the bottom. After all unbound molecules have been re-

Blood plasma containing antibodies

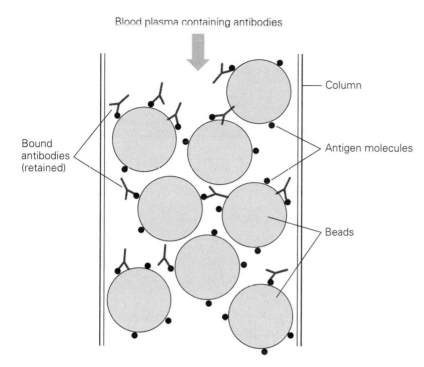

Column

Antigen molecules

Bound
antibodies
(retained)

Beads

Figure 12.2 Affinity chromatography. As practiced with plasma and antigens, affinity chromatography requires that the known antigen be chemically coupled with beads; then the antigen-coated beads are placed in a column. Upon addition of blood plasma, only those antibodies with specific antigen-binding sites will be retained—bound to the antigen.

moved, the desired proteins can be collected, usually by raising the salt concentration high enough to disrupt the ionic attractions. The bound proteins will flow out the bottom of the tube, resulting in a greatly purified protein preparation.

Assume that we are trying to isolate antibodies that will bind to the hormone insulin, which is composed entirely of amino acids. We could set up an affinity column in which insulin is attached to chromatography beads. When plasma is passed through the column, the only things that will bind to the bead-bound insulin are those antibodies that recognize it. Everything else, including all other antibodies, will pass through. After rinsing, the insulin-bound antibodies are removed from the insulin-bead complexes by passing a solution with a high salt concentration through the column. As we indicated previously, the high salt concentration interferes with the ionic interaction between the insulin antibodies and insulin, allowing the antibodies to be collected. The result is a collection of any and all antibodies that were bound to insulin. Affinity chromatography is a very useful way of separating and isolating proteins.

We call a population of antibodies separated from plasma a **polyclonal** preparation. This polyclonal mixture of antibodies is composed entirely of individual antibodies that bind to insulin. However, each individual clone actually recognizes different portions (**epitopes**) of the insulin molecule (Figure 12.3). They may also bind, although less well, to other molecules besides insulin. In general, the greater the number of different antibodies present, the greater the chance of non-specific binding, or cross-reaction.

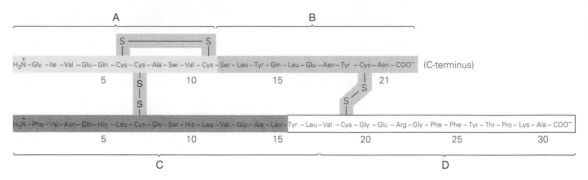

Figure 12.3 Insulin. Standard amino acid abbreviations (Glu, Ser, Lys) are used here to demonstrate the primary structure of this protein (a composite of several types of insulin). Four different (and theoretical) epitopes of the insulin molecule are shown, each capable of stimulating and binding with a different but still insulin-specific antibody.

Monoclonal Antibodies

While polyclonal preparations are useful in many ways, at times it is necessary to use only a single type of antibody so all antibodies in the preparation will recognize and be able to bind to the same epitope of the same molecule. Such a preparation is called **monoclonal**, since it contains identical antibodies. It is practically impossible to isolate monoclonal antibodies from a polyclonal collection. Nor is it always possible to produce as many polyclonal antibodies as needed, since the blood plasma of an infected animal is the original source. Fortunately, a technology has been developed for exactly this purpose—mass production of monoclonal antibodies.

We know that, individually, B cells (actually the precursors of the antibody-secreting plasma cells) produce only a single specific type of antibody singularly specific for one epitope. Could we take one B cell, a precursor of a clone of antibody-secreting plasma cells, that makes and secretes, say, one of our insulin-specific antibodies and grow this cell in the laboratory? If we could, then we could produce a large supply of the desired monoclonal antibody.

There are two problems with this idea. First, how could we select and obtain the right B cell? We've run into similar selection problems before, so we can imagine that there are ways to get around them. The second problem, on the other hand, seems more difficult to address: B cells cannot be grown for extended periods in the laboratory. When they are removed from the body, B cells survive only for a short time.

Hybridoma technology makes it possible to overcome both of these difficulties. This technology involves the fusion of two different cells to produce a single cell (called a hybridoma) that expresses properties of both originals. Figure 12.4 illustrates the process. Suppose we have B cells that produce insulin-specific antibodies and we wish to grow these cells in culture. We know that B cells won't survive long outside the body. How could we overcome this obstacle? We know that myeloma cells are cancerous forms of B cells and, as such, no longer produce antibodies, but they

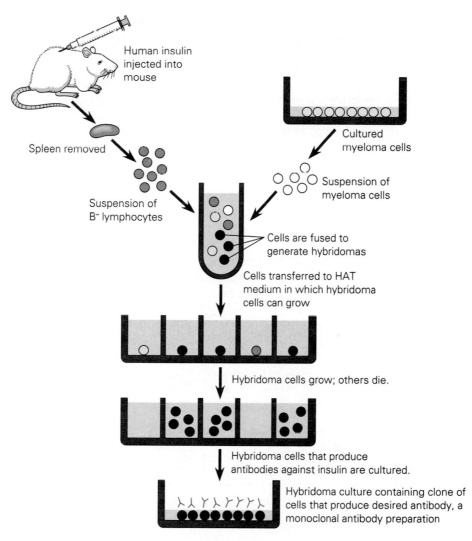

Figure 12.4 Hybridoma and monoclonal antibody formation. In the production of hybridomas, the spleen of a mouse is injected with human insulin, the antigen. Later the animal's spleen is removed, and B lymphocytes are isolated. They are fused with cancerous myeloma cells, thereby generating hybridomas. Hybridomas are separately cultured on HAT medium. Only those that produce insulin-specific antibodies multiply. Since each compartment contained just a single cell, all the progeny of that hybridoma produce a single type of antibody—a monoclonal preparation.

can be grown indefinitely in the laboratory. If two cells, an appropriate B cell and a cancerous myeloma cell, are fused, then the resulting hybridoma should display both desired traits—long-term growth outside the body and continuous production of insulin-specific antibodies.

Performing cell fusions is, in some ways, similar to constructing recombinant molecules. What will ultimately be formed is a population that contains the desired hybridoma cells as well as B cells and myeloma cells that didn't fuse. The original cells are no longer of interest. All the unfused B cells will die, so there is really no need to deal with them in any special fashion. But the unfused myeloma cells will survive, so we must devise a way to select against them.

It turns out that myeloma cells lack a particular enzyme that would allow them to detoxify a mixture of three chemicals: hypoxanthine, aminopterin, and thymine (the constituents of HAT medium). Because this mixture is not metabolized properly, that is, not detoxified, it will kill myeloma cells. Normal B cells and the hybridomas are able to metabolize this mixture, that is, to detoxify it. Thus, to select for hybridomas, we place the likely candidate cells on HAT medium and incubate them. Unfused B cells will die naturally, and myeloma cells will die because they cannot metabolize the HAT medium. Therefore, only hybridomas will survive and reproduce.

Following incubation, many different hybridomas are present since we have done nothing to specifically select a cell of interest. The selection process can be accomplished in a number of ways. One of the easiest is to separate individual candidate hybridomas and grow each of them in a separate culture dish. Antibodies made by the hybridomas will be secreted into the medium, and after some period of time we can obtain antibodies from each candidate hybridoma by sampling the medium surrounding the cells. Media samples containing antibodies can be added to individual plastic wells containing a special coating to which antibodies bind readily. The dishes will accommodate many candidates at one time, one antibody sample per well. After allowing sufficient time for the antibodies to bind to the plastic, we wash away the rest of the hybridoma growth medium. If radioactively or fluorescently labeled insulin is then added to each sample, we can relatively easily determine which samples contain antibodies that will bind insulin. Those that test positive can be tested again in order to ascertain what portion of the insulin molecule is bound. The beauty of this approach is that at all times the various hybridomas are kept separate. Once we have determined which hybridoma produces the correct antibody, we can discard the rest. Mass production of the desired antibody can be accomplished by culturing the selected hybridoma. Mass quantities of the desired antibody can then be purified from the surrounding medium.

Using this technology, we can create hybridomas that produce antibodies against almost any antigen. If a laboratory animal mounts an immune reaction against the antigen, then theoretically a monoclonal antibody preparation can be made. As with the production of proteins through recombinant DNA, most of the problems involve identifying the particular hybridoma of interest. Once it is found, it is relatively straightforward to scale up the production process to allow recovery of large quantities of the monoclonal preparation.

Note that it is possible to identify many different hybridomas that produce antibodies against a particular antigen. Each antibody probably recognizes a different epitope, or specific structural feature, of the antigen. The larger the antigen, the more different antibodies we can develop against it. Usually, the most specific, tightest-binding antibody will prove most useful.

APPLICATIONS OF MONOCLONAL ANTIBODIES

Monoclonal antibodies are used in four principal ways, all of which rely on two antibody features. The first feature is the exacting specificity of antibody binding. Just as in our bodies, antibodies used in the laboratory exhibit a keen preference for binding to one particular epitope. The use of well-characterized monoclonal antibodies allows us to determine with little uncertainty whether or not the antibody binds to the specific antigen in question. The second feature is that all antibodies of a particular class have the same general structure, including identical stem regions. The antibody stem is sufficiently long and well removed from the antigen-binding sites to allow it to be modified in a number of ways. Labels or "tags" can be added, for example, to create radioactively or fluorescently labeled antibodies. Or entire proteins can be attached to the antibody stems so that wherever the antibody goes the other protein is carried along. We will see how each of these modifications can be useful.

Medical Diagnosis—ELISA Tests

The most prominent application of monoclonal antibodies is in the detection and diagnosis of medical conditions and diseases. A particular type of test using antibodies is rapidly replacing older, less exact methods of performing medical diagnoses. We will first describe how these tests are designed and then lay out an example of the technique.

The "antibody sandwich" approach is known as an **enzyme linked immunosorbent assay**, or **ELISA** test. The basic pattern of an ELISA test is shown in Figure 12.5. A monoclonal antibody preparation that will recognize the antigen to be diagnosed is attached to the surface of a plastic dish. The solution to be tested is then

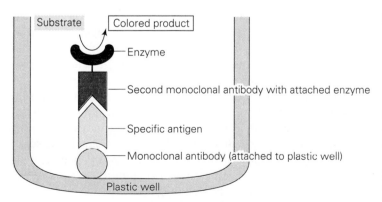

Figure 12.5 ELISA method. First, a monoclonal antibody preparation is attached to the surface of a plastic dish; antigen-containing solution is added next, and after a wash to remove all unbound materials a second monoclonal antibody (enzymatically tagged) is added. If all attachments were made, an antibody–antigen–enzyme-labeled-antibody sandwich remains. When the substrate for the enzyme is added, a blue product appears —confirming antigenic identity or establishing its presence.

applied to the dish. If the antigen is present, it will bind to the antibodies in the dish. If it is not present, no binding will occur. After a wash to remove all unbound materials, a second monoclonal antibody preparation is added that has alkaline phosphatase or another enzyme attached to its stem. If the specific antigen is attached to the first monoclonal antibody, then the second antibody will bind, effectively attaching the enzyme to the surface of the dish as an antibody–antigen–enzyme-labeled-antibody sandwich. If no antigen is present, then no enzyme will be attached. Following more washes to ensure the removal of all unbound material, the substrate for the enzyme is added. If the enzyme is still attached to the dish, then when a substrate containing a colorless substrate is added it will react and a blue product will appear. If no enzyme is present, then the substrate will not be changed and the solution will remain colorless.

The basic design of the ELISA test can be modified to diagnose the presence of almost any antigen or the presence of certain antibodies. As we will see in Chapter 13, one method of diagnosing HIV infection is to test the patient's blood to determine whether antibodies against HIV are present.

In actual practice, many different samples are tested for the presence of the same antigen. For example, an ELISA test can be used to detect the presence of a virus in the bloodstream. In a hospital laboratory, hundreds of blood samples may have to be tested daily to determine whether a particular virus is present. The virus in question bears the antigen to be detected, and monoclonal antibodies that recognize this virus are attached to the surfaces of many wells on one plate. Different blood samples are applied to each well of the plate, and the entire plate is run through the procedure at one time. At the end of the test, those blood samples that contain the virus can easily be distinguished from those that do not.

ELISA tests have rapidly become a mainstay of medical laboratory diagnosis. They are usually much faster than more traditional ways of diagnosing the presence of a virus, for example, and are at least as sensitive as most other types of tests and more sensitive than most. They are accurate, since they are based on monoclonal antibodies, and they are also simple and reliable. These virtues make them ideal for many diagnostic tests.

With the advent of monoclonal antibody technology, it has become possible to develop home pregnancy tests that are simple, rapid, and accurate. In humans, an ovum produced by the female and a sperm produced by the male fuse to form a fertilized ovum (Figure 12.6) that implants into the wall of the uterus about six days later (when it is termed an "embryo"). Implantation causes rapid, substantial hormonal changes in both the embryo and the mother—changes that prepare the mother's body for pregnancy. Ultimately, these changes cause the formation of a placenta and allow the embryo to develop into a growing fetus.

One of the earliest hormones to be produced by the embryo is **human chorionic gonadotropin** (HCG). Made following implantation, HCG is responsible for overriding the normal female hormonal cycle that would result in menstruation. So much HCG is produced that it is present in very high levels in the mother's blood, and some of the excess HCG is secreted by the kidneys into her urine. Its appearance there allows urine to be used in pregnancy testing. Any female excreting HCG in her urine has a very high likelihood (over 99 percent) of being pregnant.

Designing a home pregnancy test required being able to detect very small

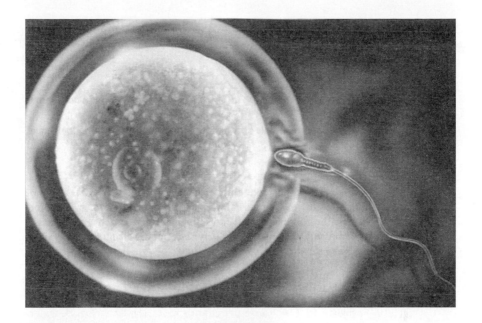

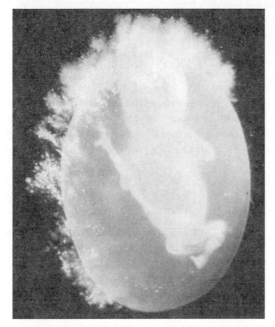

Figure 12.6 Human pregnancy.
(**a**) Conception, the moment of fertilization—a human ovum being penetrated by a sperm. (**b**) Human embryo eight weeks after conception.

amounts of HCG in urine. One commercial home pregnancy test consists of a special urine dipstick (Figure 12.7). The test area of the urine dipstick contains two different types of monoclonal antibodies specific for HCG: free antibodies unattached to the surface and "capture" antibodies attached to the surface of the stick under the test window. The free antibodies are color-labeled, while the "capture" antibodies are

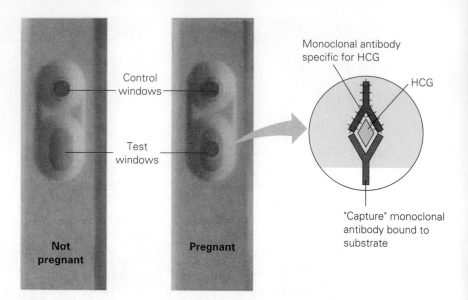

Figure 12.7 A home pregnancy test. The technology behind this test is similar to an ELISA test (Figure 12.5). It is essentially a "double-antibody sandwich" approach to identifying whether HCG is present in urine, indicating pregnancy.

specific for the HCG–free antibody complex. When urine containing HCG makes contact with the free antibodies, it binds to them, and the complex is in turn bound by the "capture" antibodies, forming a "double-antibody sandwich." Localization of the color-labeled antibody in the test window indicates pregnancy. This form of pregnancy testing is cheap, quick, accurate, and sensitive and is capable of detecting very low levels of HCG.

Monoclonal antibodies have a variety of other applications. ELISA tests, or variants of them, are used to test meat to check for the presence of pathogenic microbes (or their toxins) and verify that products labeled "beef" are in fact really 100 percent beef. Strep throat (streptococcal pharyngitis) is routinely diagnosed using an ELISA-type test that provides results in minutes instead of the several days diagnosis took only a few years ago. As we have seen with other biotechnology tools and techniques, however, the most widespread application of monoclonal antibodies is in basic research involving the detection and purification of specific molecules.

Treatment of Disease and "Magic Bullets"

Another use for monoclonal antibodies is in the treatment of disease. Again, two properties of antibodies are exploited: their specificity and the presence of stems that enable the attachment of chemical "cargo." One of the most promising applications is in the treatment of some forms of cancer (Chapter 14).

Cancer cells are cells that have lost control of their growth regulation. The majority of cells in adults grow and divide only when replacements are necessary. There are exceptions, such as skin cells, which constantly reproduce to replace cells that die on the surface and are sloughed off. For a liver cell, for example, to become cancerous, it must regain its ability to grow and divide rapidly. Several substantial changes are necessary in order for this to happen. For instance, cancerous cells are altered so they produce different surface antigens (Chapter 14). As we noted in Chapter 11, the cell-mediated immune system probably destroys many precancerous cells as they are developing. When the growth of a tumor gets past the point at which the individual's immune system can handle it, however, some form of treatment is necessary.

If we could identify a particular surface antigen produced in every case of a specific type of cancer, then we could develop monoclonal antibodies against that particular antigen. However, this antigen could not be present anywhere else in the body. Once monoclonal antibodies to the antigen are made, they could have different types of chemical cargo attached to their stems. For example, they might have a highly radioactive chemical group attached, or they might be loaded with an extremely toxic compound. With either of these alternatives, the intent is the same—to deliver the radioactive or toxic agent directly to the tumor cells.

If a preparation of modified antibodies is injected into a person's blood, they will be circulated. If they encounter their matching antigen on cells of a liver tumor, for example, they will bind there and not elsewhere. By binding to tumor cells, they bring their cargo (the toxic agent) directly to the tumor cells. If radioactive groups are attached, then the tumor cells will receive very high doses of radioactivity that will kill the cells. This is the principle of radiation therapy for cancer. However, in contrast to traditional radiation therapy, in which a regional area is bombarded with huge doses of radiation, the "magic (or smart) bullet" approach using monoclonal antibodies results in pinpoint delivery of the radiation, saving nearby tissue from exposure.

Toxic compounds carried by such antibodies will also specifically target tumor cells. Cancerous cells in humans can take up materials from the outside. Monoclonal antibodies to which toxic compounds have been attached and that subsequently become bound to the surface of tumor cells are no exception—they will be taken in by the cells, where they will release the toxic compounds. If enough toxin is delivered, the tumor cells will be killed.

Unfortunately, despite a number of laboratory successes and some clinical successes as well, application of magic bullet approaches has been difficult. Two major problems are identifying antigens that are unique to all cases of a particular form of cancer and developing a monoclonal antibody with the necessary specificity and binding strength. Many clinical trials are being performed to test the effectiveness of different magic bullets. It is hoped that some of them will prove effective in treating the less controllable forms of cancer.

Magic or smart bullets can be developed for other medical uses. We have seen published reports of successful clinical trials using monoclonal antibodies to prevent blood clotting. One series of trials involved **angioplasty**, a procedure in which a small balloon is inflated inside a clogged blood vessel and used to dislodge material from the inner wall. Ironically, during angioplasty, a process designed to open or clear a vessel, the manipulation actually induces clotting in the immediate area for a short time. Although drugs have been used to prevent postoperative clotting, a magic bullet approach has been showing some measure of success in clinical trials. In this approach, monoclonal antibodies are produced that recognize platelets, specialized cell fragments that play an important role in the clotting process. These monoclonal antibodies are modified to contain a toxin on their tail. When they are administered following angioplasty, they effectively destroy platelets and reduce the incidence of localized clotting. Treatment does not appear to be necessary for more than a few days.

Attempts have also been made to design a monoclonal antibody treatment that will dissolve blood clots after they form, perhaps during a stroke or in the obstruction of blood vessels. In this case, monoclonal antibodies will recognize proteins present in blood clots (usually fibrin, the major protein constituent of clots). Instead of a toxin being attached to these antibodies, an enzyme that can digest fibrin molecules is attached. The fibrin-dissolving enzyme can be localized at the clot, rapidly digesting it.

It is impossible to predict whether these developments will ever be commercially successful or whether they will prove to be valuable medical treatments. However, it is clear that many different approaches to treating disease can be based on monoclonal antibodies.

ESSAY: TREATING SYMPTOMS

The treatment we have described—delivering a clot-dissolving enzyme directly to a blood clot—can be used as an example of the necessary decision making involved in the use of research and development funds. Is it more effective to support the development of a treatment that will alleviate symptoms, or would the same money be better spent trying to develop a cure? Blood clots arise for many reasons, but it has become clear that life-style (diet and exercise, in particular) is a determining factor. Being able to dissolve blood clots is a wonderful treatment but does nothing to prevent future blood clots. Dissolving a blood clot in no way changes the conditions that led to the clot.

A tremendous amount of time and money has been spent on developing ways to dissolve blood clots. Not only are magic bullet approaches being developed, but recombinant DNA methods have been used to produce streptokinase and tissue plasminogen activator, products designed to dissolve blood clots. Would that money have been better spent in finding ways to reduce the number of individuals whose lives are threatened by blood clots?

This is a particularly difficult issue, for a number of reasons. Often, no way to approach finding a cure for a particular disease is obvious. For example, there is no clear-cut way to reduce atherosclerosis. The same is true regarding diabetes. One factor governing the increasing incidence of diabetes in wealthy countries is the same life-style that gives rise to atherosclerosis. Money could be spent in educational programs, to try to modify diet and exercise patterns in ways that would reduce the incidence of the disease. Such programs might show some effect but would certainly not guarantee success. Is trying to accomplish this a waste of time and money?

Policymakers must come to grips with the difficult realities of such decisions. Neglecting possible treatments that will alleviate symptoms would be inhumane. On the other hand, not to try to address the cause of the symptoms would be irrational. Where can the line be drawn? How much money should be spent in these two very different arenas?

Passive Immunization

Passive immunization, the transfer of antibodies from one organism to another, has been practiced for many years. One example involves the disease diphtheria, which was once common and deadly. To treat diphtheria, animals were injected with the bacterium that causes the disease, *Corynebacterium diphtheriae*. After a time, the **serum** (the fluid portion of blood without cells, platelets, or clotting factors) of these animals contained antibodies against both the bacterium and the toxin it produces. When the separated serum was injected into people who had recently contracted diphtheria, the animals' antibodies were often able to effectively neutralize the diphtheria toxin until the patients' own immune systems could respond. Thus, the animal antibodies acted as if they were part of an immune response even though they had simply been injected into the bloodstream.

Such a crude method of obtaining passive immunization is very dangerous. Many different substances in animal serum can cause severe allergic reactions in humans. Also, a patient's immune system might mount an attack against the animal antibodies themselves. Many cases of shock were observed when patients were administered additional doses of animal serum.

The development of monoclonal antibodies means that passive immunization may again become a viable treatment for certain diseases. Approved monoclonal antibody–based drugs are already being used to treat **septic shock**, a set of symptoms that results from the presence of a particular toxin in the cell walls of many infectious bacteria. Even after the bacteria are killed by antibiotic treatments, the possibility of sepsis remains. Sepsis results from the overstimulation of the immune system by the toxin, causing fever and lowered blood pressure. Monoclonal antibodies against the toxin have proved to be very effective in treating severe cases of septic shock.

Detection, Isolation, and Purification of Biomolecules

We have already described several examples of the use of antibodies in basic research. The most common laboratory use of antibodies is as a convenient tool for detecting specific compounds. Monoclonal antibodies are particularly useful in determining the presence and quantity of specific proteins. Figure 12.8 illustrates one method, known as Western blotting, used for this type of analysis.

1. Polyacrylamide gel electrophoresis is used to separate proteins. Each band consists of many molecules of a particular protein. The bands are not visible at this point.

Protein samples

B

Proteins from B

Polyacrylamide gel

A

Proteins from A

Larger

Smaller

2. The protein bands are transferred to a nitrocellulose filter by blotting.

Paper towels

Salt solution

Sponge

Gel

Nitrocellulose filter

3. The filter with proteins is positioned exactly as it was on the gel, then rinsed with monoclonal antibodies against a particular antigen. The antibodies are tagged with a dye so that they are visible when they combine with their specific antigen.

B

A

4. The test is read. If the tagged antibodies stick to the filter, evidence of the presence of the protein is positive.

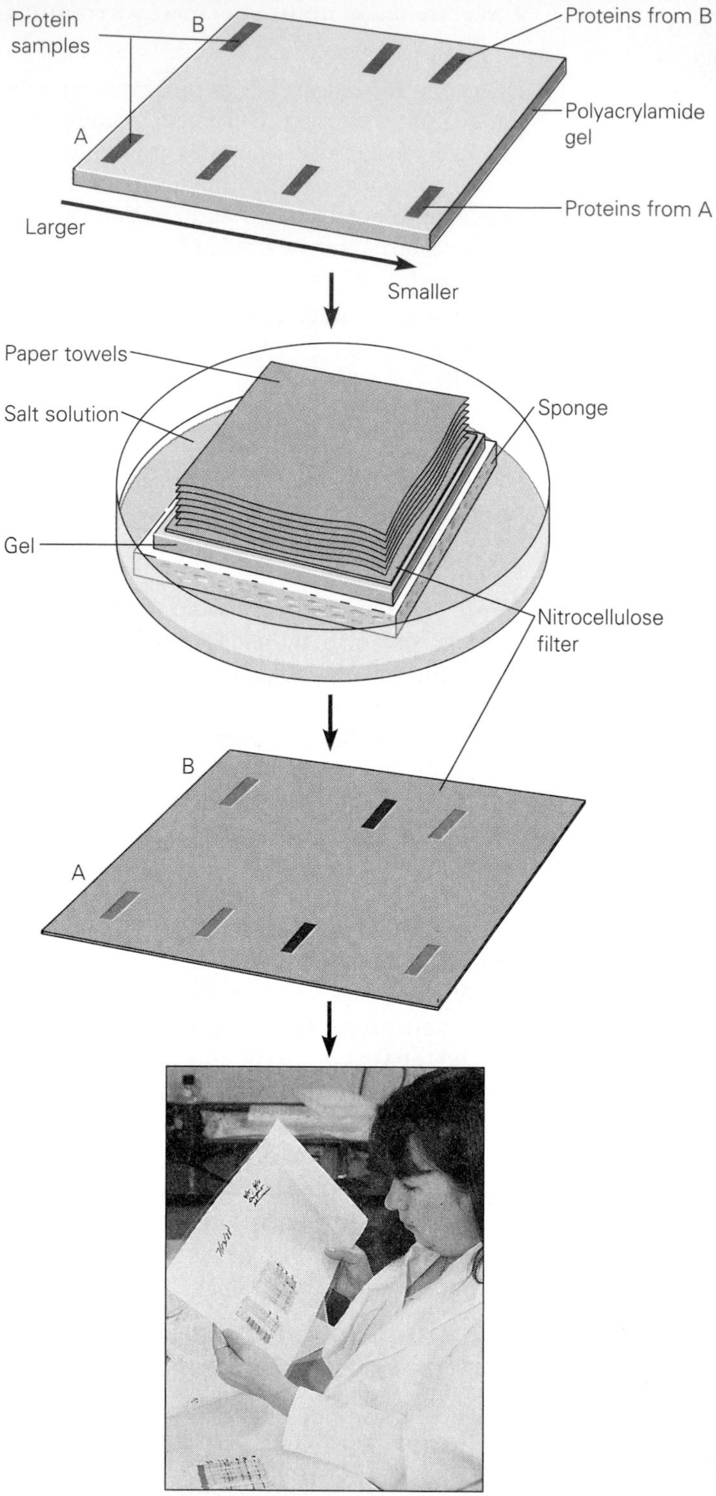

Figure 12.8 The Western blot. Specific molecules from solutions of mixtures of proteins can be separated by electrophoresis and identified by their reaction with tagged monoclonal antibodies.

Proteins from the various samples to be tested are first prepared and separated using polyacrylamide gel electrophoresis. Following this, the protein bands are transferred onto a nitrocellulose filter. The resulting **Western blot** is incubated with the primary antibody. If the specific antigen is present, then the antibody will bind. This binding will allow a secondary antibody to be used to locate the first. Typically, the secondary antibody is a preparation from a different animal that recognizes antibodies from the first animal. For example, if the primary antibody is produced in mice, the secondary antibody might be a goat preparation that recognizes and binds to all mouse antibodies.

The end result of a Western blot is a photograph showing antigen locations as dark bands. The method is quite sensitive and can be used to detect very low levels of proteins. Since samples can be displayed side by side through electrophoresis, it is also easy to perform comparative studies involving different samples.

Another very common use of monoclonal antibodies is in the isolation and purification of chemical compounds. If, for example, a particular compound needs to be separated from many other compounds, antibodies can be used in affinity purification techniques to extract the desired compound from the mixture. Such separations are very rapid and clean, and the final product is usually pure. Since monoclonal antibodies can be made that will bind to virtually any compound, such techniques have a multitude of uses, including purifying antibody preparations.

SUMMARY

This chapter introduced a new technology, monoclonal antibody production. This technology is a direct application of a natural process, the creation of antibodies. As is usual in biotechnology, the natural process can be used outside cells and organisms to accomplish specific goals, in this case, the mass production of antibodies that will recognize and bind to a single molecule. Monoclonal antibody preparations have a great variety of uses, including diagnostic and therapeutic applications. The next two chapters explore in greater detail two diseases in which monoclonal antibodies play a particularly important role.

REVIEW QUESTIONS

1. Explain how centrifugation is initially used to obtain protective antibodies from a human.

2. How is affinity chromatography used in the process of separating antibodies from blood plasma?

3. List the steps necessary to produce a specific monoclonal antibody against the hormone insulin. How is HAT medium involved in that process? What is the role of radioactively labeled or fluorescently labeled insulin?

4. How is human chorionic gonadotropin (HCG) involved in home pregnancy tests?

5. In the ELISA tests, the third solution added

contains a specific enzyme that has been chemically linked to a specific antibody. Reconsider the four steps in ELISA testing. The first solution added to the well was a specific antibody that was to have stuck to the surface of the well. Why, then, wouldn't the antibody added in step three also stick to the well?

6. What major reason is there for using an ELISA test on human blood samples to determine whether they contain HIV?

7. When the coined term "magic (or smart) bullet" is used in conjunction with chemotherapy, the implication is that the "bullet" knows

where to go, that is, it seeks specific targets. How are monoclonal antibodies used medically as "magic bullets" in chemotherapy?

8. Describe the two most significant problems with the more practical use of the magic bullet approach to chemotherapy using monoclonal antibodies.

9. Would you back research and development of a treatment of the cause of atherosclerosis (in-appropriate diet and lack of exercise) or would you be safer risking your financial resources in efforts to research and develop the means to stop clot formation (or dissolve newly formed clots)? Provide convincing arguments for either position.

10. Outline the Western Blot technique and explain why it is used.

CHAPTER 13

AIDS and HIV

Probably very few people in the United States know nothing about **AIDS (acquired immune deficiency syndrome)** and the virus that causes it, **human immunodeficiency virus (HIV)**. It is the most rapidly growing infectious disease in the world: It is estimated that some 30 to 40 million people will be infected by the year 2000. Despite increasing efforts to educate and inform people about AIDS and ways to slow its spread, however, there is still widespread ignorance about AIDS, HIV, and therapeutic agents that might prove effective in treating or preventing infection. The primary purpose of this chapter is to describe AIDS and HIV, concentrating on how we can apply the biological principles we have examined in earlier chapters to understanding this virus and the disease it causes.

You will use your knowledge of cell biology and genetics to examine how HIV infects cells and replicates. You will use your knowledge of molecular biology to discern how the genes of the virus act to allow it either to replicate or to lie dormant inside cells. Genetic engineering has already provided much of the information we have about the virus and its genes; it also provides the primary tools used to develop treatments and therapies against infection. Finally, your knowledge of the immune system will allow you to understand how the virus eventually causes AIDS, how it manages to escape destruction and elimination, and how it might be possible to develop vaccines or treatments to help prevent infection or the spread of the virus.

AIDS is one example of a class of diseases known as immune deficiency diseases. Some of these diseases are genetic, such as adenosine deaminase deficiency (ADA deficiency; Chapter 5). Others are caused by infectious agents. AIDS is the disease that results from a prolonged infection by a particular virus, HIV.

The primary effect of HIV infection is a reduction in the number of helper T cells (T_H cells) in the body. Recall from Chapter 11 that T_H cells are required for proper functioning of the immune system—they are responsible for helping each branch of the immune system become active in the presence of an antigen. With fewer T_H cells present, symptoms characteristic of immune deficiency begin to appear: Infections that the body would normally prevent appear, and tumors that the immune system would ordinarily be able to ward off arise.

HIV is a particularly insidious virus for a number of reasons. As we study the replication cycle of the virus and investigate how it interacts with some of our cells, we should keep in mind that viruses, like cellular organisms, evolve—they change and take advantage of their hosts at the same time that the hosts evolve mechanisms to prevent infection. This coevolutionary process is an interesting biological phenomenon, but unfortunately it sometimes results in a virus with particularly harmful properties.

THE HIV VIRUS

Viruses are not classified as living organisms. They possess some characteristics of living things, but they lack the ability to grow or replicate on their own. Viruses are obligate intracellular parasites, unable to replicate without the aid of a living host cell. By itself, a virus is simply a storage form of genetic material. No metabolic activity takes place within the virus, and nothing even suggests that it is alive. However, when it invades a suitable host cell, it commandeers some of the existing cellular machinery for its own use. For example, all viruses use the protein synthesizing apparatus that exists in cells in order to make additional copies of the proteins for new virus particles. The energy to accomplish these metabolic events comes from the host cell. A virus is a true parasite, taking whatever it needs from the host cell in order to replicate itself.

In general, a virus consists of two components—some sort of genetic information and a coat that encapsulates this genetic material. Some viruses are more complex, having another surrounding layer usually referred to as an envelope. Sometimes the virus carries particular enzymes or other proteins. The genetic material can be either single- or double-stranded DNA or RNA. The form of the genetic material is one way in which viruses are categorized; Table 13.1 lists the primary classes of viruses, with illustrative examples of each type.

Some viruses infect virtually every form of life, from bacteria to humans. Some are deadly, such as the Ebola virus and its relatives, the subject of popular books and movies in the mid-1990s. Others are fairly innocuous, such as the many viruses that cause the common cold. Some can result in persistent infection, such as the virus responsible for chicken pox; after childhood infection, this virus can lie dormant in the body until at some later time it becomes active and causes shingles. Some viruses can be directly associated with the development of certain tumors, such as the papilloma viruses known to cause cervical cancer. Viruses are a large and diverse group, and they present numerous and continuing challenges to our immune systems.

HIV is a member of a large family of viruses known as retroviruses (see Table 13.1). Retroviruses carry genetic material in the form of single-stranded RNA and also contain an enzyme known as reverse transcriptase, which synthesizes DNA using an RNA template (Chapter 8).

Table 13.1 Classification of Human Viruses

	Viral Family	Viral Genus (with representative species) and Unclassified Members*	Dimensions of Virion (diameter in nm)	Clinical or Special Features
Single-stranded DNA, Nonenveloped	Parvoviridae	*Dependovirus*	18–25	Depend on coinfection with adenoviruses; cause fetal death, gastroenteritis.
Double-stranded DNA, Nonenveloped	Adenoviridae	*Mastadenovirus* (adenovirus)	70–90	Medium size viruses that cause various respiratory infections in humans; some cause tumors in animals.
	Papovaviridae	*Papillomavirus* (human wart virus) *Polyomavirus*	40–57	Small viruses that induce tumors; the human wart virus (papilloma) and certain viruses that produce cancer in animals (polyoma and simian) belong to this family.
Double-stranded DNA, Enveloped	Poxviridae	*Orthopoxvirus* (vaccinia and smallpox viruses) *Molluscipoxvirus*	200–350	Very large, complex, brick-shaped viruses that cause diseases such as smallpox (variola), molluscum contagiosum (wartlike skin lesion), cowpox, and vaccinia; vaccinia virus gives immunity to smallpox.
	Herpesviridae	*Simplexvirus* (herpes simplex viruses 1 and 2) *Varicellavirus* (varicella-zoster virus) *Cytomegalovirus* *Lymphocryptovirus* (Epstein-Barr virus) human herpes virus 6	150–200	Medium-size viruses that cause various human diseases, such as fever blisters, chicken pox, shingles, and infectious mono-nucleosis; implicated in a type of human cancer called Burkitt's lymphoma.
	Hepadnaviridae	*Hepadnavirus* (hepatitis B virus)	42	After protein synthesis, hepatitis B virus uses reverse transcriptase to produce its DNA from mRNA; causes hepatitis B and liver tumors.
Single-stranded RNA, Nonenveloped + Strand	Picornaviridae	*Enterovirus* *Rhinovirus* (common cold virus) hepatitis A virus	28–30	At least 70 human enteroviruses are known, including the polio-, coxsackie-, and echoviruses; more than 100 rhinoviruses exist and are the most common cause of colds.

(continued)

Table 13.1 (cont.)

Viral Family	Viral Genus (with representative species) and Unclassified Members*	Dimensions of Virion (diameter in nm)	Clinical or Special Features
Single-stranded RNA, Enveloped + Strand			
Togaviridae	*Alphavirus* *Rubivirus* (rubella virus)	60–70	Included are many viruses transmitted by arthropods (*Alphavirus*); diseases include eastern equine encephalitis (EEE) and western equine encephalitis (WEE). Rubella virus is transmitted by the respiratory route.
Flaviviridae	*Flavivirus* *Pestivirus* hepatitis C virus	40–50	Can replicate in arthropods that transmit them; diseases include yellow fever, dengue, St. Louis encephalitis, and Japanese encephalitis. The unclassified hepatitis C virus is most likely in this family.
Coronaviridae	*Coronavirus*	80–160	Associated with upper respiratory tract infections and the common cold.
− Strand, One Strand of RNA Rhabdoviridae	*Vesiculovirus* (vesicular stomatitis virus) *Lyssavirus* (rabies virus)	70–180	Bullet-shaped viruses with a spiked envelope; cause rabies and numerous animal diseases.
Filoviridae	*Filovirus*	80–14,000	Enveloped, helical viruses; Ebola and Marburg viruses are filoviruses.
Paramyxoviridae	*Paramyxovirus* *Morbillivirus* (measles virus)	150–300	Paramyxoviruses cause parainfluenza, mumps, and Newcastle disease in chickens.

(continued)

Table 13.1 (cont.)

Viral Family	Viral Genus (with representative species) and Unclassified Members*	Dimensions of Virion (diameter in nm)	Clinical or Special Features
– Strand, Multiple Strands of RNA Orthmyxoviridae	*Influenzavirus* (influenza viruses A and B) influenza C virus	80–200	Envelope spikes can agglutinate red blood cells.
Bunyaviridae	*Bunyavirus* (California encephalitis virus) *Hantavirus*	90–120	Hantaviruses cause hemorrhagic fevers such as Korean hemorrhagic fever and *Hantavirus* pulmonary syndrome; associated with rodents.
Arenaviridae	*Arenavirus*	50–300	Helical capsids contain RNA-containing granules; cause lymphocytic choriomeningitis and hemorrhagic fevers.
Produce DNA Retroviridae	oncoviruses *Lentivirus* (HIV)	100–120	Includes all RNA tumor viruses and double-stranded RNA viruses. Oncoviruses cause leukemia and tumors in animals; the *Lentivirus* HIV causes AIDS.
Double-stranded RNA, Nonenveloped Reoviridae	*Reovirus* Colorado tick fever virus	60–80	Involved in mild respiratory infections and infantile gastroenteritis; an unclassified species causes Colorado tick fever.

*Unclassified viruses have not been assigned to genera; therefore, only their common names are listed.

HIV Replication Cycle

Figure 13.1 illustrates the rather complex replication cycle of the HIV virus. As we consider the details of this cycle, we will see why HIV is so difficult for our immune system to defend against and why it is proving difficult to find effective treatments, cures, and vaccines.

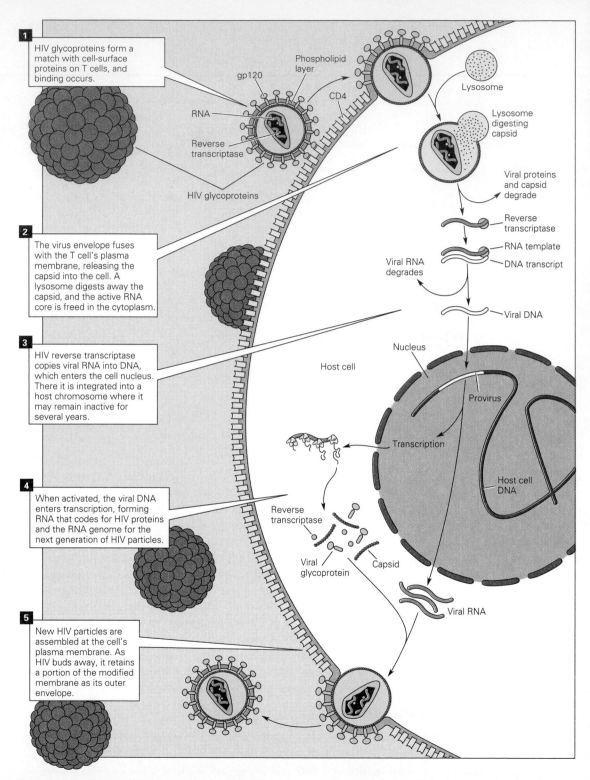

Figure 13.1 Replication cycle of the HIV virus. HIV displays a replication cycle typical of retroviruses. The various stages are described more completely in the text.

Infection of a Host Cell Once an HIV particle enters the body, it circulates through the blood and lymph system like any other foreign body. The virus contains a number of proteins in its envelope, including one called gp120 (a glycoprotein, or gp, of a particular size). The protein gp120 binds to proteins called CD4 receptors (CD stands for cell differentiation antigen) that are expressed on the surface of particular cells. Once this interaction occurs, the virus is tightly bound to the cell. The envelope of the virus is similar in structure to the cell membrane and fuses with the cell membrane, much as two soap bubbles fuse when they come into contact on the surface of water. This fusion allows the virus to enter the cell.

Inside the cell, the viral coat is stripped off by enzymes in the cell (a process often termed uncoating) that normally take part in the degradation of cellular material. In this case, they attack the viral coat, degrading it as if it were cellular material to be recycled. Once the coat is removed, the contents of the viral particle—including the RNA genome and reverse transcriptase—are released.

Although it is possible for HIV to infect cells that do not contain CD4 receptors, these cells are the primary targets of HIV. Cells that express CD4 receptors ($CD4^+$ cells) include T_H cells and macrophages, as well as a few others. Note that the primary cellular targets of HIV are two of the more important cells in the immune system.

Reverse Transcription Recall from Chapter 8 that reverse transcriptase is an enzyme that constructs a strand of DNA using RNA as a template. Once viral reverse transcriptase is released inside an infected cell, it begins synthesizing a double-stranded DNA molecule, called a provirus, that contains the same genetic information as the viral RNA. The provirus is then transported into the nucleus, where it is integrated into a chromosome of the cell.

Production of New Virus Particles The provirus, now located within a host cell's chromosome, can do one of two things. It can lie dormant (a process we will describe shortly), or it can begin to produce viral RNA and new viral proteins. In the latter case, the viral DNA functions no differently than any other set of genes in the DNA: mRNA is produced and then translated into proteins. These proteins make up new virus particles—new coat and envelope proteins, for example. As RNA copies of the viral genes are made, they travel to the cytoplasm of the cell. At the same time, the viral proteins that make up the coat are accumulating in the cytoplasm. When all these components are produced in appropriate amounts, they begin to spontaneously self-assemble into new virus particles, each containing viral RNA, reverse transcriptase, and other necessary proteins within the coat.

Release of New Virus Particles from the Cell The final site of assembly of a new virus is the inner surface of the cell membrane. The assembled coat interacts with the viral proteins on the inner surface of the cell membrane, proteins previously synthesized under the direction of the viral genes. In this way, particular viral membrane proteins are gathered in the membrane in the region of the virus coat, initiating the budding of this region of the membrane and enclosing the virus within a membrane envelope. Although this envelope is derived from the cell, it has the characteristics of the viral envelope, since viral proteins (such as gp120) were collected in the region before the virus assembled.

Latent Form of the Virus If the provirus incorporated into a chromosome does not begin directing the production of new virus particles, it enters a **latent** state, in which it is not active at all but instead is simply carried along with the other DNA present in the chromosomes. Whether a particular viral DNA becomes latent is apparently determined by whether the infected T_H cell is activated. If it is, the viral DNA is active as well and begins directing the production of new viruses. If the cell is not active, the viral genes remain latent.

If an inactive T_H cell harboring a latent infection becomes activated by a totally unrelated antigen, the viral genes become active as well and begin the process of replicating new virus particles. Since T_H cells often exist in an inactive state for years, a latent infection can also last for years, becoming active only when that particular cell becomes active. This may partly explain the long time that often passes before the onset of full-blown AIDS.

Why Does HIV Infection Result in AIDS?

The replication cycle described may appear to do no great harm to the infected cells or to the individual. Unfortunately, this is not the case. Although in its latent form the virus does no immediate harm, during the course of HIV infection large numbers of activated cells are producing new virus particles. Any parasite that is using a cell for its own purposes will eventually create problems within the cell. T_H cells are no different—they eventually succumb to the presence of so many new virus particles. The synthesis of new viruses siphons off so much energy and nutrients that the cell eventually dies. Before it dies, however, the cell produces and releases many new virus particles, which are carried passively through the body and infect other susceptible cells.

Thus, over the course of time, the viral infection kills most or all T_H cells. Recall from Chapter 11 that these cells serve a vital function within the immune system: They are responsible for overall regulation of both the antibody-mediated (humoral) and cell-mediated immune responses. Without T_H cells, the rest of the immune system is unable to react to the presence of foreign invaders. This situation results in the immune system deficiency characteristic of AIDS.

The symptoms of AIDS are not directly caused by the HIV infection. Rather, they arise because the loss of T_H cells inhibits the ability of the immune system to provide a normal defense against microorganisms and cancerous cells. Because the immune system is compromised, common opportunistic viruses, bacteria, fungi, and protozoans that are normally harmless inhabitants of the human body become pathogenic. Initial symptoms often involve persistent yeast (*Candida albicans*) infections. Later, as more T_H cells are lost, other opportunistic infections become common, including a particularly rare form of pneumonia caused by the fungus *Pneumocystis carinii*. Other viruses in the patient's cells may begin to express themselves, often resulting in shingles (caused by herpes zoster virus) and the eye infection retinitis (caused by cytomegalovirus, CMV). Tumor cells that would normally be destroyed by the cell-mediated immune response now escape destruction and give rise to tumors, including the tumor most common in AIDS patients, a skin tumor known as Kaposi's sarcoma.

The types of infections and other diseases that AIDS patients exhibit are related to the degree of HIV infection, which is reflected by the number of T_H cells remaining in the body. As the population of T_H cells decreases, patients become increasingly susceptible to opportunistic infections and tumors. Patients do not generally die as a direct result of HIV infection. Rather, they eventually succumb to a continuous on-slaught of opportunistic infections and other maladies that result from severe immune deficiency.

HIV AND THE IMMUNE SYSTEM

We have seen how HIV results in the decreased effectiveness of a patient's immune system, which eventually allows lethal damage. But why doesn't the immune system of an infected individual destroy the HIV virus? There are a number of reasons, some of which we encountered in our examination of other infectious agents, such as the parasite that causes malaria (Chapter 11). Other reasons, however, are unique to HIV.

Initial Response to HIV Infection

When an individual is infected by HIV, an immune reaction is triggered. Generally, within a few weeks after infection, it is possible to detect HIV-specific antibodies in the blood of an infected individual, indicating that the immune system has reacted against the virus. However, the fact that the virus eventually persists and causes AIDS means that the immune system is not able to completely rid the body of the infection.

Figure 13.2 shows a typical pattern of HIV infection and the subsequent loss of T_H cells. Within the first six months of infection, the number of HIV viruses in the body increases as the virus multiplies. The immune system mounts a response, however, and dramatically reduces the number of infectious HIV particles. At this time, the victim undergoes **seroconversion**, meaning that HIV antibodies are detectable in the blood. This initial infection is sometimes accompanied by flulike symptoms (slight fever, swollen lymph nodes), but it might be symptom-free.

How HIV Evades the Immune System

HIV manages to evade the immune system in a number of ways. As with most viruses, once HIV particles infect cells, they are no longer subject to an antibody-mediated immune response since they are no longer circulating in the bloodstream. The cell-mediated immune response is the only defense system that can destroy the infected cells. In order to accomplish this, cytotoxic T cells (T_C cells) must be able to recognize viral antigens that are processed and displayed on the outer surface of the infected cell.

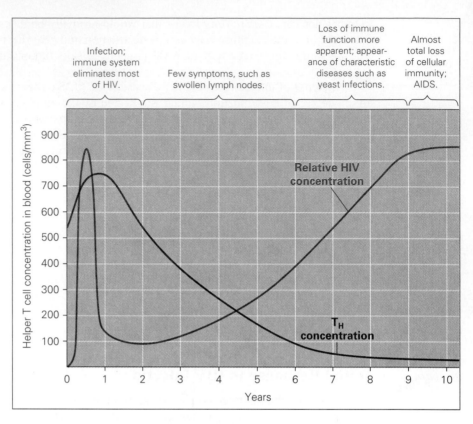

Infection; immune system eliminates most of HIV.

Few symptoms, such as swollen lymph nodes.

Loss of immune function more apparent; appearance of characteristic diseases such as yeast infections.

Almost total loss of cellular immunity; AIDS.

Figure 13.2 The progression of HIV infection. HIV infection can be characterized by four stages. Initially, there is a burst of viral replication, as with most viral infections. During this time, the immune system mounts a defense and eliminates most of the virus particles. During the second stage, the number of T_H cells begins to decrease, allowing the production of more HIV particles; some symptoms may appear at this point. The third stage is characterized by an apparent loss of immune system functions: The T_H cell count is very low, and opportunistic infections begin to appear. By the fourth stage, full-blown AIDS, there is almost a complete loss of immune system activity.

But this does not always happen. Many virus particles that infect cells enter latency, and thus the infected cells do not display viral antigens. Such cells appear normal to the immune system, despite the fact that they contain viral genes. Cells harboring latent viruses or proviruses can allow the virus to become active, resulting in small bursts of virus production throughout the ensuing years.

HIV poses another problem, that of antigenic variation. We encountered this with malarial parasites (Chapter 11). The viral genes responsible for the production of viral proteins are subject to very high mutation rates. In other words, the viral genes that direct the production of the components of new virus particles mutate frequently, so the viral antigens show great variation. This makes it particularly difficult for the immune system to eliminate all virus particles and infected cells—some of them will require different antibodies and T_C cells in order to be eliminated. In this way, at least some virus particles manage to escape the immune system. It is also possible for HIV to pass directly from one cell to another without first being released from the infected cell, thus avoiding exposure to the circulating antibodies that normally bind to and destroy virus particles.

Thus, HIV manages to evade complete eradication from the body. Some viruses remain latent and become active only years later. Others manage to mutate in such a way that they escape immediate destruction. During this cat-and-mouse interaction, the surviving viruses continue to replicate and destroy the T_H-cell population of the infected individual. As these cells decrease in number, the immune system becomes less effective. More and more viruses escape destruction, eventually destroying more and more T_H cells. Eventually, the immune system is so compromised that other microorganisms are able to establish infections. The onset of full-blown AIDS results from a very long, drawn-out battle between HIV, which is able to avoid destruction and therefore maximize its chances of replicating, and the immune defenses, which ordinarily work so well that we are unaware of the constant bombardment by infectious microorganisms.

DIAGNOSING HIV INFECTION

The first step in treating any disease is proper diagnosis, which is often based on the appearance of particular symptoms. But in the case of HIV infection, symptoms are at first so mild, or even nonexistent, that it is impossible to diagnose based on symptoms. Thus, early diagnosis of HIV infection must rely on other methods.

Biotechnology, through the development of monoclonal antibodies and the polymerase chain reaction, has made available several different ways of testing for HIV infection. These tests are very accurate and, at least in the case of the ELISA test, are rapid and inexpensive. Note that the principles behind these tests are equally applicable to the diagnosis of many other types of infectious disease.

Testing for the Presence of Antibodies Against HIV

Most individuals who are HIV positive develop antibodies against the virus. The two most common diagnostic tools presently used both rely on the detection of antibodies as evidence of active HIV infection.

ELISA Test In the last chapter, we discussed the development and use of ELISA tests as rapid diagnostic tools. They are particularly useful in determining whether a specific molecule is present, and this is exactly how they are applied in tests for HIV antibodies.

Figure 13.3 illustrates how an ELISA test is organized to detect the presence of antibodies against HIV. Like most ELISA tests, it is rapid, accurate, and sensitive. However, also like other ELISA tests, it is only as good as the monoclonal antibody that goes into making it. The key to the development of a test lies in the selection of

Figure 13.3 An ELISA test for detecting HIV antibodies in blood. In order to detect HIV antibodies in blood samples, an ELISA test must be designed using purified HIV protein attached to test dishes. Into these dishes are placed samples of blood. If antibodies to HIV are present, they will bind to the HIV protein attached to the dish. In order to detect the bound antibodies, a second antibody is added that recognizes human antibodies and carries an enzyme. Subsequent addition of the enzyme substrate will result in either the production of color (indicating the presence of HIV antibodies) or the absence of color (indicating the absence of HIV antibodies).

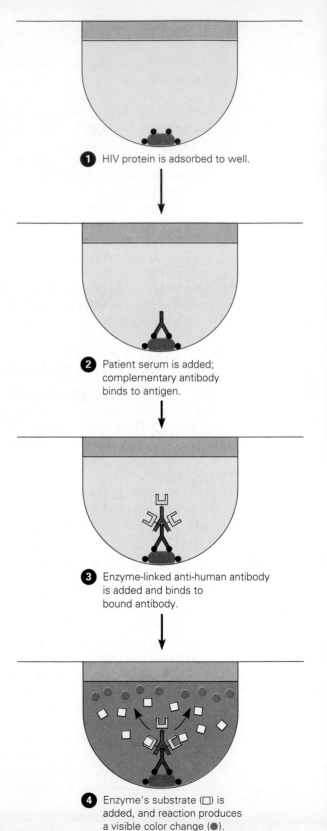

1 HIV protein is adsorbed to well.

2 Patient serum is added; complementary antibody binds to antigen.

3 Enzyme-linked anti-human antibody is added and binds to bound antibody.

4 Enzyme's substrate (□) is added, and reaction produces a visible color change (●).

the correct monoclonal antibody. The population of hybridoma cells from which the monoclonal antibodies will be produced will undoubtedly contain cells that produce antibodies of differing effectiveness. Some antibodies will bind specifically to a particular antigen, and others might bind less well to that antigen while perhaps also binding to variants of it.

While the test may be very effective, there is room for error. A false positive results from a test that gives a positive result, indicating the presence of HIV antibodies, when there really are none. False positives are often identified by repetition of the test using the same blood sample. If the results are still positive, a different type of test, which we will describe shortly, is used in order to gather independent evidence of the presence of HIV antibodies.

Although false positive tests do occur, they are rare and usually fairly easy to identify. More problematic are results indicating no HIV antibodies when, in reality, they are present. The aim of diagnostic testing is to completely avoid these false negatives, even at the expense of incurring more false positives. Usually, during the design and testing of a diagnostic procedure, it is possible to change the test and alter its specificity, either through the use of a different monoclonal antibody or through a change in conditions that allows the test to detect samples that might otherwise have gone undetected. Since accurate diagnosis of HIV infection is so important to the individual (and others), tests have been purposefully designed to err on the safe side, in other words, to give false positives rather than risk false negative results. In this way, it is hoped that the diagnostic process will avoid missing anyone who is actually infected. False positives, while perhaps greatly increasing anxiety, can almost always be identified through further testing.

Western Blot Test A somewhat different diagnostic test, based on the same principle as the ELISA test, is used as a second way of screening for HIV antibodies—the Western blot, described in Chapter 13. These tests are slightly better than ELISA tests, but because they are much more expensive, time-consuming, and labor-intensive they are not used for routine screening. Instead, they are often employed as a second, independent way to assess the presence of HIV antibodies in a particular blood sample. Any sample that consistently tests positive for HIV by ELISA tests is usually subjected to a Western blot for confirmation.

Testing for the Presence of HIV Genes: PCR Test

A third HIV test, based on a different technique, is now available. It is much more expensive than either ELISA tests or Western blots; however, it provides another independent source of information and can be used to help determine whether infection has occurred in individuals who do not develop antibodies against HIV. This test is based on PCR (polymerase chain reaction; Chapter 9), the technique used to amplify DNA.

Since the sequence of the HIV genome is known, it is possible to design PCR primers that will result in the specific amplification of HIV genes should they be present. The test consists of drawing a sample of blood, extracting DNA from the cells in

Figure 13.4 PCR test to detect HIV DNA. This test is a straightforward application of PCR technology. DNA is isolated from cells derived from blood. PCR is performed using primers specific for HIV DNA. If HIV DNA is present, it will be amplified. Gel electrophoresis of the resulting DNA molecules allows visual confirmation of the presence or absence of HIV DNA.

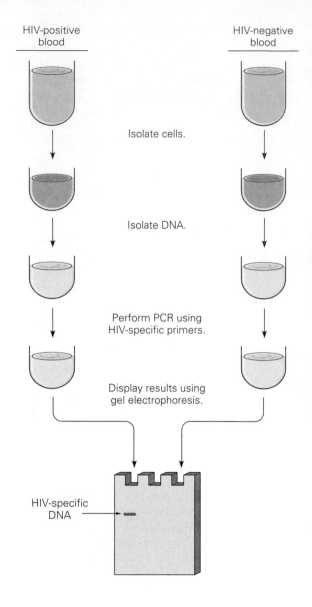

the sample, and then using PCR to try to amplify HIV-specific DNA. If amplified DNA results, then HIV DNA must have been present in the sample. If there is no amplified DNA, then either there was a technical problem with the PCR reaction or no HIV DNA was present in the sample. Technical difficulties can almost always be tracked down by performing appropriate control experiments. Figure 13.4 illustrates the principle of PCR diagnosis, as well as what the results of such tests look like.

TREATING HIV INFECTION

As of this writing, there are no effective treatments for HIV infection and only a handful of treatments that provide marginal benefit. Perhaps, as we consider the available treatments, it will become clear why HIV infection is so difficult to treat.

What general approaches might prove useful in treating HIV infection? There are at least two: Alleviate the symptoms of infection, and try to slow or prevent the spread of the virus within the patient's cells. Unfortunately, the symptoms of HIV infection often don't appear until long after the infection is firmly established. Treatment of AIDS consists primarily of trying to prevent or control the multitude of infections and complications that result from the loss of immune system activity. Little can be done for a patient suffering from AIDS that in any way affects the HIV infection. Rather, treatment consists of controlling the secondary infections in any way possible.

Efforts to Boost the Immune System

As we mentioned, almost all HIV-infected individuals develop antibodies against HIV. Even if HIV cannot be eliminated through immune system attack, it is possible that, by somehow boosting the immune system or providing a stronger response, the spread of HIV could be held in check or at least slowed considerably. If such a boost could be attained shortly after infection, when few viruses are present in the body, then such a treatment might prevent the onset of AIDS.

A number of ways to try to fight HIV through an enhanced immune reaction are possible. One of the simplest is to provide monoclonal antibodies against HIV to an infected person. Another approach is to stimulate the body to produce T_H cells faster than they are depleted. If the production of T_H cells could be maintained at a high enough level, it might forestall the onset of AIDS. A number of treatments involving growth factors and interleukins are also being pursued.

However, it is not likely that HIV infection can be held in check simply by increasing the immune response. As we have seen, HIV has many strategies for avoiding the immune system. Most likely, HIV will ultimately overcome any efforts to fight it by boosting the immune response. We must seek other ways of preventing the spread of the virus.

Preventing the Spread of HIV Throughout the Body

Many strategies can be used to prevent the spread of HIV in a patient. Figure 13.5 illustrates approaches that may prove to be effective. Success with any of these approaches would represent a significant breakthrough in the treatment of HIV infection. In theory, any treatment or action that prevents (1) attachment of HIV to uninfected cells, (2) integration of the HIV genome into that of the host cell,

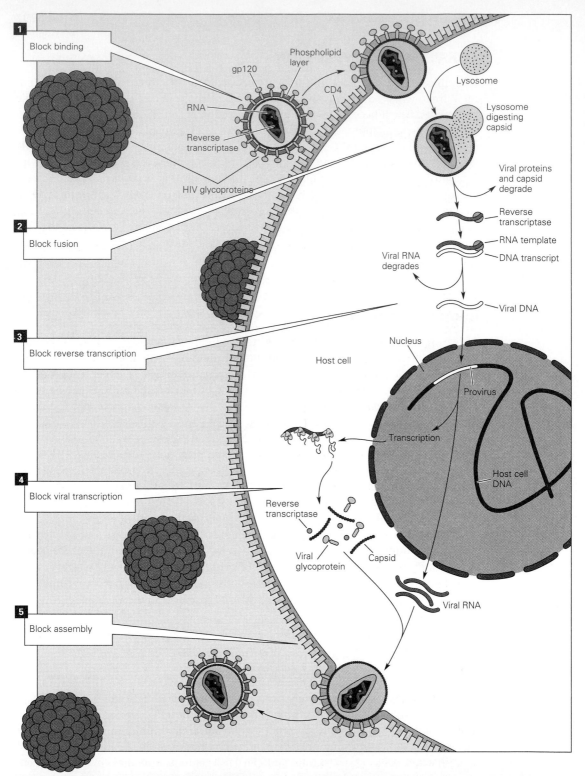

Figure 13.5 Potential steps at which HIV infection or replication might be blocked. Superimposed on the HIV replication cycle are some of the steps being targeted for drug treatment. If any step can be slowed or prevented, then the spread of HIV in the body will be slowed as well.

298

(3) replication of the virus, or (4) assembly of new virus particles would be an effective treatment. The pace of research into such treatments is exceedingly rapid. Therefore, we can make only general comments about each approach.

Preventing the Entry of HIV into Uninfected Cells Recall that an association between gp120 on the viral envelope and CD4 on the cell surface is required for HIV to enter a cell. Anything that prevents this interaction would be an effective treatment. A number of approaches are being tried, most of which involve trying to "overload" the system, in essence trying to block access to CD4.

One way to accomplish this is to add antibodies that recognize and bind to CD4 molecules on the surfaces of cells. If the antibodies bind to CD4, an HIV particle will not be able to bind, since two entities cannot occupy the same physical space, and the entry of the virus will be thwarted. Of course, to be effective as a treatment, such antibodies would need to be present at all times—or at least throughout the time that HIV particles are moving between cells. It is difficult to imagine being able to create such a situation long-term, for several reasons. First, covering all CD4 cells with antibodies, even artificial ones, might trigger an immune response—perhaps even resulting in the destruction of the CD4 cells, the very cells that are the object of the treatment. Further, generating an immune response might result in severe shock, a potentially lethal response.

Second, binding of antibodies to CD4 molecules would very likely interfere with the normal function of these molecules. Although we don't know exactly what CD4 does in normal cells, it may serve a function that should not be compromised by such a brute-force approach. Third, there is a continual turnover of molecules on cells in the body, and CD4 is no exception. Newly made CD4 receptor molecules would have to be coated with antibodies, necessitating a constant supply of new antibodies entering the bloodstream. Despite such drawbacks, a number of CD4-related approaches are undergoing experimentation. Even if they are not perfect, they may prove to be valuable in slowing the spread of HIV.

Another approach to preventing HIV entry into cells is to try to coat gp120 molecules in the viral envelope to prevent them from binding to CD4 molecules. One way to accomplish this is to prepare a portion of the CD4 molecule that binds to gp120 and inject it into the bloodstream (Figure 13.6). If sufficient numbers of CD4 fragments are present, then HIV particles in the bloodstream will bind to them instead of to the CD4 receptors on host cells. Therefore, the virus would be blocked from entering potential host cells.

This approach has been shown to provide protection to cells that are artificially exposed to HIV in laboratory cultures. However, it requires large quantities of CD4 fragments in the bloodstream, and such large quantities might have adverse side effects, perhaps interfering with the normal function of CD4 receptors. Whether such a system can be developed and applied as a therapy remains to be seen.

Preventing Reverse Transcription One of the most effective approaches thus far involves ways of slowing or preventing the reverse transcription of HIV RNA. The logic behind this approach is straightforward. HIV is a retrovirus and contains reverse transcriptase, which is required for the replication of HIV genes. If reverse transcription

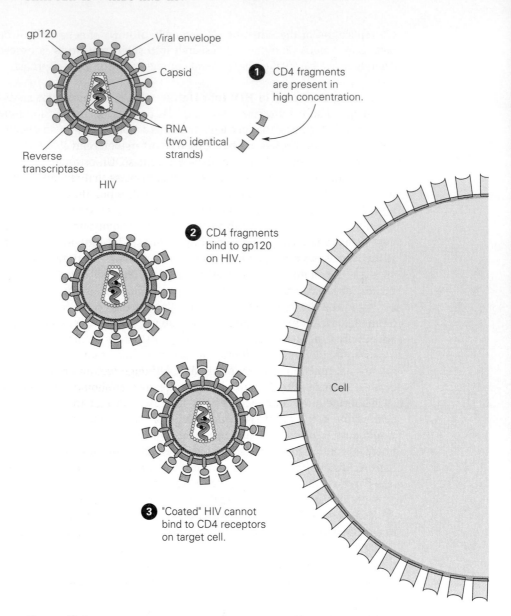

Figure 13.6 Blocking infection by preventing binding of HIV to target cells. If CD4 fragments are present in high enough concentrations, they can bind to the gp120 molecules on the surface of HIV particles. This binding will prevent the gp120 molecules, and hence the HIV particles, from interacting with the CD4 receptors on the surface of the target cell, effectively blocking infection.

were blocked, there would be no productive infection of cells by HIV. Since normal cells do not have reverse transcriptase, it might be possible to find chemicals or treatments that would preferentially block it and have little or no effect on other cellular processes.

Fortunately, we know something about reverse transcriptase. Recall its function—to synthesize a copy of DNA using RNA as a template. In this regard, it is similar to other enzymes that synthesize DNA and RNA, such as DNA polymerase and RNA polymerase. These enzymes are well understood, and knowledge about them has proved useful in studies of reverse transcriptase. Unfortunately, it is also true that at least some attempts to block reverse transcriptase activity would most likely interfere with DNA and RNA polymerases as well. Since these enzymes are critical, such interference would likely be harmful to reproducing cells.

Several drugs inhibit reverse transcriptase activity, including azidothymidine (AZT, also known as zidovudine or Retrovir) and dideoxyinosine (ddI). Figure 13.7 shows how these **nucleotide analogs** work. In essence, they provide a substitute DNA base that, once incorporated into a growing chain, prevents further synthesis of DNA. AZT and ddI are fairly well known compounds, although other similar compounds, such as dideoxycytidine (ddC), have also been tested. While many studies indicate some benefits from use of these and similar drugs, much controversy surrounds their use. They exhibit adverse effects and may do more harm than good during early stages of HIV infection.

Another method of preventing HIV from integrating and replicating that is being investigated takes advantage of the ability of complementary single-stranded nucleic acid molecules to hybridize to each other. The fact that the HIV genome exists as single-stranded RNA allows the application of hybridization technology to block reverse transcription.

The principle involves putting a specific RNA molecule that is complementary to the HIV genome into cells. Since this complementary, or *antisense*, strand of RNA does not code for any proteins (it is the complement of the coding strand), it would seem to serve no function within the cell—that is, until the cell becomes infected with HIV. When HIV RNA is released into the cell, the antisense RNA should bind to it and form a stable double-stranded RNA molecule. This should effectively block the reverse transcription step, preventing the HIV genome from being turned into a provirus that can integrate into the chromosome.

Gene therapy (Chapter 10) can be used to provide the antisense RNA molecules. An artificial gene encoding the antisense RNA is put into the target cells (Figure 13.8). When this gene is expressed, an antisense RNA molecule is produced. Using gene therapy to provide the antisense RNA has merit. The most important cells to protect are T_H cells. Recall that these cells are produced in bone marrow. One of the most successful ways of introducing foreign genes into human cells is by inserting them into **stem cells** (cells that give rise to nearly all blood cells) obtained from bone marrow. The stem cells are removed from the patient, modified in the laboratory to contain the antisense gene, and then returned to the patient. Clinical trials have shown that this approach works; research is proceeding to determine whether or not gene therapy and antisense technology can slow the spread of HIV. One question still to be asked is whether the presence of antisense RNA in normal cells will alter their function. Such a question may prove very difficult to answer.

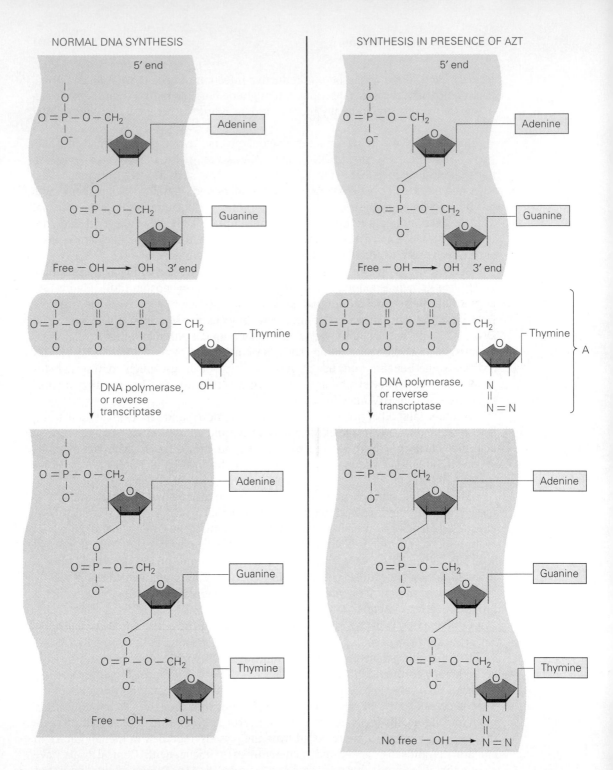

Figure 13.7 How AZT inhibits reverse transcription. AZT is chemically similar to thymidine, except that it lacks a hydroxyl (OH) group on the 3' carbon of the ribose. When thymidine is added to a growing chain of DNA (left), it results in a molecule that still has a 3'OH group, allowing subsequent nucleotides to be added. In contrast, when AZT is incorporated into a growing chain (right), there is no 3'OH on the chain terminus. Since 3'OH is required for the addition of subsequent nucleotides, the incorporation of AZT blocks further DNA synthesis.

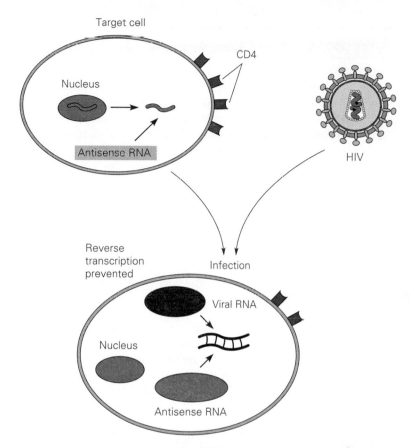

Figure 13.8 Preventing HIV infection by using antisense RNA. Cells that are targets for HIV infection could be genetically engineered to contain a gene that expresses a complementary copy of the HIV genome. If this antisense gene is expressed, then the antisense RNA will be in the cytoplasm of the target cell. If the cell is infected by HIV, the antisense RNA will bind to the viral RNA when it is released into the cytoplasm, forming a double-stranded RNA-DNA hybrid molecule. Such double-stranded molecules cannot be used by reverse transcriptase. Thus, a DNA copy of the HIV genome is never made and cannot be incorporated into the chromosome of the cell.

ISSUE: VACCINES AGAINST HIV

Since we have already examined the development of vaccines against a number of different diseases (Chapter 11), we can appreciate why developing a vaccine against HIV is a daunting task. All the methods HIV uses to effectively evade the immune system make development of a vaccine difficult. Nevertheless, tremendous research efforts are under way to find a vaccine against HIV.

The development of an HIV vaccine must take into account all the challenges that HIV presents as an infectious agent, including its ability to hide from the immune system in a latent state, pass directly from cell to cell without passing through the bloodstream, and kill those cells most central to the immune system. While it is unlikely that any potent vaccine will be developed in the very near future that will prevent HIV infection, it is possible that one will be introduced that substantially reduces the rate of HIV replication and greatly reduces the risk of transfer from one individual to another. Even a partially effective vaccine would prove very useful in reducing the incidence of AIDS.

All the usual approaches to developing vaccines are under investigation. Killed whole-virus vaccines, subunit vaccines, and even vaccination through DNA injection are all being tested clinically. So far, no vaccines have been even marginally effective in clinical trials. This result is not surprising; we have known of HIV for less than two decades, and we have learned that developing a vaccine is extremely difficult. Perhaps we should be amazed at how far we've come.

ESSAY: THE ECONOMICS OF ILLNESS

Where does the money to develop and test potential AIDS treatments and vaccines come from? Certainly, the federal government, through the usual means of funding scientific research (National Science Foundation, National Institutes of Health, National Cancer Institute, and so on), provides substantial funding for AIDS and HIV research. Nonprofit foundations also provide research funds. We should note that a tremendous amount of money is spent by pharmaceutical and biotechnology companies in efforts to develop effective vaccines and treatments. The impetus for such corporate spending is that a tremendous amount of money will be made by those who develop a vaccine or treatment, even a partially effective one. Probably no other new vaccine or drug would hit the market with such fury.

This is almost always the case with medical research—huge profits are made because of people's illnesses. The cost of developing and testing new drugs is immense, especially considering that very few tested drugs turn out to be marketable. (Probably fewer than one in 10,000 potential drugs ever make it to market.) Obviously, if the financial rewards were not tremendous, the incentive to try to develop such drugs would not exist. Our patent system helps in recouping research and development costs by protecting new developments for a period of time before other companies can reap the financial rewards.

There is no doubt that being ill is costly. Our medical care system allows sky-rocketing costs and treats them as normal and expected. Ultimately, research and development costs are passed on to consumers, mainly through increased insurance premiums. The United States is struggling with our existing system of medical care. The struggle should not be unexpected given that a significant number of wealthy and powerful individuals believe their wealth and power will be threatened by changes.

The current system of financing medical research may work as well as any other. If there is a demand for a product, increased efforts will usually bring it to market more rapidly—a desirable outcome. We all share the costs of developing new drugs by paying higher prices for all drugs. Should the cost of medical care be more tightly controlled? The socialist goal of spreading the costs of medical care throughout the population are in large part met by the capitalistic nature of medical research and development. Should this system be changed? If it is, how can the incentive of healthy competition necessary to develop new drugs be maintained? On the other hand, many of us cannot afford to continue to pay exorbitant amounts for health care. Unfortunately, all the high technology now being applied to the fight against HIV, as well as cancer and many other diseases, only serves to further increase the cost of being sick—or rather the cost of remaining healthy.

Preventing Assembly of New Virus Particles Another promising approach to drug development has been to block one of the steps necessary for virus assembly. One viral enzyme in particular, HIV protease, has been the target of a number of new drugs. HIV protease processes specific viral proteins before they can properly assemble into finished particles. Blocking this protease, therefore, prevents the proper formation of new virus particles. Some of these drugs are currently being used in treatment efforts. Often, they are more effective when they are combined with other drugs, for example, the nucleotide analogs described earlier.

Animal Models of HIV Infection

One of the primary challenges in performing research on HIV and AIDS is that AIDS is strictly a human disease. It has counterparts in other animals, but these diseases are not exactly the same as AIDS in humans. HIV has a very narrow host range—it infects only humans, at least with the devastating results it exhibits. Consequently, HIV infection must be studied primarily in human cells that are maintained in the laboratory, not actually within individuals.

Viruses very similar to HIV infect other animals, most notably simian immunodeficiency virus (SIV), which infects certain primates. Many properties of viral infection can be studied in such animals, but always with the understanding that what we learn might not be directly applicable or transferable to HIV infection in humans.

Using animals for research purposes poses a number of problems. First, many people oppose the use of animals for any type of research even though most medical advances have relied on research involving laboratory animals. Since AIDS is a very human-specific disease, those animals most closely related to humans (other primates) are used in AIDS studies.

Second, the use of large animals in research is expensive. The primates used in AIDS research come from Africa or other tropical locations and must be fed and maintained under highly controlled conditions. Significant resources are involved in simply maintaining these animals.

Third, some of the most suitable primates for AIDS research are now endangered species, or at least very difficult to obtain. This situation heightens the dilemma of whether to use them in great numbers.

Fourth, importing suitable primates into the United States (or other countries where they are not normally found) poses certain health hazards. Occasionally, animals have carried with them viruses or other infectious agents that can infect humans, some with devastating results. The movie *Outbreak* and the nonfiction book *The Hot Zone*, by Richard Preston, focus on such events. One appropriate caution we should consider is that, if we must use lab animals, we must take great care in obtaining, maintaining, and using them for research.

Testing New Therapeutic Agents

We now pause to consider one of the basic processes involved in developing any new therapeutic agent for commercial use. Superficially, the necessary steps are to make sure the new drug does what it is supposed to do and to make sure it has no serious side effects. Depending on the nature of the disease and treatment, certain undesirable side effects might be tolerable—for example, drowsiness in the case of cold medications or, in a more extreme case, hair loss during some anticancer chemotherapies. Unfortunately, the testing of a new therapeutic agent is seldom simple and straightforward.

The Food and Drug Administration (FDA), the regulatory body that governs the introduction of any new therapeutic agent to the market, has established guidelines that specify how new agents should be tested. The first step is to test the agent as thoroughly as possible in any animals or other nonhuman organisms that are appropriate. Once such tests are performed and the agent's mode of action and immediate side effects in animals are known, approval is sought to proceed with *clinical trials*, or tests involving humans. There are three phases to clinical testing. Although the details of testing might vary with each particular disease and agent, the ideas behind them are the same.

Phase I Clinical Trials The purpose of phase I trials is to determine whether any immediate, serious side effects appear in humans that did not appear during animal testing. As we have noted, animals respond similarly, but not identically, to humans, and thus new and different side effects may show up in humans. Phase I trials involve a small number of volunteers who do not suffer from or are not at great risk of contracting the disease in question. The test is designed simply to determine whether there are adverse side effects in normal, healthy individuals. Phase I trials may be completed quickly (perhaps in six months).

Phase II Clinical Trials The goal of phase II trials is twofold. Primarily, phase II is designed to determine the dosage or length of treatment necessary in order to elicit the desired response. While some information about dosage is developed during animal testing, more precise tests are necessary to determine the appropriate dosage in humans. Again, a small number of individuals are involved, and they are normal, healthy, low-risk volunteers. Phase II trials typically last longer than phase I trials in order to allow the course of treatment to be followed more completely. The second purpose of phase II trials is to continue to check for side effects. Since the tests run longer and involve more individuals, it is possible to more completely describe any undesirable side effects. Assuming that no serious side effects appear and that a proper dosage or course of treatment can be developed, testing will move into the third phase.

Phase III Clinical Trials This phase is the major test of a new therapeutic agent. The primary goal of phase III trials is to determine the efficacy of the treatment, or how well it works. How can we develop a suitable test that measures this?

The ideal process seems to be straightforward. Two groups of individuals, all of whom are healthy, are assembled. The first group is given the therapeutic treatment and then exposed to the disease. In order to determine how well the treatment works, the results are compared to those obtained from the second group, which is exposed to the disease and not given the treatment. The relative number of individuals who contract the disease in group I compared to the number in group II will give an indication of the effectiveness of the treatment.

Two additional groups are also used in the test. Individuals in group III are given no treatment and are not exposed to the disease. This allows a determination of the random incidence of the disease. Group IV is given the treatment but is not exposed to the disease. This helps determine whether the treatment itself causes any incidence of the disease, a possibility if, for example, the new treatment is a vaccine developed through inactivation of whole organisms.

To make matters even more complicated, each treated group is usually split into two smaller subgroups, one of which is given the real treatment and the other of which is given a **placebo**. The tests are conducted completely at random so that they will not be influenced either by those performing the test or by those receiving treatment.

Phase III trials necessitate the testing of a large number of individuals from different social groups and geographic regions. The trials may take a long period of time, perhaps a number of years, depending on the nature of the disease and the course of treatment. During this time, additional data concerning the side effects of the treatment can also be gathered.

How Must These Tests Be Modified for HIV Infection? You may be wondering about some potential difficulties associated with testing anti-HIV drugs and AIDS therapies. The most serious difficulty is that the ideal test would involve exposing uninfected people to HIV. In other words, it would involve purposely infecting individuals (groups I and II, as described earlier). Even though group I is given the treatment prior to exposure, there is no concrete evidence that the treatment actually prevents the disease—this is why the test is being done in the first place. Isn't it unethical or immoral to conduct a clinical test involving such a deadly infectious agent?

Since testing cannot involve the deliberate exposure of individuals to HIV, the tests must be modified. The test groups can include large numbers of high-risk individuals. Because these individuals are aready at high risk of contracting the disease, the odds are that a significant number of them will become infected without being deliberately exposed. If enough individuals are involved, then it is possible to determine statistically whether the treatment has any effect by noting the difference in the number of infected individuals from the group of high-risk people who receive no treatment and from the group of high-risk people who do receive treatment. The necessity of conducting clinical tests in this manner complicates analysis of the test results and often extends the time of testing required to obtain statistically significant results.

A further complication in testing HIV-related therapies is that a particular treatment might provide protection in several different ways. For example, a potential treatment might prevent initial infection by HIV, or it might slow the spread of an already existing infection. Both are desirable effects, but they must be tested and studied differently.

Another major difficulty in anti-AIDS testing is that the normal course of the disease is very slow. Remember that, on average, ten or more years may pass before an infected individual develops clinical AIDS. Does this mean that tests should run at least ten years in order to determine the effectiveness of a new drug? The spread of HIV can be measured in other ways, but it is clear that determining the efficacy of drugs meant to prevent or slow HIV infection requires long-term tests.

At least as important as determining the parameters of clinical tests is establishing who is responsible for controlling and approving them. Who has the authority to decide whether to approve a particular test? What if a company or corporation presents falsified results or doesn't report all test results in order to make a stronger case for approval? There have been instances of such practices—imagine the economic rewards of being the first to develop an effective vaccine against HIV. Would the economic rewards be worth the potential losses in lawsuits that might follow? Should all tests be repeated by independent investigators in order to establish their validity? Consider the extra time and money this would involve.

The FDA is under tremendous pressure to relax its rules regarding drug approval for agents that might prevent or slow HIV infection. How much should such rules be relaxed—and who should decide?

ESSAY: SHOULD PROMISING DRUGS BE APPROVED FOR USE BEFORE TESTING IS COMPLETE?

What if, during the early part of a phase III trial, a drug seems to show great promise in slowing the spread of HIV? Should the test be stopped and the drug approved for immediate use? Ethically, should an experimental drug be withheld if it appears to help individuals who have AIDS?

Aside from asking these questions in relation to all HIV-infected individuals, we also need to ask them regarding the members of the phase III experimental groups. If a treatment is approved for immediate use, then presumably the high-risk individuals not receiving treatment (those in group II) would want the treatment, and completing the test as originally conceived would be difficult if not impossible. Side effects or complications might appear later in many more individuals as a result of shortening the testing procedure. On the other hand, many more individuals might be helped if the drug is indeed effective.

Should individuals have the right to request and receive experimental treatments if they want them? The FDA establishes which drugs will be available under what circumstances. Sometimes, drugs become available in countries where guidelines for their use might be quite different. In this case, an economic component is clearly involved, since only those who can afford the cost of the drug and its procurement can obtain treatment.

The more serious the disease in question, the more difficult it is to withhold drugs from general use. Yet clearly some method must be established for determining drug efficacy and safety. This is another example in which the best decision to make for society as a whole (complete the tests and establish the safety of the drug) is in direct conflict with the best decision that might be made for the benefit of individuals.

....

ESSAY: HOW SAFE SHOULD THERAPEUTIC AGENTS BE?

How safe does a drug need to be in order to garner approval from the FDA? What is an acceptable risk for a medicine? These questions are terribly difficult to answer, yet they are precisely the questions we must address as new drugs are introduced.

Entire fields of study are devoted to risk assessment, and we cannot hope to cover all there is to know about the subject in a few paragraphs. But we can identify several important ideas.

There are at least two different ways of questioning the safety of medical products. First, what are the known immediate risks to the particular individual about to receive treatment? Usually, such risks are well documented by the treatment provider. For example, a particular drug might cause 16 percent of users to experience drowsiness, 13 percent to experience headaches, and so forth. When children are immunized, parents receive information about the possible risks of such treatment. There may be a one in 10,000 chance of developing symptoms of the disease, if not the disease itself, as a result of the vaccination. Such risks are usually well known and statistically determined. Whether an individual receiving treatment has full knowledge of the risks is another question. However, it seems clear that, in matters of personal safety, informed consent requires as complete a knowledge as possible of potential side effects. Individuals should and can make their own decisions.

A second way to question safety is to ask what unknown risks might exist for the individual receiving treatment or for others. For example, what if a particular drug increases the incidence of colon cancer, but only after prolonged use? Long-term effects such as these are usually not—and cannot be—studied as well as short-term effects, at least when a drug is first developed. No one would argue that drugs should undergo forty years of testing before becoming available to the public in order to assess the long-term effects. It seems reasonable to allow use of a drug after much shorter testing, but a long-term study should be conducted in order to more accurately determine the long-term effects. Numerous drugs have been withdrawn from the market after long-term studies have revealed previously unknown side effects.

Can drugs affect individuals besides those taking them? Two ways come to mind. First, fetuses may be exposed to drugs taken by pregnant mothers. Testing should have been done to predict such potential effects, but often there are complications. The effect of thalidomide on unborn children is one of the most dramatic examples of this danger. Second, at least some drugs might cause behavioral changes or modifications that affect others. Even cold medicines, some of which induce drowsiness, can affect individuals other than those taking the medicine if the patient happens to fall asleep while driving. Warnings are provided to try to limit such occurrences, but the fact is that individuals taking the medicine must show some responsibility in order to avoid potential hazards to others.

Clearly, we should all be concerned about the safety of drugs and should choose wisely what medications we take. At some point, however, a decision must be made about whether a particular drug is safe enough to allow on the market. Arriving at such a decision is not an easy task. Minor side effects such as drowsiness or headaches might be quite acceptable compared to hair loss or increased risk of heart disease. The key is to establish a reasonable balance between the effectiveness of a particular treatment and the risks it entails. The FDA, in conjunction with the maker and tester of the drug, must set these guidelines. They will of necessity differ for different drugs, and it is impossible to make many generalizations. In essence, the ultimate safety of any drug is subject to consumer approval—if the inherent risks are too high, then the drug will not be used. However, this market control is almost always in the hands of physicians, not the final consumers.

It will be interesting to follow the development of new drugs and therapies for HIV treatment in order to see how risk assessment applies to a disease that is nearly 100 percent fatal. HIV-positive individuals are quite worried about their futures, and they should have available to them an entire arsenal of experimental drugs. But it is also important to remember that virtually all drugs and therapies have some side effects and that these often are not well known until long-term studies have been completed.

THE ROLE OF BIOTECHNOLOGY IN HIV AND AIDS RESEARCH

The number of individuals infected by HIV continues to grow worldwide. Clearly, its spread demands that a tremendous effort be made to develop treatments, cures, and vaccines against HIV. But before such treatments can be effectively designed and developed, we must first understand the basic biology of HIV more completely.

Major gaps in our knowledge of how HIV infects cells and how it escapes the immune system still exist. However, based on what we do know, a wide variety of treatments might be effective. Many of these treatments are in various stages of clinical trials. While none have proved to be consistently effective, remember that we have only been studying HIV and AIDS for less than twenty years and that the progress we have made thus far is nothing short of phenomenal. It seems likely that effective treatments and perhaps even vaccines will be developed soon. Unfortunately, this wishful thinking and well-deserved pride in the progress of scientific endeavor are of little solace to those already suffering from the disease.

Biotechnology occupies a central role in HIV and AIDS research. Our knowledge of basic cellular biology helps us better understand HIV. Our understanding of molecular biology and immunology is necessary in order to understand how HIV causes disease. Ultimately, this knowledge of molecular biology and immunology, as well as our continued development of new technology, will allow us to develop more effective treatments, cures, and preventive vaccines. Last, HIV and AIDS research efforts have been spurred and financed by a great number of biotechnology businesses and pharmaceutical companies that have added a tremendous capability to the research effort, in conjunction with government- and foundation-sponsored activities.

Unusual challenges are associated with designing and conducting tests of potential therapies directed against HIV infection. Some of these challenges are purely technical, while others involve making ethical and moral decisions. Unfortunately, many of these questions have no easy answers, resulting in unhappiness in some circles. By learning more about the basic biology of HIV and the difficulties inherent in developing and testing potential therapeutic agents, we can make more informed decisions not only about testing procedures, but also about life-styles, the avoidance of risk, and the practice of safer sex.

SUMMARY

AIDS has captured global attention like no other disease, and the causative agent, HIV, has been studied more intensively than any other pathogen in history. In this chapter, we have seen that HIV infects cells by binding to the CD4 receptor of susceptible cells and that the progression of AIDS is marked by the gradual depletion of helper T cells. Without T_H cells, neither the cell-mediated nor the antibody-mediated immune response is fully activated; hence, opportunistic infections and cancers can kill afflicted individuals. We have also examined approaches to prevent and/or treat HIV infection, including diagnosis, efforts to boost the immune system, preventing HIV entry into cells, preventing reverse transcription, and preventing the assembly of new viruses. Finally, we considered the testing of therapeutic agents and the role of biotechnology in AIDS research.

R E V I E W   Q U E S T I O N S

1. How do viruses, including HIV, differ from prokaryotic organisms (bacteria)?

2. Describe how HIV is replicated in T_H cells.

3. How does HIV manage to escape and evade the human immune system?

4. How can viruses, claimed by most scientists to

be nonliving entities, attack and actually kill living cells?

5. How do gp120 and CD4 receptors interact to direct specific HIV attacks?

6. Explain the truth behind this quote: "Almost no one dies of AIDS."

7. What is the role of the provirus in HIV infections?

8. Give several reasons for the unusual difficulty pharmaceutical companies are having designing effective AIDS vaccines or chemotherapeutic treatments.

9. How are ELISA tests, Western blots, and PCR used in cases of suspected AIDS or HIV infection?

10. List several different means currently used in the treatment of full-blown AIDS.

11. Explain how AZT and ddI function in the treatment of AIDS patients.

12. Clinical trials are generally conducted in three phases. Briefly explain the purpose of each phase and note the special problems posed by AIDS.

CHAPTER 14

Cancer

Nowhere is the debate over basic versus applied research more apparent than in the attempt to find treatments and cures for all types of cancer. Next to AIDS, cancer may be the most feared disease in the United States. Cancer is responsible for about 500,000 deaths each year in the United States, and the current odds are that one in four individuals will develop some form of cancer during his or her life.

Certainly, we know more today than we did twenty years ago about the changes that a cell has to undergo to initiate a tumor. Yet despite our basic understanding of what happens at the cellular and molecular levels, the 1990s have seen only a slight reduction in the mortality due to cancer. So it is not surprising to find the following question being asked: Why should basic research be supported if it doesn't yield positive results in the search for better treatments and therapies?

In defense of scientific inquiry in general and the research basis of biotechnology more specifically, we should point out that both basic and applied research have contributed greatly to our knowledge of cancer. At the same time, we admit that much remains to be discovered. In Chapter 13, we outlined a few applications of recombinant DNA and immunology in the struggle against AIDS. Many of those same ideas and techniques are also being used to study cancer. We should not be surprised to learn that much of what we have learned about cancer has come as a direct result of applying recombinant DNA and immunologic methods in cancer research.

The purpose of this chapter is to highlight what we know about cancer, focusing on molecular and genetic changes that turn a normal cell into a cancerous cell. Then we will describe ways of treating the disease and assess the status of the fight against cancer.

WHAT IS CANCER?

At the outset, we need to distinguish between *cancer* and *tumor*. While we often use the words interchangeably, they refer to different clinical forms of disease. A **tumor** (neoplasm) is a localized, abnormal, growing mass of tissue that serves no normal

Table 14.1 Types of Tumors

Tissue of Origin	Name of Tumor
Carcinomas (about 90% of all tumors; arise from cells known as epithelia, which are on the surface of tissues or organs)	
skin	squamous cell carcinoma
lung	pulmonary adenocarcinoma
breast	mammary adenocarcinoma
stomach	gastric adenocarcinoma
colon	colon adenocarcinoma
uterus	uterine endometrial carcinoma
prostate	prostatic adenocarcinoma
ovary	ovarian adenocarcinoma
pancreas	pancreatic adenocarcinoma
urinary bladder	urinary bladder adenocarcinoma
liver	hepatocarcinoma
Sarcomas (about 5% of all cancers; arise from any primary tissue other than surface, glandular, and parenchymatic epithelium)	
bone	osteosarcoma
cartilage	chondrosarcoma
fat	liposarcoma
smooth muscle	leiomyosarcoma
skeletal muscle	rhabdomyosarcoma
connective tissue	fibrosarcoma
blood vessels	hemangiosarcoma
nerve sheaths	neurogenic sarcoma
meninges	meningiosarcoma
Lymphomas/leukemias (about 5% of all cancers; arise from cells of the blood and lymph)	
red blood cells	erythrocytic leukemia
bone marrow cells	myeloma or myelocytic leukemia
white blood cells	lymphoma or lymphocytic leukemia

metabolic purpose. Tumors can arise in nearly all types of tissues in the body. Table 14.1 lists the types of tumors and their origins.

Some tumor cells move from their place of origin in the body to another location where they take up residence and cause new tumorous growth—they **metastasize**. When this happens, the disease becomes known as **cancer**.

Why do tumors develop? Under normal circumstances, the cells in all the different tissues of the body are in a precise balance. Consider your skin. Individual skin cells have a preprogrammed lifetime. After a given period of time, they die and are sloughed off. New cells are constantly being made in lower layers of the skin to replace those that are lost. One reason for the constant replacement of lost skin cells is that the skin, as our outermost layer of protection, is constantly bombarded by all sorts of noxious agents that can cause mutations and other alterations within the cells. By constant replacement of these old, external cells from beneath, potential long-term damage can be avoided.

Balanced cell turnover involves reproducing the same number of new skin cells as the number of cells that die. Cell division and differentiation must be in balance with cell death in order to maintain the integrity of the skin. For this particular tissue, the turnover is quite rapid. Contrast this with what occurs in your heart muscle cells. They are not constantly reproducing. New heart muscle cells are required much less frequently than are new skin cells. Therefore, much less cell division takes place in heart muscles.

Numerous and presumably complex control systems must be operating in all of our cells to regulate division and differentiation. If these controls are not operating properly, one result may be uncontrolled growth—a hallmark of cancer. As a starting point, we can visualize a tumorous cell as one that has lost its normal regulation of cell division and is generally dividing more rapidly than normal.

COMMON TRAITS OF TUMOR CELLS

Tumor cells share a number of traits no matter where they originate. Appreciating these similarities will allow us to better understand how tumor cells arise.

1. *Tumor cells have a distinctive appearance.* Tumor cells usually appear distinct from the cells surrounding them. If normal cells are oriented in a particular direction, tumor cells usually lose that orientation. Tumor cells appear less differentiated, often with a different morphology. A common example of this distinctive appearance and how we can use it is in Pap (for Papanicolaou) smears, in which cervical cells are removed and examined microscopically (Figure 14.1). Identifying tumor cells within such a sample is fairly easy, thereby allowing us to catch cervical cancer early. Early detection is a primary weapon in the war against individual cancers.

2. *Tumor cells are immortal in culture.* Most cells in the body have finite lives and do not usually reproduce once they reach maturity. Mature cells are also finite when they are removed from the body and grown in laboratory culture. In fact, many do not survive for very long at all. In sharp contrast, tumor cells are often referred to as immortal, because they can be maintained in culture for a great many generations, perhaps indefinitely. Not only do they survive, but they also reproduce. This is the primary hallmark of a tumor cell—uninhibited and generally quite rapid growth.

3. *Tumor cells interact with their neighboring cells quite differently from normal cells.* Imagine what happens during liver growth and development.

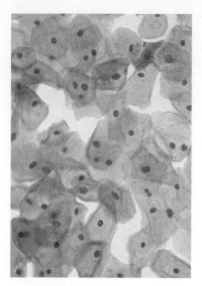 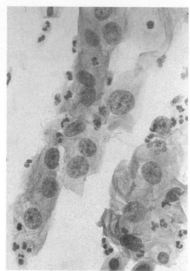 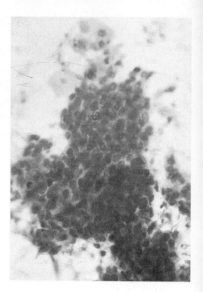

Figure 14.1 Appearance of cells found in Pap smears. In Pap smears, cells are scraped from the cervix and viewed under high magnification. When abnormal, cancerous cells are observed, their early detection allows treatment to be started before the actual appearance of a tumor. (**a**) Normal cervical cells are large, with nuclei well defined and small. (**b**) Abnormal cells show various states of differentiation and are not yet cancerous. (**c**) Invasive cancerous cells have little cytoplasm, large nuclei.

Some cell is initially instructed to become the first liver cell. This cell then reproduces, producing more liver cells, until the liver is as large as necessary, and then most liver cells stop reproducing. The same behavior occurs in most normal cells grown outside the body—they can usually reproduce or be induced to reproduce until they reach a certain density. When the surface of the container in which they are growing becomes completely covered with cells, for example, the cells often stop reproducing. This trait is sometimes referred to as *contact inhibition*, because it was once thought that contact with multiple neighboring cells was the signal to stop reproducing. Whether or not this is the case for all cells is not clear, so this process is now called *density-dependent growth inhibition*. No matter what it is called, tumor cells usually do not display this property. They can grow to a much higher density, not stopping at a single layer of cells. Tumor cells have lost contact or density-dependent growth inhibition.

4. *Tumor cells do not attach themselves to surfaces.* Again, we'll use normal liver cells as an example. When these cells are removed from the body, they require a particular type of surface on which to grow. Liver cells, and most other normal cells, fasten themselves to the surface; without attachment they will not grow properly. Attachment probably simulates the interactions the cells would have had within the body, either with other cells

or with some sort of connective tissue. Tumor cells have lost the necessity for anchorage and reproduce without attaching to the surface of the culture dish.

5. *Tumor cells have altered cell surfaces.* We now understand that our cell surfaces are important in a number of ways, particularly in establishing proper communications between cells and in the immune system functions. Tumor cells express, present, or display different cell surface proteins, making them appear different to both their neighboring cells and to the immune system. Altered surface appearance is valuable, because many developing tumor cells are recognized as "abnormal" by the immune system and are destroyed long before they become damaging. It is only when such cells escape destruction by the immune system that they become a serious threat.

6. *Tumor cells may contain chromosomal alterations.* Very early in our study of cellular reproduction, we found that chromosomes must be replicated very precisely during reproduction in order to produce exact, or nearly exact, copies. We also noted that certain defects in this process caused a variety of different diseases. Tumor cells sometimes display chromosomal abnormalities such as breaks or shuffled pieces. Sometimes, a particular tumor displays a very specific type of chromosomal alteration (Figure 14.2).

7. *Tumor cells secrete proteins.* Some of these proteins affect the circulatory system in the tumor itself. As a tumor mass develops, it needs increased circulation to provide nutrients for the developing cell mass. To meet these nutritional needs, tumor cells secrete proteins, called *growth factors*, that stimulate the development of more blood vessels in and surrounding the tumor. Other proteins, including proteases and embryonic proteins that are secreted into the blood, can aid in cancer diagnosis.

There are other similarities among tumors; we have listed the most apparent. Keep in mind, however, that not all cancers share all these traits. There are many types of tumor cells, each characterized by a particular set of properties.

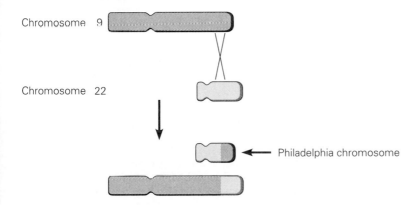

Chromosome 9

Chromosome 22

Philadelphia chromosome

Figure 14.2 Chromosomal abnormality associated with a specific cancer.
When a portion of the long arm of human chromosome 22 is exchanged with the long arm of chromosome 9 by a reciprocal translocation, the result is the so-called Philadelphia chromosome—associated with a cancer called chronic myelogenous leukemia.

Benign and Malignant Tumors

A tumor cell begins as a normal part of the tissue from which it arises, among neighboring cells that are presumably still normal. As the cell becomes abnormal and grows, it produces a mass of reproducing cells characteristic of a tumor. Sometimes, all the cells of this growing mass stay together, never moving elsewhere in the body. Such a tumor is called **benign**. It is often possible to surgically remove entire benign tumors.

In other cases, or perhaps in more advanced cases, cells of the tumor may be released, travel through the body, and establish residence in some other location (Figure 14.3). This is the process of metastasis. Tumors that have metastasized are said to be **malignant**.

Tumors and the Immune System

Given that tumor cells express distinctive proteins on their cell surfaces, how can they escape detection and destruction by the immune system? There are several possibilities. As we will see shortly, the development of a tumor cell is an accumulation of changes, not a single, sudden conversion. Intermediate cells—those on their way to becoming tumorous—may or may not exhibit significant changes. That is, early in the process, radically different cell surfaces may not be displayed, so the immune system does not react or reacts only weakly.

It may be that the changes in cell surface proteins are initially so insignificant that they cause no immune reaction. In other words, even though differences might exist in cell surface proteins, the altered proteins may still be similar enough

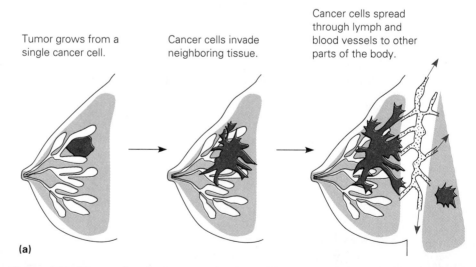

Tumor grows from a single cancer cell.

Cancer cells invade neighboring tissue.

Cancer cells spread through lymph and blood vessels to other parts of the body.

(a)

Figure 14.3 Tumor development and metastases. (**a**) Development and metastasis of a malignant breast cancer. As a tumor grows, the cells become invasive, breaking through layers of tissue until they work their way through the walls of capillaries and circulate throughout the body. Finally, they find a location on the inner wall of the capillary, reproduce, again break through the capillary wall, and invade the surrounding tissue—initiating cancer at the new location.

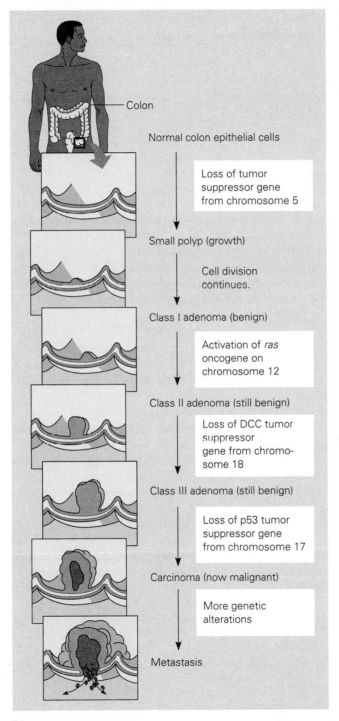

Colon

Normal colon epithelial cells

Loss of tumor suppressor gene from chromosome 5

Small polyp (growth)

Cell division continues.

Class I adenoma (benign)

Activation of *ras* oncogene on chromosome 12

Class II adenoma (still benign)

Loss of DCC tumor suppressor gene from chromosome 18

Class III adenoma (still benign)

Loss of p53 tumor suppressor gene from chromosome 17

Carcinoma (now malignant)

More genetic alterations

Metastasis

(b)

Figure 14.3 (b) Multistep development of colorectal cancer. Changes in the tumor parallel three genetic alterations in tumor suppressor genes and the mutation of one oncogene. In each of the four cases, the mutations are in genes that normally regulate cell growth or division, causing a breakdown in regulation.

to normal ones that the immune system does not recognize them as foreign and target them for destruction.

Since tumor cells often have radically different cell structure and organization, masses of them often have different characteristics as well. For example, most tumors do not have the normal organization of blood vessels, and circulation to the interior of a tumor is sometimes limited. In such cases, it is extremely difficult for the specific antibodies or sensitized T cells to gain entrance to the tumor.

For whatever reasons, some tumor cells can easily evade the immune system. Still, throughout our entire lifetime, our immune system protects us from developing tumor cells. It is difficult to estimate how many tumorous cells the immune system destroys, but the number is probably quite large.

CANCER AS A GENETIC DISEASE

In Chapter 5, we highlighted several diseases, including some genetic diseases. At that time, we concentrated on genetic abnormalities that could be passed from generation to generation—hereditary genetic diseases. All these diseases are characterized by particular mutations within the reproductive cells that are passed to offspring.

Cancer is a different sort of genetic disease. Although it is ultimately caused by mutations or other chromosomal alterations, these changes are not present in all cells of the body. In fact, they arise first in a single cell that subsequently develops into a tumor. Only the cells of the tumor, those derived from the original altered cell, contain the genetic changes. These changes are usually not present within the reproductive cells, and thus cancer is not generally a heritable disease. (We will have much to say about the inheritance of predispositions to cancer later in the chapter.)

One serious implication of the fact that cancer is a genetic disease (but not heritable) is that cancer, like the mutations that cause it, is inevitable. Anything that causes mutations can cause cancer, including exposure to carcinogenic chemicals, some forms of radiation, and even certain viral infections. Mutations also occur spontaneously. As we will see, cancer results from an accumulation of mutations. Thus, exposure to mutagens early in life may cause mutations that become detrimental only much later, after other mutations have occurred. Cancer seems to be one result of cumulative alterations in our genetic material over the course of time.

It has not always been clear that cancer results from genetic changes. By examining one avenue of cancer research, we will be able to gain a fuller appreciation of how rapidly our understanding of cancer has progressed and at the same time begin to lay out some of the basic changes that arise in tumor cells.

 ## Cancer and Viruses

Viruses were first proposed as a cause of cancer in the early 1900s. However, this idea was not widely accepted, since cancer does not generally have the properties of an infectious disease. There are exceptions to this—for example, feline leukemia. This disease is caused by a specific virus, the feline leukemia virus, and epidemics of

Table 14.2 Tumor Viruses

Virus Class	Examples	Tumors Induced
DNA Viruses		
herpesviruses	Epstein-Barr virus	Burkitt's lymphoma, nasopharyngeal carcinoma
papovaviruses	human papillomaviruses	cervical cancer
RNA Viruses		
C-type viruses	Rous sarcoma virus	sarcomas
	human T-cell leukemia virus	leukemias/lymphomas

feline leukemia have resulted from viral spread. Some other viruses were shown to cause specific cancers or were identified as at least being associated with tumors (Table 14.2). Most are specific to a particular host organism and, further, to particular tissues. Curiously, few viruses have been positively associated with human cancers. This is probably an example of our inability to manipulate human experimental subjects in the same way as laboratory animals. Providing evidence that a virus causes cancer requires deliberate infection and development of cancer in test subjects, experiments that we cannot and do not wish to perform on humans. Many human cancers may have viruses associated with them, but we simply have no evidence.

We can make some observations concerning two viral associates of human cancers. Epstein-Barr virus (named for its discoverers) was first identified as the agent that causes Burkitt's lymphoma, a massive jaw tumor found in areas of Africa where mosquito-transmitted infections are common. Epstein-Barr virus (EBV) has also been found to cause a different type of cancer, nasopharyngeal carcinoma, in southern China. However, EBV infections in the United States result in infectious mononucleosis, not cancer. Mononucleosis is a disease that results in proliferation of lymphocytes, but the proliferation is limited, not cancerous. We still don't know why the same viral agent causes three completely different diseases in different human populations.

Human T-cell leukemia virus (HTLV) causes acute T cell leukemia in humans. It infects lymphocytes and causes them to proliferate (causing cancer). Perhaps coincidentally, HTLV is structurally similar to HIV-I and -II (the AIDS viruses). HTLV and the AIDS viruses both infect lymphocytes, but the former results in cell proliferation of infected cells while the latter results in cell death.

The cancer-causing capacity of some viruses results from the insertion of viral genes into the chromosomes of the host cell. Studies of these oncogenic viruses have identified a number of particular genes (**oncogenes**) that cause transformation, or the taking on of tumorous cell characteristics. The term *transformation* in this context refers to a cell in which growth has been deregulated; in most instances, the rate of cell proliferation has increased. Following the first oncogene discovery in the mid-1980s (the *src* gene, found in the Rous sarcoma virus), many others were identified. Their discovery caused great excitement at the time, because many believed we would soon find all types of cancer to be caused by viruses. That excitement, however, was short-lived, since no associated viruses have been found for many forms of cancer.

Proto-oncogenes and Tumor Induction

The study of oncogenes in viruses led to a general understanding of the role of altered genes in cancer. Once viral genes were identified as oncogenes, molecular biologists applied techniques of hybridization to determine whether oncogenes were unique to viruses. It turns out that viral oncogenes are very similar to, but generally not identical with, genes found in normal, healthy cells. Sometimes, the difference between the viral oncogene and the cellular gene is just one base (although sometimes the differences are greater). The surprise was that cellular analogs, or counterparts, of viral oncogenes even existed.

Are the cellular analogs also oncogenes? No. As it turns out, they are normal genes that typically produce proteins involved in signaling within cells. In fact, such cellular genes are often important to normal cell functioning. When these genes become mutated, they can cause cancer. Prior to mutation, cellular genes are known as **proto-oncogenes** to distinguish these normal gene copies from oncogenes.

The next major step in understanding oncogenes came as a result of further study of tumors that had no viral associates. Did these tumors also contain oncogenes? Examination of many of the tumors revealed that many oncogenes are present in various tumor cells, some of which do not originate from viruses. Nonviral oncogenes apparently arise through mutation of existing proto-oncogenes. We shouldn't be surprised. Since carcinogens and radiation can induce tumor formation, it is logical that changes in otherwise normal cellular genes might result in tumor formation without viral assistance.

We can now outline a general scenario for the induction of tumor formation (Figure 14.4). While it may contain errors, the basic principles involved in tumor formation and genetic behavior are well established. Tumor cells arise from the addition of oncogenes or the alteration of normal cellular genes involved in regulation of growth and cell-cell interactions. The addition of oncogenes can occur most readily through viral infection, when the viruses insert oncogenes into the chromosomes of the host cell. The subsequent expression of these oncogenes alters normal cell physiology, resulting in at least some of the properties of a tumor cell. Alternatively, proto-oncogenes found normally within the chromosomes of a cell can mutate through a variety of means. If they mutate into oncogenic forms, then they will alter the metabolism of the cell, allowing it to become tumorous.

Transformation: What Does It Take?

Recall that many cellular properties are altered when they are transformed into tumor cells. Presumably, even more properties are altered when these same cells become metastatic. It is difficult to conceive of a single mutation or alteration in the genes of a cell that would result in such dramatic and diverse changes in cell function. Rather, it has been known for quite some time that cellular transformation results from collections of several mutations, which provide the basis for a normal

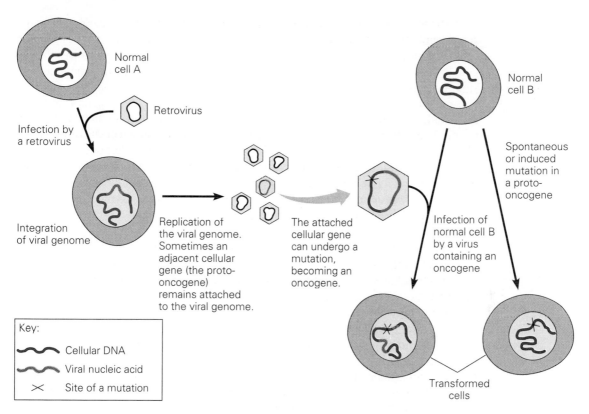

Figure 14.4 General schemes for accumulating cancerous mutations. Cells can accumulate cancerous mutations by at least two different means. Viruses carrying oncogenes can infect cells, after which the viral oncogenes may be integrated into the cells' chromosomes. If these integrated oncogenes are expressed, they will transform the cell. Alternatively, mutations in proto-oncogenes can result in oncogenes that can cause transformation. Such mutations may arise spontaneously, through exposure to radiation or chemical carcinogens, or as a result of chromosomal rearrangements.

cell to become tumorous. We still do not know the exact progression of tumor induction, but we have learned more about the genes involved in abnormal growth. Several observations at this point will help crystallize our understanding.

What Properties Are Altered in Transformed Cells? We've described some of the general traits of tumor cells: loss of growth control, loss of contact inhibition, and immortality in culture. Some of these traits are easily understood in terms of the proto-oncogenes and oncogenes and their signaling roles within cells. For example, it is possible to see how the loss of contact inhibition might prevent cells in contact with each other from signaling properly. If the tumor cell does not perceive that it is in fact surrounded by other cells, it will continue to grow and divide.

It is much more difficult to imagine how signal alterations result in the greater proliferation of tumor cells. However, our understanding of the processes that regulate cellular reproduction has advanced greatly and now allows us to see how genes other than the oncogenes are involved in the development of tumor cells.

One analogy is to think of driving a car. You can make the car move or stop by stepping on one of two pedals. You, the driver, are responsible for integrating traffic lights and signs, other cars, pedestrians, weather conditions, the funny noises the engine is making—and arriving at a decision about whether to continue moving or to stop. Let's assume that the normal state of the car is motion, and that a positive action must be taken in order to make it stop (hit the brakes!). As long as everything is normal, the multitude of signals will be processed and the result will be a decision to allow the car to continue moving. But if anything goes wrong, the decision will be made to stop.

Cells operate in an opposite manner. The "normal state" of a mature, fully differentiated cell is not dividing, but numerous signals control whether or not the cell will divide. Internal and external cellular sensors are probably setting up questions: Is everything OK on the outside? Is the cell surrounded by its appropriate neighbors? Are food and water available in sufficient quantities? Is the DNA damaged, or is an important component missing? If all the signals are still correct, then the do-not-divide "decision," the normal state, continues. If something is wrong, for example, if adjacent cells have been damaged, then the divide "decision" will be made.

Oncogenes operate by allowing the cell to divide despite other signals saying "stop." Functional oncogenes are said to be *gain-of-function mutations*. These mutations provide cells with properties they did not have before, resulting in the occurrence of specific cellular processes that would not take place under normal conditions.

To continue the analogy, specific genes produce proteins analogous to automobile brakes. These proteins prevent cells from dividing until the cell is ready. If such genes are lost, or mutate such that they no longer function as brakes, cells begin to proliferate abnormally. These genes are known as **tumor suppressor genes**, since in their normal state they suppress the proliferation of normal cells. The mutation of tumor suppressor genes is also implicated in the development of tumors. These mutations, however, are not gain-of-function mutations as with oncogenes.

Inheritance of Predispositions to Cancer Our knowledge of tumor suppressor genes allows us to understand how predispositions to cancer might be inherited. In order for a tumor suppressor mutation to allow proliferation of cells, the function associated with that gene must be lost, or at least greatly reduced or altered. Since we are diploid organisms, we carry with us two copies of each tumor suppressor gene. It is likely that both of these genes must be mutated or lost in order to allow the sort of free, rapid proliferation that some tumors exhibit. This means that, most likely, two distinct mutagenic events must occur within that cell in order to alter both copies of the gene. Clearly, the probability of two such events occurring is much lower than the probability of only one occurring. Contrast this with the gain-of-function oncogenetic mutations, in which a mutation in a single copy of the gene can alter the product so that it performs a transforming function.

Assume that one parent has a mutation in one copy of a tumor suppressor gene. This mutation might not be expressed, since it is likely that the other copy of the gene provides sufficient function to result in healthy cells. The mutated tumor suppressor gene will be passed along, and some children will be born with it. The odds of receiving an additional single mutation that will destroy the sole remaining copy of the normal tumor suppressor gene are far greater than the odds of receiving two mutations independently, making it more likely that the child will develop the particular cancer associated with a loss of the tumor suppressor gene. Thus, a predisposition to develop cancer is based on the inheritance of mutated tumor suppressor genes.

In a sense, then, we know that at least two classes of genes are involved in the transformation of a normal cell. Oncogenes provide the cell with new or enhanced functions, allowing it to do things it was not able to do previously, or at least to do them at different times. Mutations within tumor suppressor genes allow cells to overcome the normal barriers to rapid growth, resulting in uncontrolled cellular proliferation. One strong likelihood is that, in all tumors, both oncogenic and tumor suppressor mutations together result in transformation.

Why Is Cancer Predominantly a Disease of the Elderly? Cancer seems to result from an accumulation of mutations that alter a particular cell in specific ways, providing all the characteristics necessary to begin tumor formation. A cell accumulates mutations over the course of its lifetime and inches closer to becoming tumorous. Clearly, the longer a human lives and the greater the incidence of mutation, the more likely tumor formation will occur.

DIAGNOSIS AND TREATMENT OF CANCER

Our next consideration is how this basic knowledge, gained over the past twenty years, can be used to develop better diagnostic procedures, treatments, and possibly even cures for specific cancers. We will consider both diagnosis and treatment of cancer, since biotechnology contributes significantly to both.

Diagnosis

The diagnosis of cancer has undergone vast improvement in recent years as a result of new techniques, including new biotechnology techniques. There are a variety of ways to identify abnormal growth; some of these aim at directly observing abnormal tissue, while others search indirectly. Pap smears are a good example of one method of direct detection: microscopically viewing cervical cells to determine whether abnormal cells are present. As we indicated earlier, the technique relies on the observer's ability to distinguish visually between normal and cancerous cells.

Numerous blood tests have been developed to detect proteins secreted by tumors. The detection of embryonic proteins or proteases in blood is often diagnostic of particular cancers. Sometimes, it is possible to identify circulating cancerous cells in the blood—a tumor that is metastasizing may release millions of tumor cells into the blood daily. Most of these cells will die before they find another location suitable for growth, but while circulating they can be used diagnostically.

Biopsies of tissues suspected of being cancerous involve removing cells and examining them microscopically. As with Pap smears, it is sometimes possible to identify abnormal cells and determine how far along the path toward tumor formation they are. Many skin cancers are routinely diagnosed in this way. Many people have had moles removed that might be indicative of cancer; once removed, part of the mole is examined to determine the state of cellular transformation, information that may be exceedingly valuable in the diagnosis of melanoma.

Another way to identify cancer is to use recombinant DNA technology to perform genetic diagnosis. It is now possible to genetically screen individuals or even individual cells for the presence of particular oncogenic or tumor suppressor mutations. This screening helps assess the state of developing cancers, as well as providing information about any predispositions an individual might have.

One technical advance that has helped in the diagnosis of cancer is **magnetic resonance imaging (MRI).** With this specialized technique, we can examine internal organs without doing exploratory surgery. MRI technology has greatly aided in the diagnosis of brain cancers, which are often particularly difficult to identify.

Traditional Treatments and Therapies

We are familiar with the three traditional treatments for cancer: chemotherapy, radiation therapy, and surgery. Vast improvements have been made in these procedures in recent years, with the result that tumors can be more safely removed and the side effects of radiation and chemotherapy are reduced. But, fundamentally, no new widespread, effective treatment of cancer has been introduced in quite some time. Treatments may be effective; often they are not. Some people have said rather cruelly that the medical role in cancer treatment consists of cutting, poisoning, burning, and hoping.

Surgical Removal The most common way to halt the growth of a tumor is to physically remove it from the surrounding nontumorous tissue. It may be possible to completely rid the body of benign tumors in this way.

However, not all tumors are benign or can be surgically removed. The tumor must be a single, isolated mass, or at worst a localized collection of a relatively small number of identifiable masses. Tumors that have metastasized are singularly poor candidates for surgical removal, since they have spread to other locations in the body. But surgical removal of benign tumors is relatively straightforward, unless the tumor happens to be in a location in the body not surgically accessible. Brain tumors frequently fall in this category—undergoing brain surgery is often riskier than having the tumor.

Sometimes, tumors are found and identified only after they have grown to such size that permanent damage has been done to an organ or surgery will inflict damage. Thus, early detection of tumor masses is of the utmost importance, from two points of view. First, treatment can be initiated before the tumor grows too large to be accommodated within the confines of the tissue. Second, treatment can perhaps be initiated before a single tumor mass metastasizes. The success of treating malignancies is generally much lower than the success of treating isolated tumors.

Chemotherapy Another way to try to rid the body of tumors and cancers is to kill the offending cells using toxic chemicals. How can we kill the cancer cells without also killing healthy cells? The drugs used in chemotherapy must preferentially kill cancerous cells. We've encountered several situations in which we needed to identify properties of certain cells. What properties of tumor cells could we exploit to develop such chemotherapeutic agents?

One significant difference between most tumor cells and normal cells is their rate of growth. The generally higher growth rate of tumor cells means a much more rapid metabolism, specifically, higher rates of DNA replication. Therefore, most chemotherapeutic agents target important metabolic steps such as DNA replication or cell division. Many drugs induce mutations in the cells as well, since inducing mutations in tumor cells will frequently cause them to die. The rapid replication rate of tumor cells makes them much more susceptible to mutation, and, conversely, normal (nondividing) cells should be less affected. Chemotherapy usually consists of brief periods of exposure to high levels of the drugs, with time between treatments for the healthy cells to recover from side effects. In this way, prolonged exposure to toxic drugs is avoided. Table 14.3 lists a number of common chemotherapeutic drugs and the processes they target.

The primary, immediate side effects of chemotherapy—nausea, anemia, and hair loss—result directly or indirectly from the preferential effect the treatment has on any rapidly dividing cells. The cells lining the stomach and intestine are in a constant cycle of replacement. Thus, the drugs used in chemotherapy will necessarily target them as well as tumor cells, causing the lining of the digestive system to become impaired and leading to nausea and probably vomiting. Cells of the blood, particularly red blood cells, are also targeted because they are constantly being formed. As they

Table 14.3 **Some Drugs Used in Cancer Chemotherapy**

Class	Examples	Mechanism of Action
antimetabolites	methotrexate 5-fluorouracil 6-mercaptopurine	inhibit pathways for biosynthesis of nucleic acids by substituting for normal compounds
antibiotics	actinomycin D adriamycin daunorubicin	bind to DNA
alkylating agents	nitrogen mustard chlorambucil cyclophosphamide imidazole carboximide	chemically modify DNA
mitotic inhibitors	vincristine vinblastine taxol	interfere with mitosis
hormones	estrogen (for prostate cancer) cortisone progesterone androgens	inhibit growth of particular hormone-sensitive cells
others	asparaginase	degrade the amino acid asparagine

are killed by chemotherapeutic drugs, anemia results. Loss of hair results from the preferential effect chemotherapeutic drugs have on the actively metabolizing cells that make hair. Fortunately, if the treatments are not too severe or too prolonged, the effects on the digestive system, blood, and hair may be reversed.

One of the greatest challenges of chemotherapy is that the drugs that are used to kill tumor cells must be administered in ways that minimize the adverse effects on noncancerous cells of the body. There is still a significant amount of trial and error involved in chemotherapy, since each patient—and each tumor—responds differently to drugs.

One advantage of chemotherapy over surgery is that the drugs, because they are typically delivered through the bloodstream, circulate throughout the entire body. Cancers that have metastasized can thus be treated the same way as an isolated tumor. Neither surgery nor radiation therapy, discussed next, can be used effectively to treat tumors that have spread throughout the body.

Radiation Therapy The third traditional cancer treatment works on the same principle as chemotherapy. We know that radiation can damage cells, primarily by causing mutations. Mutations are more likely to kill cells that are replicating quickly; thus, radiation preferentially kills rapidly growing cells such as tumor cells.

But, like chemotherapy, radiation will damage normal, healthy cells as well as tumor cells. Unlike chemotherapy, radiation treatment can often be localized at the exact site of a tumor. The entire body is not exposed to radiation as it is to chemotherapeutic drugs. This means that the location of a tumor must be clearly defined before treatment can begin. It also means that radiation therapy generally cannot be used on metastatic tumors, since they are present in different locations.

Ironically, two of the primary weapons in the fight against cancer are external agents actually known to cause cancer. Many of the drugs used in chemotherapy are carcinogenic, and radiation surely causes cancer. How is it possible to use agents that cause the disease to treat the same disease?

A tumor is generally an immediate health threat and must be treated as quickly as possible. Carcinogens and radiation may in fact induce tumors, but probably only after the passage of time as "normal" cells exposed to the treatments accumulate the mutations necessary to develop into separate tumors. In the meantime, the treatments may rid the body of the cancer. Until treatments with less severe side effects are developed, the simple fact is that both chemotherapy and radiation therapy have to be applied to destroy tumors, which pose the immediate threat to the patient's health.

New Approaches to Cancer Therapy

A tremendous amount of research has been and is being performed with the goal of developing better treatments and potential cures for cancer. Many research programs are based directly on the biotechnology we have discussed in this book—recombinant DNA and immunology. Describing all the efforts being made in this area would be impossible, so instead we will concentrate on some of the basic biotechnological approaches that are being applied to cancer.

Recombinant DNA Approaches Since cancer is a genetic disease, caused by mutations or other changes in genes, recombinant DNA techniques can be applied to affect the disease's progress. The most promising procedures being tested are gene therapy approaches. The intent is not usually to correct the genes that have gone awry but rather to deliver genes to tumor cells that will result in their death. Such "killer genes" can code for toxic proteins or for proteins that block a vital metabolic reaction. Or the genes may direct the synthesis of products that interfere with some other process in the cell.

One of the many challenges we face in developing a recombinant DNA approach is that of delivering the killer genes only to the tumor cells, and not to normal cells. Selective delivery is one of the more difficult technical aspects of gene therapy. Most successful gene therapies employ an *ex vivo* approach, in which cells are removed from the body, manipulated, and returned. An *ex vivo* approach cannot be used against tumor cells, because they cannot be removed from the body. (If we could remove them, patients would be instantly freed of cancer.) For a recombinant DNA approach to succeed, we must perfect methods that will allow *in vivo* delivery of genes to targeted tumor cells—a tall order.

One promising idea is illustrated in Figure 14.5. Here, a retrovirus is used as a vector to transport the killer gene to tumor cells. Recall that retroviruses enter cells through an interaction between a specific viral envelope protein and some receptor on the cell surface. Also recall that tumor cells generally have distinct sets of cell surface proteins compared to normal cells. In order to target the retroviral vectors carrying killer genes to specific tumor cells, we modify the viral envelope protein so it will be recognized only by tumor-specific receptors. This process is difficult but theoretically possible. First, a specific receptor on the tumor cell must be identified and studied. Once we know the structure of this molecule, we can determine what structure the viral envelope protein would need in order to interact specifically with the tumor cell receptor. We can then create envelope proteins containing that structure by altering the genes coding for the viral envelope genes, using recombinant DNA techniques. As we have seen repeatedly, not only do cells take great advantage of the ability of different molecules to interact very specifically, but biotechnologists also take great advantage of such interactions.

Another recombinant DNA approach involves the antisense technology described in Chapter 13. If specific genes are responsible for causing cells to lose their normal growth control, then preventing the expression of these genes might be sufficient to prevent the cells from proliferating. Antisense therapy could be used to accomplish this. The general idea is to introduce into a cancerous cell an antisense gene that produces an mRNA molecule complementary to the mRNA encoded by an oncogene, thereby decreasing or preventing expression of the oncogene. The complementary mRNAs would bind to each other and prevent translation of the oncogenic mRNA. Again, one significant challenge inherent in this approach is delivering an antisense gene to the target cells.

A different approach relies on the ability of certain DNA molecules to form triple-helix structures, instead of the normal double helix. We have known that, theoretically, DNA strands could be combined in a triple helix. However, only within the past decade have researchers been able to make this happen in a variety of DNA molecules

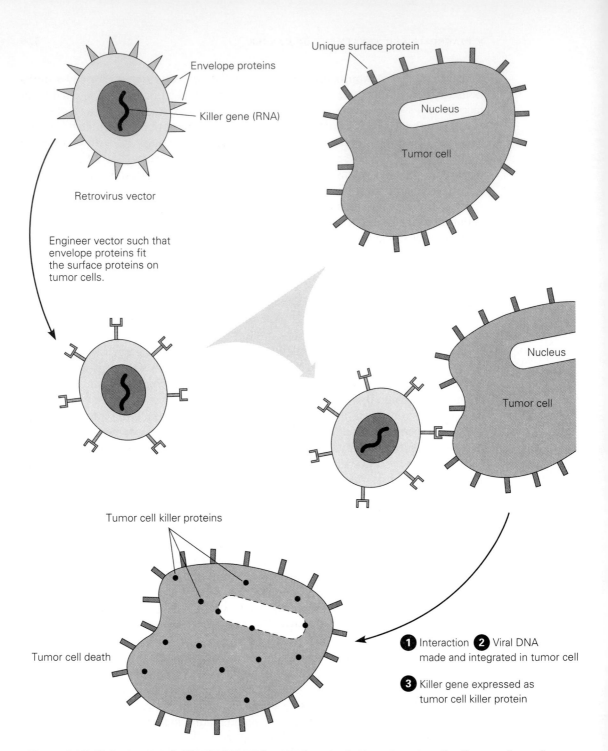

Figure 14.5 Using a genetically engineered retrovirus to destroy tumor cells. Tumor cells usually have unique cell surface proteins. By identifying them and determining their structures, we could genetically engineer the envelope proteins of a retrovirus to bind only to those specific tumor cell surface proteins. Once engineered, the modified retrovirus, carrying "killer genes" (RNA form), can be injected into the bloodstream. It will attach to the tumor cell and infect it. After the killer genes (DNA form) are incorporated into the tumor cell chromosomes, they will be expressed, resulting in the production of cytotoxic proteins. The tumor cell will be preferentially killed.

under conditions that might be found in the cell. One observation has been that a rather small strand of DNA, if constructed properly, can bind to a very specific sequence of DNA and cause triplex formation. The triplex structure so alters the normal structure of the DNA that other molecules that might have bound to that region can no longer do so. For example, if a triplex is formed in the promoter region of an active oncogene, then that promoter will no longer be functional and the oncogene will not be expressed. Although triplex therapy is probably still in the future, the trials that have been performed thus far have been encouraging.

Not only is recombinant DNA technology responsible for much of our knowledge about tumor cells and the genes that cause cancer, but we can also use it directly to develop new therapies and cures. As we discover more about the molecules that function within our cells, we also discover ways of manipulating them—in this case, preventing them from causing cancer.

Immunological Approaches The immune system is well suited to meet the threat posed by tumors. What better way to combat such abnormal cells than through the immune system? Cells that begin to develop abnormally are almost always detected by the immune system and destroyed. We know that tumor cells contain abnormal cell surface proteins and are thus self-identifying. Therefore, T_C cells should be able to recognize them as foreign and destroy them, much as they recognize and destroy otherwise normal cells that become infected by viruses.

Apparently, our immune system does a good job of rendering potential tumor cells ineffective. One way to determine this is to note that patients suffering from immune deficiency diseases frequently develop some form of cancer, although patients appear to suffer from infectious diseases more frequently because these develop much more rapidly.

How do some tumorous cells avoid being identified and destroyed by the immune system? Perhaps the cell surface antigens on precancerous cells are not different enough to stimulate a strong immune response, or perhaps there is not a strong enough interaction between a T-cell receptor and a cell surface antigen. Either event would not allow the immune response to be initiated.

During investigations of a number of different tumor types, scientists have found evidence of types of T cells that can invade and destroy tumor cells. One type, called **tumor infiltrating lymphocytes** (TILs), has allowed development of a treatment known as **adoptive immunotherapy**. The principle resembles one of the potential treatments for HIV infection. Blood cells are removed from the body, and the immune cells are separated from other cells. Those that are particularly active against tumor cells are then identified and isolated. They are treated with interleukins to cause them to proliferate and returned to the body. In this way, the particular cells that react against the tumor are increased in number to fight the tumor more effectively. This is what the immune system should be doing on its own and in fact does most of the time: creating lots of copies of the particular cells that fight disease agents or toxins. Adoptive immunotherapy seeks to accomplish what the immune system cannot against a particular tumor mass.

The ability to perform adoptive immunotherapy depends first on being able to identify and isolate the desired cells, which is done by determining whether the immune cells interact with antigens from the tumor cells. This, of course, requires that the antigens be identified and often purified—by biotechnology methods that we have covered earlier in the book. Once potential TIL cells have been identified and isolated, they need to be induced to proliferate. Immunologists know of various cytokines, such as interleukins, that induce the proliferation of immune cells. In some instances, including adoptive immunotherapy, scientists have been able to reproduce this effect in the laboratory, leading to the production of many copies of cells that can potentially attack tumors. Sometimes, injecting interleukins directly into the tumor or surrounding region stimulates whatever TIL cells or other immune system cells might already be present.

Another major immunological approach to treating cancer is through magic bullets, the monoclonal antibodies described in Chapter 12. Recall the basic principle: Monoclonal antibodies that react specifically with tumor cell surface antigens are produced and modified to carry a toxin or highly radioactive molecule directly to the tumor cells. This approach, like most immunological methods, relies on the ability to identify and isolate a unique antigen present on tumor cell surfaces. The antigen is used to produce monoclonal antibodies that react against it. After being modified to carry harmful cargo and injected into the bloodstream, the monoclonal antibodies will circulate until they happen on the antigen to which they bind—which by design is found only on tumor cells. Once they bind, the cargo has been delivered—the toxin or radioactive entity is now localized directly at the site of the tumor. Antibodies that do not bind to tumor cells clear the body in a relatively short period of time, minimizing potential damage to other cells and tissues.

The greatest advantage of the magic bullet approach is that it is highly specific. It overcomes some of the difficulties associated with traditional chemotherapy and radiation therapy by avoiding overexposure of healthy cells to noxious agents. On the other hand, it has at least one significant drawback—in order for it to work, unique cancerous antigens have to be known. This is very difficult for even one type of cancer, let alone for the many different tumor cells known to exist. Despite its difficulties, however, the magic bullet approach has shown its potential effectiveness, at least in certain circumstances. With continued study, it may be possible to more fully develop this method of treatment.

ISSUES: THE WAR AGAINST CANCER

In 1971, President Richard Nixon officially began a new chapter in medical history by declaring war on cancer. The proclamation was more than simply symbolic. He signed into legislation the National Cancer Act, establishing a new government research center, the National Cancer Institute (NCI), to investigate all scientific leads that might be used to combat cancer. Presently, NCI has an annual budget of over $2.1 billion.

How has the war on cancer been progressing? Based on the records of some other government medical institutes, there is reason to expect great success. For example, the National Heart, Lung, and Blood Institute has reported a reduced incidence of heart disease of some 30 percent in the last two decades, primarily through improved treatments and life-style education. Compare that reduction with the 1.2 percent per year *increase* in cancers from 1973 to 1990 and the increased mortality rate due to all types of cancer of 6 percent. In March 1998, a combined report from the American Cancer Society (ACS), the National Cancer Institute (NCI), and the Centers for Disease Control and Prevention (CDC) provided the first really good news: The incidence rate of all types of cancer declined 0.7 percent per year, while the overall death rate dropped 0.5 percent from 1990 through 1995. In that period of time, the four leading cancer sites as well as the leading cancer killers were lung, prostate, breast, and colon-rectum.

Consider how many deaths occurred from cancer in the United States in 1996. In that year, the age-adjusted death rate was 129 deaths per 100,000 people, for a total of more than 540,000 fatalities. Globally, the total is over 6 million fatalities, or more than 11 percent of deaths due to all causes. In the United States, this second-place killer, behind heart disease, was responsible for 23.4 percent of the 2.3 million deaths that year.

These statistics do not seem to represent a tremendously effective campaign against the disease. Looking at the data more closely proves interesting and leads to more difficult questions. Lung cancer decreased 1.1 percent per year from 1990 to 1995, the leukemia rate declined 0.3 percent per year, and colon and rectum cancer declined 1.5 percent. On the other hand, non-Hodgkin's lymphoma (1.9 percent) and melanoma (2.5 percent) have increased. As we can see, while the trend in all types of cancer in the early 1990s was toward reduced incidence and death rates, the trend was not unidirectional.

Interpreting statistics can become even blurrier when we consider the incidence of cancer, or the number of new cases diagnosed each year. Detecting and diagnosing cancer is now much easier and more accurate than it was in the early 1970s, so increased rates or the actual number of confirmed cases, by themselves, do not mean that more people are getting cancer. They may simply mean that more people are appropriately diagnosed as having cancer. It has also been suggested that, in our present litigious (suit-happy) society, doctors may diagnose cancer rather freely simply to avoid a potential malpractice suit if the patient later develops cancer. Such societal influences make it very difficult to accurately ascertain how successful we have been in the battle against cancer.

●●●

SUMMARY

A tumor or neoplasm is an abnormal, growing cellular mass that has no normal metabolic purpose. When tumor cells move from their place of origin to another location, take up residence, and cause new tumorous growth, they have metastasized. When this happens, the disease is known as cancer and the tumor is termed malignant.

Cancer is a genetic disease caused by mutations that arise first in a single cell. Tumor suppressor genes undergo loss-of-function mutations and allow tumor development. Proto-oncogenes have gain-of-function mutations and form oncogenes (or oncogenes are brought into cells by viruses). Their subsequent expression alters normal cells such that they (1) have a different physical appearance, (2) are immortal in culture, (3) interact differently with neighboring cells, (4) do not attach themselves to surfaces, (5) have chemically altered cell membranes, (6) may contain chromosomal alterations, and (7) secrete different proteins.

Early diagnosis is key to successful anticancer treatment. Biotechnology focuses on several oncogenic changes and is employed in a number of blood tests for specific proteins secreted by tumor cells. Recombinant DNA technology is used to genetically screen individuals for the presence of specific oncogenic or tumor suppressor mutations. Not only do these tests assist in the initial diagnosis of cancer, but they also help assess the state of cancer development and provide information concerning cancer predisposition.

Traditional treatments for cancer include chemotherapy, radiation therapy, and surgery. Today, biotechnology and immunology offer several alternative means of treatment. One biotechnological approach is to use gene therapy. The central idea is to deliver "killer genes" to tumor cells and only tumor cells. Another recombinant DNA approach is to introduce an antisense gene that produces an mRNA molecule complementary to the mRNA encoded by the cancer-inducing oncogene. The complementary mRNAs bind to each other and prevent translation of the oncogenic mRNA.

Immunological approaches to cancer therapy include the enhancement of tumor infiltrating lymphocytes (TILs) *in vitro* and their subsequent use in adoptive immunotherapy. An equally opportunistic approach to treating cancer involves monoclonal antibodies that are attracted specifically to the abnormal antigens of tumor cells. The monoclonal antibodies are altered to carry cytotoxins or radioactive molecules that will dispatch only the cancer cells, not the adjacent normal tissue.

REVIEW QUESTIONS

1. Distinguish between *tumor* and *cancer*.

2. Provide several reasons why there is no real cure for cancer.

3. Most tumorous cells have seven traits in common. List them and indicate how each trait sets tumor cells apart from normal cells.

4. Explain how some cancers may be considered genetic diseases.

5. Few cancers are positively associated with viral agents, even though both virologists and oncologists believe it is likely that many more types are at least partially initiated by viruses. Why are viral agents of any type of human cancer difficult to pin down? (*Hint:* What experiments would be absolutely necessary?)

6. Explain why oncogenes are considered gain-of-function mutations, whereas mutations in tumor suppression genes are considered loss-of-function mutations.

7. Why are most cancers found in elderly people?

8. Why can't more cancers be successfully treated by surgical removal?

9. Anticancer chemotherapy has what significant advantage over surgical removal?

10. How can retroviruses be used in *in vivo* approaches to cancer treatment?

11. How is antisense therapy intended for use against cancer?

12. Describe the general theory behind the use of triplex formation in cancer treatment.

13. Explain how TILs are supposed to be used in adoptive immunotherapy.

14. One immunologic approach to cancer treatment is the use of magic bullets composed of monoclonal antibodies attached to either a toxic molecule or one that is radioactive. How is this approach supposed to work?

15. List two or three statements that back the idea that some tumorous cells are neither identified nor destroyed by human immune responses.

16. Why do many patients experience nausea, vomiting and hair loss under many current chemotherapeutic regimes against cancer?

17. What is the single, most significant advantage of chemotherapy over either surgery or radiation in cases of metastatic cancer?

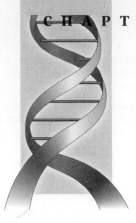

CHAPTER 15

The Business of Biotechnology in the United States

In 1980, a normally shy, reserved microbiologist by the name of Ananda Mohan Chakrabarty (Figure 15.1) reportedly shouted at the top of his lungs, "I won!" This out-of-character behavior was due to the news that the U.S. Supreme Court had just upheld a ruling awarding him a patent for the development of the first oil-eating bacterial strain. It didn't matter that the strain had already been found to be too fragile for use in the wild. In many ways, this court decision marked the real beginning of the commercialization of biotechnology. Because biotechnology involves the use of living organisms, cells, or tissues—or materials derived from them—the notion of ownership and exploitation of resources for profit has been and will continue to be of great importance to the development of the industry. The Chakrabarty decision allowed a living organism to be patented for the first time.

The business of inventing and discovering new products and procedures is time-consuming and often risky. Early on, the U.S. government recognized the value of rewarding those who make technological improvements and discoveries. The patent system in the United States was established in 1793 to provide some protection for those willing to take the risks involved in invention and development of new products. A patent granted by the government protects inventors or developers for a certain number of years (usually seventeen) by allowing them to be the sole owners, producers, and sellers of the product for that time. During this time, the costs of development and research can presumably be recovered in the market. At the conclusion of the patent period, other manufacturers can utilize the information to produce and market identical products. The availability of generic drugs and pharmaceuticals is an example of the process. When a new drug is developed, it is patented, and all rights to the production of the drug belong to the holder of the patent. At the end of the patent period, however, the product is fair game. The usual result is a drastic drop in price to the consumer as generic equivalents are produced.

The patent system provides a legal means to ensure economic reward for creativity, research, and development. This chapter introduces commercial biotechnol-

Figure 15.1 Ananda Chakrabarty. Chakrabarty was awarded a patent in a landmark U.S. Supreme Court case, the first to allow a living organism to be patented.

ogy. How are the products and processes of biotechnology developed for market? Who pays for this development, and how is it protected? How does the patent system function to encourage business ventures in biotechnology?

THE CHOICE OF PRODUCT OR SERVICE

The primary purpose of any for-profit corporation is to make money. This means that companies must choose wisely in deciding what products they will offer. Products must be competitive within existing markets and be valuable enough to ensure a profit. Given the abundance of potential products and processes available to biotechnologists, how do they make decisions about what to develop for commercial production?

Ideas and Research

The entire process begins with an idea—usually related to a valuable biochemical or pharmaceutical compound. Many of the products on which biotechnology companies initially concentrated were involved in the treatment of cancer or in the development of vaccines and antiviral drugs. Such compounds tap into an immense market and thus can be expected to provide an unusually good return on investment should they prove successful. Generally, these compounds are available only in minute quantities through traditional means. Interferons and interleukins are good examples. When these compounds were first discovered, they were shown to play some role in the growth of cells. They quickly became targets for commercial application, since any agent that affects the growth of cells might have anticancer or antiviral use. However, in the case of interferons and interleukins, isolating needed amounts for testing was prohibitively costly until methods were developed to produce them through genetic engineering (Figure 15.2).

Figure 15.2 *E. coli* genetically engineered to produce gamma interferon, a human protein that promotes an immune response.

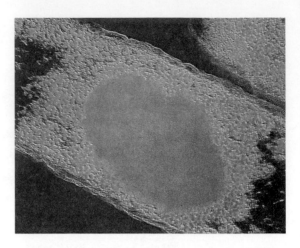

With the expansion of our biotechnological capabilities, enormous opportunities have opened up for potential commercialization. Once ideas become available, those ideas with the best possibility for commercial success must be identified. This process involves a combination of scientific research to determine how difficult it may be to engineer the product and commercial or market research to establish the likely demand for the product (Figure 15.3). Biotechnology organizations that can combine smooth scientific development with accurate market research will stand the best chance of successfully generating profits.

Typical Sequence of Events

The commercialization of biotechnology is no different from other business ventures. Of course, there are particular nuances and unique approaches in biotechnology, but the same sequence of events takes place as in other businesses. Five basic steps are involved:

1. Ideas.
2. Scientific development and market research.
3. First production capabilities.
4. Testing and approval.
5. Marketing and final production.

Some of these steps have already been explored, for example, in our consideration of drug testing in Chapter 13. Others, such as the first step, have been the subject of much of this book. However, several aspects of biotechnology development should be explored further.

Market Research Economic forecasting is tricky. Despite this, it is a fundamentally important corporate tool. Usually, before any business venture is attempted, substantial research is done to assess the feasibility of the proposed service or product.

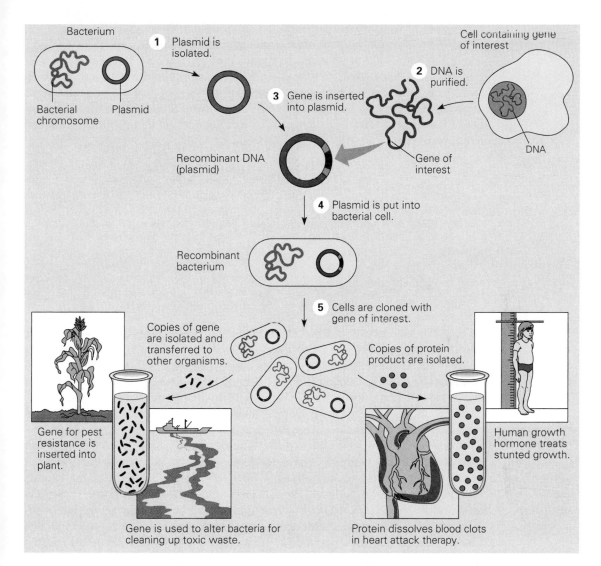

Figure 15.3 A typical genetic engineering procedure, with examples of applications.

Is there a market? How much will the market generate in sales? Is the market likely to grow or shrink? Making economic forecasts that address questions like these is central to business planning.

Forecasting becomes especially difficult when the projections have no historical basis. Before the advent of recombinant DNA, no one wondered at all about the prospects of gene therapy or the availability of laboratories that would specialize in forensic DNA testing. Such activities were still in the realm of science fiction.

Once the technology became real, however, an almost immediate need for biotechnology forecasting arose. Many biotechnological products and services are linked to the medical industry, so general trends operating in that industry might be expected to be mirrored in medical biotechnology. On the other hand, many biotechnological products and services might significantly alter or even replace some forms of traditional medicine and therefore alter forecasts based on them. In the very early days, the potential of the biotechnology industry was such that many people went into business strictly as biotechnology forecasters. John Elkington, editor of *Biotechnology Bulletin*, wrote about forecasting the growth of the fledgling biotechnology industry:

> Think of a moderately high number. Feed it into your calculator and multiply it by a million or, if you are feeling bullish, by an American billion. Dress up the resulting figure in some breathless text and publish. If you can find the nerve to charge several thousand dollars for the resulting report, so much the better. The market for surveys of the developing market for biotechnology products is itself booming.

However, he went on to say:

> Yet one conclusion which can have escaped few of those who actually read the results is that the forecasts are so wildly different that only a small minority are likely to be near the mark. The problem is identifying which these might be.

The range of "educated estimates" was exceptionally broad, from a market growth of over 10,000-fold to the more modest projection of less than 50-fold in twenty years.

Why was there such a plethora of market analyses and surveys? The simple answer is that whenever a new technology is in its early, developmental stages it has a tremendous potential for economic reward. Thus, many investors were interested. The incredibly rapid growth of biotechnology throughout the late 1970s and the 1980s was unprecedented—and resulted in many fortunes being made and lost.

Investment Capital Before a corporation can embark on an entirely new research program, it must secure funding for the work. Facilities must be built, individuals must be hired to do the work, and specialized equipment must be available—to name just a few of the usual expenses. If a long time will pass before any return can be realized from the market, a considerable amount of money must be invested initially. Where does this money come from? It often comes from venture capitalists, people with the monetary resources who are willing to invest deeply in new, speculative technology. They often provide money to support the initial formation of a corporation as well as research and development work through the first phase of the company's existence.

The initial investors can realize a profit in one of two ways. If the products are developed and brought to market, the investors share in the profits. Alternatively, the corporation can be sold to other investors at a profit or can "go public" by offering public stock. Offering public stock is one way of gathering investment money. Shares of the corporation are offered for sale, producing the necessary operating capital to fund the corporation. Venture capitalists can often make large amounts of money at this stage, because they typically own a significant number of shares of the stock. The basic principle is that the venture capitalists, who put up the initial funding for a speculative operation, are the ones who will either gain or suffer as the company progresses.

The first biotech corporation to offer shares of stock to the public was Genentech. On October 14, 1980, Genentech's stock was initially offered at $35 per share. Within twenty minutes, the share price soared to $89. Other public offerings by biotechnology companies have yielded similar gains. The offering by Cetus Corporation, which occurred shortly after that of Genentech, raised about $115 million for the corporation.

Why have some biotechnology companies performed so well in the stock market? The answer is the high expectations that were built up both by those involved in the industry and by outsiders. The promise of biotechnology had been instilled in many people, and investors were absolutely ready to make money from this new technology. The glut of positive market surveys certainly did not hurt the promise of the industry. The offering of Genentech stock was perhaps the single most anticipated stock offering in history.

Acquiring Expertise Ideas and money provide some of the necessary raw materials, but ultimately suitable scientific expertise must be available for any biotechnology venture to succeed. Where can such expertise be acquired?

As with most new technologies, the development of the basic processes of biotechnology occurs at academic institutions, where scientists are generally free to pursue activities without regard to market value and commercial potential. Restriction enzymes and the procedures used to clone and analyze DNA were developed in academic laboratories (Figure 15.4).

However, the growth of the private biotechnology sector has required that expertise be hired. Initially, all of the expertise came from academic institutions. Early on, experts were hired as consultants. Lucrative consulting arrangements were offered to prominent scientists in order to obtain their services. Competition for scientists became intense. Many academic scientists found that this competition put their positions in distress. Academic research flourishes through the free exchange of ideas, and this freedom and sharing allowed much of biotechnology (particularly genetic engineering) to develop rapidly. However, once contracted as a consultant to a biotechnology firm, an academic scientist is no longer able to freely discuss or share all information. Proprietary rights force the scientist to behave differently within the academic workplace. Some thought the trend of hiring scientists as consultants would spell disaster for science education in this country, but such concerns have not materialized.

Instead, the work force has adjusted. Enough qualified scientists are now available to occupy both traditional academic and commercial positions. At present, between the relatively slow growth of biotechnology firms and the rapidly increasing number of qualified scientists, there is actually a personnel surplus.

Risk and Reward

The potential reward for business investors serves to drive the commercial development of any new industry. Such rewards do not come without significant risks, however. For instance, the investment necessary to test, produce, and market new medical products is enormous. It is advantageous to allow corporations time to

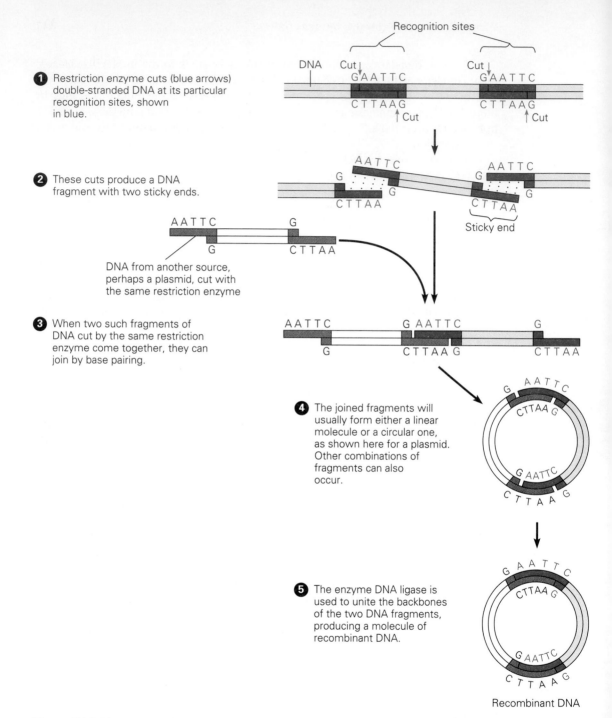

Figure 15.4 How restriction enzymes are used in constructing recombinant DNA molecules. The numbered steps depict the steps involved in cloning a gene in bacterial cells.

recover their costs when they introduce newly developed products into the marketplace. Without such assurances, companies would have far less incentive to invest in new products in the first place. These assurances are the underlying reason for the existence of patents.

PATENTS AND THE PROTECTION OF IDEAS

Millions of patents have been issued by the U.S. government, many of which protect products or processes that never materialized or reached commercial status. However, it is safe to say that very few inventions or technological breakthroughs have not enjoyed the protection of patents.

The products and processes of biotechnology are no different from any other technological innovation in this regard. Discovery and development are still risky, and recovery of research costs must be realized in the marketplace. Therefore, the patent system should operate to protect developments in biotechnology as it does other products and processes. However, biotechnology raises a number of issues that could never have been foreseen by the designers of the patent system.

Patenting Life

Many biotechnology patents have been awarded, a few of which have come to be seen as landmark awards because they set precedents. The first of these precedent-setting events occurred in 1971, when Ananda Chakrabarty, then a microbiologist working for General Electric, filed a patent application for a genetically altered bacterium. Chakrabarty was the first scientist to genetically alter bacteria to use crude oil as their energy source. His oil-eating bacteria were produced by transferring a number of plasmids known to encode genes responsible for the breakdown of oil products into a single strain of *Pseudomonas* (Figure 15.5). His work did not involve genetic engineering per se, since it was done before restriction enzymes had been discovered and so no new DNA molecules were created, but it did involve the movement of genetic material.

For almost ten years the patent office and the courts fought his application. The U.S. Patent and Trademark Office initially rejected the patent on the basis that living organisms were not patentable. This decision was appealed to the Court of Customs and Patent Appeals, which reversed the decision and held that Chakrabarty could be awarded a patent for a living organism. The court stated, "The fact that microorganisms . . . are alive . . . is without legal significance." The court further justified its decision by claiming that microorganisms are patentable because they are "more akin to inanimate chemical compositions such as reactants, reagents, and catalysts than they are to horses and honeybees or raspberries and roses."

The Patent and Trademark Office, still believing that living organisms should not be patented, appealed the decision to the U.S. Supreme Court. The Supreme Court instructed the Court of Customs and Patent Appeals, which had ruled in favor of the patent, to reexamine its decision in the light of *Parker vs. Flook*, a Supreme Court decision that stated there should be no expansion of patent coverage without a clear

Figure 15.5 *Pseudomonas* at work. *Pseudomonas* bacteria increased through nitrogen and phosphorus plant fertilizers were useful in cleaning Alaskan beaches following the *Exxon Valdez* oil spill. Chakrabarty filed a patent for a strain of *Pseudomonas.*

and certain signal from Congress. Clearly, at this time the Supreme Court did not favor the patenting of life without congressional direction to do so.

Ultimately, both the Court of Customs and Patent Appeals and the Patent and Trademark Office held to their original decisions, forcing the Supreme Court to issue a ruling. In what was a surprise for many, the Supreme Court upheld the decision of the Court of Customs and Patent Appeals and ruled that Chakrabarty could indeed receive a patent for his development of an oil-consuming bacterium. The microorganism was viewed not as a product of nature, but as a product of Chakrabarty's invention, making it patentable.

Aside from the ethical dilemmas presented by a decision that living organisms could be patented, the effect on the biotechnology industry cannot be underestimated. The awarding of a patent means clear ownership of the right to a product or process, establishing who will be allowed to make money from the commercial development of the idea. Since so much biotechnology involves the use of living organisms—as sources of raw material, hosts for cloning experiments, or eventual producers of genetically engineered products—the Chakrabarty decision became a landmark that allowed businesses the freedom to experiment. The promise of commercial freedom for biotechnology that this decision heralded is in part responsible for the dramatic performance of corporations offering public stock soon after.

But the Supreme Court intended the Chakrabarty decision to be very narrow. All the justices, whether for or against, agreed that the decision should not affect the future of scientific research and that congressional action would be necessary to further expand patent law to allow the routine patenting of living organisms. However, in 1988, the Patent and Trademark Office issued its first patent for a living animal, a transgenic mouse containing various oncogenes originally found in other species. Although the mouse in question was labeled "the Harvard mouse" because it was created by researchers at Harvard University, the actual patent rights were held by DuPont Company, the firm that financed the Harvard research. The patent awarded to DuPont was extremely broad, covering not only a specific transgenic mouse, but also almost any species of nonhuman mammal that contained recombinant oncogenes and their offspring. DuPont now markets this mouse as it markets chemicals, with the trade name OncoMouse. Other animals have since been patented, and hundreds of patent applications have been filed claiming rights to various genetically engineered fish, cows, hogs, and sheep (Figure 15.6).

The changes in patent rulings that led to the patenting of animals included some very specific exclusions. For example, human beings could not be patented. This exclusion was based on interpretation of the Thirteenth Amendment of the Constitution, which prohibits ownership of a human being. However, human cells and tissues, along with embryos and fetuses, were not specifically excluded from the rulings and therefore might still be patentable.

What were the effects of extending patents to animals? As we have stated, it provided a solid boost to the commercial development of biotechnology. Animals and organisms that serve as research tools can now be patented, allowing companies to recover costs associated with the development of new technologies. However, a more philosophical argument can also be made that patenting living organisms turns the entire living community of the earth into a preserve of potential research tools. President George Bush was severely criticized for not signing an international agreement on species protection, biodiversity, and biotechnology during the Earth Summit of 1992. Part of his reluctance to do so was based on pressure to allow U.S. firms to retain the ability to patent any species of the world that might prove beneficial to biotechnology, such as a particular plant in a tropical region that might contain an anticancer drug.

Figure 15.6 Genetic engineering in mice. The "supermouse" on the left, much larger than its littermate, is an example of a genetically engineered mouse.

Despite restrictions prohibiting the patenting of humans, numerous applications received by the U.S. Patent and Trademark Office and its European equivalents anticipate a change in policy. For example, it is possible to genetically engineer certain animals to make potentially valuable pharmaceuticals in their mammary glands. It is conceivable, then, that human females could be engineered to make a desired pharmaceutical in their mammary glands to achieve the fabrication of a wholly human material. Patents for such processes involving humans have already been filed—even though they are technically not yet feasible or allowed under current interpretations of the law. Researchers appear to be ready to take the technology in this direction when it does become possible.

All of this information concerning patents highlights two important ideas. First, having a system that rewards innovation and provides protection of ideas and products is an extremely potent stimulus for business. Without the potential for patents, biotechnology would not be where it is today. Second, somewhat straightforward concepts about ideas, products, and processes are not so straightforward when they are applied to biotechnology developments. In some ways, living organisms are like chemicals and therefore should be patentable. Yet never before have individuals or corporations attempted to be granted the right to control entire species of animals. Patenting living organisms is not as simple as owning a herd of cattle; it is more like owning all cattle.

Other Patents

Patents on living organisms have not been the only controversial patents being pursued. One of the most forward-thinking patent ideas was filed in 1974 by Stanley Cohen and Herbert Boyer (Figure 15.7) on behalf of Stanford University and the University of California. This patent application covered the use of restriction enzymes to insert foreign genes into plasmids. Cohen and Boyer, along with at least two other individuals, had published their experimental results and thought they should patent

Figure 15.7 Herbert Boyer. Herbert Boyer, one of the founders of Genentech and one of the first scientists to apply for a patent on genetic material.

the new technology. Following numerous difficulties and several reformulations, the patent was awarded to the two universities in 1980.

Stanford immediately set up a licensing arrangement. Companies were required to pay a $10,000 licensing fee and an annual fee of $10,000 for use of the techniques in research and development, but academic researchers were exempt from these obligations. Stanford also claimed royalties on the sale of all products resulting from this technology, initially set at 1 percent but falling to 0.5 percent for sales over $10 million. Within two years, the two universities had collected well over $2 million as a result of this patent, all of which became part of the schools' operating budgets.

One distressing aspect of this patent was that two scientists, and their respective universities, thought it appropriate to take decades of work on genetic engineering and claim exclusive rights for completion of the final stages. Not only was the entire field ignored, but individuals named on the initial publication were not allowed a part in the ownership rights. Many academicians were upset by what they viewed as an arrogant move on the part of two scientists and thought the situation might create an unworkable relationship between academic science and biotechnology businesses.

In 1991, biotechnology patenting went in a new direction when Craig Venter, a scientist working in the National Institutes of Health (NIH), filed a patent application on behalf of NIH to secure the right to individual human genes. His initial application covered 337 genes found within the human brain. He quickly followed with another patent application covering another 2,000 genes. If awarded, the two patents together would have given him and NIH rights to over 5 percent of human genetic material. For the first time, a request was made to give legal rights of ownership to naturally occurring human genetic material. These patents were rejected, as have been all others involving naturally occurring human genes. Again, many in the scientific community thought Venter and NIH were presumptuous in thinking they had rights to natural genes. However, a number of scientists thought genes ought to be patentable. Many individuals involved in the Human Genome Project believe that, ultimately, a corporation should be formed as part of the project in order to oversee the patenting of the enormous number of genes that will be described. Even though no known function will be associated with most of these genes in the foreseeable future, they reason that if even a single gene out of this group proves to be important commercially, the profits to be earned by its exploitation will be worth patenting it now. These patent applications attempt to control the rights to naturally occurring genetic material, not living organisms. A subset of the human body is being claimed.

Other subsets of human tissue have also been the subject of patent battles. The most famous case involved John Moore, a patient diagnosed with a rare form of cancer known as hairy-cell leukemia. In Chapter 14, we noted that cancer involves the altered regulation of a large number of cells. Sometimes, the altered cells produce biochemicals that are desirable and difficult to produce in other ways. Thus, it is fairly common practice to remove tissue, particularly from the immune system, of patients with cancer to try to grow cells from this tissue in the laboratory. Rarely, **cell lines**—cells that will live under laboratory conditions—can be established. These cell lines sometimes prove extremely valuable for testing and research or as sources of scarce materials.

When John Moore's diagnosis was confirmed, his spleen had enlarged to such a degree that it was considered wise to remove it. Following removal of his spleen, portions of it were sent to labs in order to try to produce cell lines. One of those established, known as Mo cell line, was patented by the doctors and researchers involved and their associated universities in 1984. John Moore was not a part of the patent. The patent covered both the cell line itself and nine products that could be derived from it. Even before the awarding of the patent, the universities and individuals involved had received significant payments from biotechnology firms in anticipation of the licensing fees that would arise from the issuance of a patent.

John Moore immediately filed suit to win the right to share in the profits derived from his own tissue. Initially, courts rejected the claim that he had any right to part ownership in his own tissue, but upon appeal the decision was modified. John Moore was given no right to ownership of the cell line derived from his tissue. The court did allow, however, that the doctors and holders of the patents had not fulfilled their obligations since they had not informed Moore of the potential reward for their work. Eventually, Moore was allowed to share in the profits.

From the commercial standpoint, such a ruling was a blessing, since it set the precedent that individuals had no legal rights to ownership of material derived from their bodies. In other words, parts of human bodies and cells derived from them could be owned wholly by corporate interests, to be bought and sold like any other commodity. From the individual's standpoint, while it did not support direct property rights, the ruling at least ensured some degree of financial reward.

In 1991, another step was taken in the patenting of human tissue. Systemix, Inc., was granted a patent covering a method of isolating very pure stem cells. These cells are the progenitors of nearly all blood cells. They serve as research tools because they retain the ability to divide and differentiate into many different types of cells. Not only did the patent cover the method of isolation, but it also gave Systemix, Inc., the rights to the stem cells themselves, as the product. This was the first instance of any normal, nonengineered human cell or tissue being patented.

Although this patent is generally viewed as suspect and court battles over it continue, the stage has been set for significant corporate ownership of various parts of the human body. The examples we have cited, including cell lines, tissues, and even genes, confirm that tremendous scientific resources are available within the human body and other living organisms. As these scientific resources are converted into commercial products and processes, ownership and financial interest must be properly served. But how are we to do this? Certainly, the originators of the patent process could never have foreseen such difficulties, and, as usual, our ability to deal with the effects of new technology on society lags behind our ability to develop the technology itself.

BIOTECHNOLOGY IN THE UNITED STATES: A STATUS REPORT

Just two decades following the public launching of Genentech, over 1,300 biotechnology companies in the United States now employ more than 150,000 people and earn about $13 billion annually. Consider the following governmental estimates of the amount of money that will be spent on different areas of biotechnology by the year 2000:

Energy	$15,392,000,000
Foods	$11,912,000,000
Chemicals	$9,936,000,000
Pharmaceuticals	$8,544,000,000
Agriculture	$8,048,000,000
Metal recovery	$4,304,000,000
Pollution control	$96,000,000
Total	$58,232,000,000

As we reflect on the total global impact and consider that areas of biotechnology have been growing rapidly for only two decades, we should not be surprised to learn that many interested groups have joined together in associations. For example, the Biotechnology Industry Organization (BIO) represents 600 companies, academic institutions, and state biotechnology centers plus other organizations in forty-seven states and more than twenty countries. Member organizations are engaged in research and development of biotechnological products related to health care, agriculture, and the environment. The diversity of areas presently being fueled by biotechnology products and techniques is astonishing. However, the areas of health care, agriculture, and environmental biotechnology are clearly in the lead.

Biotechnology in Health Care

In the health care area alone, an estimated 60 million people have been helped by the more than forty biotech drugs and vaccines approved by the U.S. Food and Drug Administration (FDA). Biotech medicines are presently being used to treat anemia, cystic fibrosis, diabetes, dwarfism, multiple sclerosis, hemophilia, leukemia, heart attacks, hepatitis, genital warts, and tissue or organ transplant rejection. A biotech-derived vaccine for hepatitis B has been used effectively for several years. Because this vaccine is produced by modified yeast cells, it is entirely safe. A number of other disease-preventing vaccines are being researched, including vaccines for influenza, AIDS, Rocky Mountain spotted fever, cholera, and several severe types of diarrheal disease. More than 270 biotech drugs are undergoing clinical trials or are in development for treatment of various cancers, Alzheimer's disease, obesity, and other conditions. The president of BIO noted in 1995 that a drug that would simply delay by five years the average age at which Alzheimer's disease strikes would save $50 billion annually; the same delay in the onset of Parkinson's disease would save $3 billion, and delaying the onset of cardiovascular disease and strokes by five years would save $67 billion. It is easy to see that economics could be a substantial driving force in health care industries nationwide.

Another aspect of health care is diagnostic testing. Emphasis is on accuracy and swift delivery of results. Biotechnology is used in hundreds of routine medical diagnostic tests, including several home pregnancy tests.

Two interrelated issues facing pharmaceutical biotechnology today are the lengthy time required for development and approval (nearly seven years for biotech drugs) and the high cost of development—$125–$359 million per drug. Given these figures, we need to answer two questions: Why does the process of research, testing, and approval take so long? and Why is it so expensive?

One problem is that many genetically engineered medicines are proteins. After a potentially valuable candidate has been identified, it may take years to learn how it works. Then the likelihood of commercialization has to be determined.

Also, in order to study any new drug, researchers need large quantities of the pure compound. This is where genetic engineers exercise their skills—by isolating the specific gene (and its expression-controlling site) that tells cells how to make the product. The isolated genetic information is spliced into the DNA of a cell or microbe, and then, if everything works properly, the new drug is produced for extensive testing.

Availability of larger amounts of the new drug allows preclinical testing. Laboratory studies and animal trials are used to determine whether there is sufficient biological activity to warrant further testing and what adverse side effects are likely when the drug is finally administered to people. If we assume a high degree of biological activity and minimal adverse side effects, the process toward commercialization enters a three-stage clinical evaluation. The last stage may require thousands of patients and is extremely expensive. Only after the third phase has been completed and all the test results have been compiled is a product license application submitted to the FDA.

The FDA has six months to act on the product license application. It may approve the product, approve the product on condition that further clinical trials be accomplished, or disapprove. The average time required for the approval process is generally about one year.

Biotechnology in Agriculture

On the agricultural front, biotech efforts have already provided consumers with more flavorful vegetables, such as carrots and peppers, and tomatoes that actually taste vine-ripened. In addition, the extensive dependence on chemical herbicides and pesticides has been somewhat reduced with the adoption of **biopesticides** by backyard gardeners and commercial growers alike. There are eleven biopesticides in use today that are toxic only to targeted pests and do not harm nontargeted organisms. For instance, we have had excellent results with the insect-specific toxic proteins produced by the bacterium *Bacillus thuringiensis*. A logical next step would be to introduce the bacterial toxin gene into crop plants so that, if expressed, the toxin would kill any insects that began to feed on the plant. This would surely lessen our dependence on chemical insecticides.

One of the more controversial biotech products released in the recent past was the **bovine growth hormone** (**BGH**), or bovine somatotropin. Although BGH injections were shown to increase both growth rate and milk production in cattle, a storm of protest arose from those who pointed out that the United States already had an abundant milk supply and further increases would more than likely lead to reduced payments to dairy operators. This economic difficulty was coupled with an indication that increasing BGH levels tended to shorten a cow's life and also made individual animals more susceptible to injury.

In the near future, in addition to nearly seed-free green peppers, tastier and fresher tomatoes, and whole baby carrots, we will likely see longer-lasting strawberries, raspberries, broccoli, and cauliflower; sweeter snap peas; and potatoes that absorb less cooking oil. In addition, due to biotechnology, many fruits and vegetables will be insect- and virus-resistant. The FDA has recently approved squash with natural resistance to viruses, cotton plants that require less chemical herbicide, and corn with engineered resistance to the devastating European corn borer. Many more advances are on the horizon.

Biotechnology may well have a greater global impact on agriculture than the so-called "green revolution" of a few years ago. Large areas of many developing nations are not prime agricultural land, and weather patterns are often unfavorable for crop production. Biotechnologists can alter the genetic makeup of many plants to allow them to grow more abundantly whether wet or dry, in intense heat or chilling cold.

Biotechnology and the Environment

Environmental biotechnology includes the use of microbes for pollution control and for preventing or limiting environmental damage. Genetically modified microbes can more efficiently clean up hazardous wastes and reduce our dependence on methods such as incineration or hazardous waste dumps. How does this process work? Microorganisms take in nutrients and produce wastes. Different microbes "feed" on a wide variety of materials—some even consume toxic materials such as PCBs and methylene chloride. Cleaning up these materials and a host of other hazardous chemicals has been approached in two different ways: by adding nutrients to the sites to be cleaned such that the naturally occurring microbes are stimulated to grow (**biostimulation**), or by adding living microbes (**bioaugmentation**) that "eat" the hazardous wastes and render them harmless (**biodestruction**). Two other types of bioremediation are noteworthy: **phytoremediation**, or the use of plants or fungi to take up metals or waste materials, and **biofiltration**, or using microbes to remove complex pollutants from manufacturing or sewage discharges.

SUMMARY

Modern biotechnology has been developing commercially for about twenty years. As in the early 1980s, the business is still risky for those on the cutting edge of research as well as for the venture capitalists and stock purchasers who provide the financial backing. Ideas spring from creative minds, and these ideas need to be researched and developed. Extensive market research is ideally accomplished before production is initiated. Also, critical testing and regulatory agency approval must be received prior to commercial production and marketing. Persons or corporations engaged in the research and development of biotechnological products are especially challenged, because the risk is unusually great and the ability to forecast is hindered by the lack of historical precedents. In the United States, our patent system operates protectively in biotechnology. However, many previously unheard of issues have arisen, perhaps the most complex of which concerns the patenting of living organisms.

As we have seen, practical applications of biotechnology abound. They extend from such widely separated fields as energy, food, and pharmaceuticals to controlling environmental pollution. Alcohol as a renewable fuel, tomatoes that hold the peak of ripeness longer, human insulin made by bacteria, and engineered microbes that clean up spilled crude oil—all express the breadth of modern biotechnology.

REVIEW QUESTIONS

1. Why was the 1980 Chakrabarty decision of the U.S. Supreme Court especially significant for biotechnology?

2. Explain how the U.S. patent system serves to reward scientific research and development.

3. Biotechnological commercialization typically involves five basic steps. List them, in order from the first.

4. Provide examples of the general questions that market research attempts to answer.

5. What questions would a sharp, wealthy venture capitalist pose for scientists presenting a new biotechnological product for market development?

6. What conflicts did academic scientists face when they became active participants in commercial biotechnology?

7. Provide outlines of arguments for and against the patenting of living organisms.

8. Explain how the following are interrelated: hairy-cell leukemia, Mo cells, spleen, John Moore, financial reward, patents of cell lines and products, and ownership.

9. What arguments are made in support of experiments that might lead to preventing the onset of serious genetic disorders? To whom would these "money saving" arguments be made?

10. Describe several recent agricultural advances due to biotechnological methods and experimentation.

11. Define the following with respect to environmental biotechnology: biostimulation, bioaugmentation, biodestruction, phytoremediation, and biofiltration.

12. Describe the two sides of the BGH, or bovine somatotropin, controversy. Should the program be considered a success or a failure? Is it likely to be repeated?

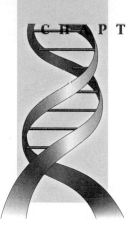

Biotechnology in the Developing World

Biotechnology has become an incredible technology not only in the United States and other industrialized nations but in many developing countries as well. Some countries have made biotechnology a prime national interest—Malaysia, for example, has touted biotechnology as one of the three developing technologies that will lead the nation into a new economic era.

Why is biotechnology of such intense interest to developing countries? Surely, the answer must lie in economic concerns. Since its inception, biotechnology has been projected to bring fantastic advances along with huge economic rewards. For example, one of the greatest advertisements for biotechnology is that it will revolutionize agriculture by allowing fewer crops to be used to provide more raw materials and better-quality food. Also projected are improved plant strains that will be resistant to pests and herbicides, thereby increasing yields. In short, biotechnology promises more for less: larger yields from limited resources. Since developing countries tend to have growing populations but scarce resources, such claims naturally appeal to those in power.

But national economic gain may not be the primary, or even a significant, interest. Developing countries recognize their need to participate in the global economy, which generally means the production of exportable goods. If biotechnology could be applied to some unique national resource, opportunities would arise for the development of new, valuable goods. Countries fostering biotechnology also probably believe that it will allow them to get in on the ground floor of the next technical revolution. Just as some countries developed strong capabilities in the electronics industry as it was expanding, other countries are now trying to capitalize on the prospect that biotechnology may become equally weighty in the global economy.

Are these motives appropriate? How would biotechnology benefit the developing world? How can biotechnology become a part of a national scientific endeavor? What is the relationship between countries with strong, active commercial biotechnology, such as the United States, and smaller, poorer countries trying to find or create a niche in world markets? We will explore these questions in this chapter, not

353

necessarily with the goal of developing exact or absolute answers but as a way of becoming informed as to how biotechnology may be applied in several political, economic, and social environments.

BLUEPRINTS FOR SUCCESS

At the outset, understand that the goals of different countries are likely to be quite different. One country might be most interested in biotechnology as a way of fostering overall improvement in medical services, enhancing agriculture, and raising the standard of living. Another country might be particularly interested in developing products based on unique natural resources, probably for export rather than internal consumption. Still other countries, including the most industrialized nations, might be interested in developing biotechnology in third-world countries as a means of gaining access to their raw materials or to sources of relatively cheap labor. Regardless of the national or corporate goal or from which standpoint we view the issue, biotechnology has much to offer. The question is, Will it provide necessary economic relief, or will it simply exacerbate existing inequalities?

The answer to this question depends on many factors. How much research and development already occur within the country? An established research community is a likely key to the success of new scientific ventures. Imagine how difficult it would be to write papers for classes without access to computers and libraries. Conducting scientific research in some developing countries can suffer from similar constraints. The better established research activities are, the better the chances for the rapid success of any new venture. Other key questions involve markets. For instance, can regional, national, or international markets for particular biotechnological products be developed within the country? The people of most developing countries would benefit from increased food supplies, but perhaps they would not benefit as much from improvements in mining technologies. On the other hand, an exploitable export market may influence the decision to pursue a particular biotechnological program.

The question of available money is critical. How much will a particular investment in biotechnology cost? Are sources of investment capital available? Will there be sufficient funding to carry a particular research and development project through to fruition? Would the project's failure be devastating? The United States and its corporations can readily absorb the failure of numerous biotechnology efforts simply because of the availability of both expertise and capital investments for speculative business ventures. However, most developing countries are not able to pursue many programs at once and must be extremely careful when they select products for potential commercialization.

One highly critical consideration is whether a particular nation has a unique and potentially valuable natural resource. Many developing countries, particularly those in the tropics, harbor incredibly diverse plant and animal life. Investigations of such natural treasures are presently being pursued, often in search of potent new drugs. Biotechnology increases the number of possibilities and the rate at which they can be explored.

The success of biotechnology in developing countries is chiefly a matter of careful choice and logical implementation. The great advantage of biotechnology is that it can be applied to many different economic sectors at many levels. The choice might be between developing a multimillion-dollar-per-year recombinant DNA research laboratory and building a much less expensive production facility for high-protein animal feed from microorganisms.

We cannot consider the total effect of biotechnology development without considering the unique and varied economic and social changes that must accompany it. Recall the example of Malaysia. We must also consider how biotechnology will affect the population of Malaysia, one of the world's largest producers of natural rubber from trees. If rubber trees are genetically engineered to be much more productive, far fewer trees would be required unless demand increased comparably. The rising unemployment that would, in all likelihood, follow might be viewed as unacceptable or perhaps even threatening to the well-being of the country, as well as to individuals. On a more positive note, plantations of vastly more productive genetically engineered rubber trees might free a large segment of the population to seek new and perhaps more financially rewarding endeavors.

There are no easy answers—simply great possibilities and potential disasters. Our strategy in this chapter is to provide a brief introduction to the kinds of biotechnology that are most applicable in less wealthy countries. We will primarily consider two products: ethanol in Brazil and palm kernel oil in Malaysia. These cases highlight a number of issues that can be extended to other situations, although each is unique in its own right. What succeeds in one location at one time under a particular set of circumstances may not necessarily be applicable to other situations.

THE PROMISES AND POTENTIAL
OF BIOTECHNOLOGY

To begin, let's take a larger view of biotechnology and examine some applications that may be better suited to developing countries. This is by no means a complete survey but rather a representative sample of the kinds of tools and techniques that might be applied in a variety of circumstances.

Agriculture

Tremendous success has been achieved in the years since World War II through the development of higher-yield crops. Particularly for wheat and rice, the improvements in yield have been dramatic and have helped feed an ever-growing world population. The first wave of such high-yield crops came about through traditional breeding practices—using Mendelian genetics to identify and select strains of crops that exhibit particularly valuable traits. However, there are inherent disadvantages in the use of such high-yield strains. They generally require more fertilizer, more pesticide and herbicide treatments, and greater irrigation—all of which are costly. In addition, as we pointed out in Chapter 6, the intensive use of highly inbred strains results in a lack of genetic diversity, which can have dire consequences.

Biotechnology is currently catalyzing another green revolution by allowing

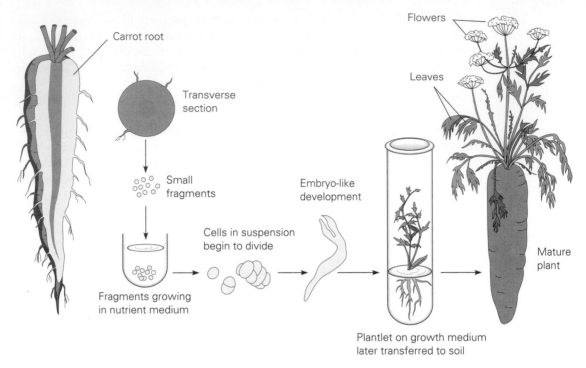

Figure 16.1 Plant tissue culture. A classic example of plant tissue culture: cloning carrots. Small carrot fragments are placed into a nutrient medium. Cells divide and, with proper hormone treatments, will produce new mature plants, all of which are genetically identical to the original plant.

crops to be identified, selected, and even created that are more resistant to pests and drought and, perhaps at the same time, require less fertilizer. Obviously, the introduction of such modified crops would enable more people to be fed using fewer resources, a goal of considerable interest in most developing nations. On the other hand, new and improved crops usually have indirect costs associated with them. Although it is impossible to know exactly what costs might be associated with bioengineered crops, as in traditional breeding there will undoubtedly be some.

We can use biotechnology not only to create crops with extremely valuable traits, but also to develop tools and techniques of great benefit agriculturally. One of the more beneficial techniques is **tissue culture** (Figure 16.1). Through the controlled use of plant hormones, it is possible to regenerate whole plants from individual cells that are maintained in culture. Often, it is easier to maintain and culture plant tissues than to raise mature plants from seed. The ability to generate whole plants on command makes a theoretically endless supply of plant material available, of the best possible strains. By using tissue culture techniques, for example, it is possible to isolate and maintain strains of crops that are virus-free. Many potato plants used in the world today originated in this manner, as did a great number of ornamental flowers. Because tissue culture is an inexpensive form of biotechnology that does not require extensive equipment or training, it is one of the more common forms of biotechnology found in developing nations. A number of economically important products are propagated, at least in part, through tissue culture, including potatoes, palm oil, coconuts, coffee, and sugar cane.

Implementation of a newly successful agricultural practice generally means that others (farmers, counties, states, or countries) will follow suit. Thus, a particular crop strain that performs well will be used by more and more farmers, a practice that tends to quickly limit the number of varieties of that particular crop used overall. Astonishingly, of the 50,000 edible plant species on earth, just 29 account for more than 90 percent of the caloric value of foods produced. Unfortunately, the success of these particular crops means less exploration of other potential crops, although many more food crops could certainly be used.

Food Products

A number of biotechniques are used in producing food. Some are ancient practices, such as fermentation. Others, like the production of highly nutritive foods from microbes, are relatively new. But whatever the process, new and better ways are available for preparing food for consumption.

Fermentation is a process by which microorganisms alter the relative nutritional qualities of foods. Everyone is familiar with brewing, in which carbohydrates derived from a variety of agricultural products are subjected to fermentation (usually by yeast) in order to produce alcohol. But many other forms of fermentation are in use, some of which are at least as economically important as producing alcoholic beverages.

Soy sauce has been produced by microbial fermentation of soybeans for hundreds of years. In 1999 the Indiana Soybean Board indicated on their web pages that there are three major commercial types of soy sauce: *shoyu, tamari,* and *teriyaki* sauce. Shoyu and teriyaki sauces are blends of fermented soybeans and wheat whereas tamari is made only from soybeans. These sauces are consistent additives in many Far Eastern diets. There are many other fermented soybean foods. For instance, *tempeh,* a traditional Indonesian food, is a chunky soybean cake with a smoky or nutty flavor. *Tofu,* or soybean curd, is a cheese-like food made from curdled soymilk. Tofu easily absorbs the flavors of other ingredients with which it is cooked and serves as a source of high-quality protein and B-vitamins. *Miso* is a rich, salty condiment that is the essence of Japanese cooking. The Japanese make miso soup and use it as a flavoring. *Natto* is made of fermented, cooked whole soybeans and has a sticky, viscous coating with a cheesy texture. In many Asian countries natto is served as a topping for rice, in miso soups and with vegetables. As these foods are dietary staples, improvements in fermentation technology, perhaps the oldest form of biotechnology, are being actively explored by developing nations, particularly in the Far East.

A related technology is the large-scale culture of microorganisms for use as a direct source of high-protein food known as **single-cell protein** (SCP; Figure 16.2). The

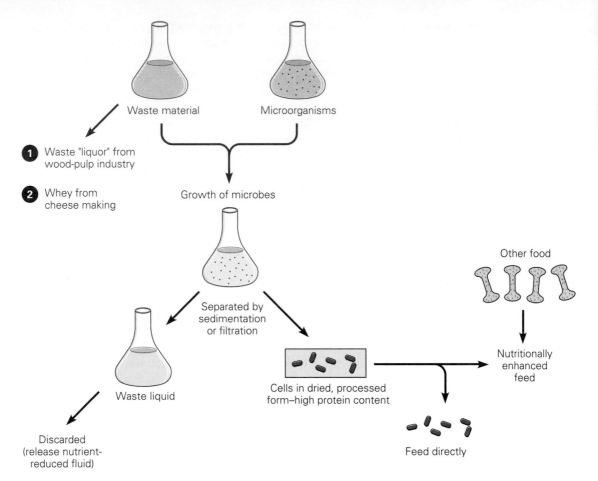

Figure 16.2 Single-cell protein. Microbes grown on organic waste fluids can be recovered and either used directly as food or added to other feeds for nutritional enhancement. In cases where waste pulp liquor or whey serves as the microbial nutrient, the fluid may be returned to rivers or streams with little further treatment after SCP has been produced and separated from the fluid, because the nutrients have been reduced and will not stimulate microbial (algal) growth (an algal bloom).

technology leading to SCP is relatively simple and can be combined with fermentative processes to produce a number of substances. SCP was considered as a short-term solution to the shortage of food in Germany during World War I. However, only in the 1960s were real efforts made to investigate SCP as potential human food.

At that time, interest was focused on growing microorganisms on natural gas and its by-products. Although such materials are useless as food for complex organisms, they can be utilized by microbes as a source of carbon and energy. The result is that the gas products are turned into microbial proteins. The recovery of the cells following fermentation allows for production of a high-protein solid material.

SCP has not yet been an economic success, for a variety of reasons. First, throughout the 1970s, the much-anticipated food shortages never materialized; in fact, there were often huge surpluses of grains. Second, oil embargoes and other adjustments in the flow of hydrocarbons meant that, at times, fewer resources were available at low cost for the production of SCP. The SCP that has been produced has been used primarily in animal feeds, usually mixed with grains and other by-products. Despite limited economic success, however, the technology is available and is neither complex nor costly. An obvious use of such technology is in processing by-products (think waste products) from other agricultural activities. For example, in Cuba huge quantities of fibrous material remain following sugar-cane processing. Using these waste fibers as a growth medium for microorganisms could transform them into usable feed.

Livestock Breeding and Animal Health

Animals, like plants, play a critical role in all cultures. Livestock has traditionally been subjected to selective breeding practices similar to those used for plants, although the time periods involved may be considerably longer. Ways to more effectively and efficiently select new animal breeds are constantly being sought, as are ways to make them grow more rapidly. Also, biotechnological advances in human health care have, in some cases, been applied to veterinary care.

Embryo Transfer The most important biotechnology in animal husbandry is the ability to remove and manipulate cattle embryos (Figure 16.3). Such embryo transfer techniques are now relatively common. A female is first induced to superovulate, producing many more eggs than normal. Following artificial insemination, the many fertilized embryos are washed from the uterus of the female and collected. They are frozen in liquid nitrogen ($-179°C$), where they can be maintained indefinitely. Since it is not possible to induce animal cells to regenerate back into an animal (as with plants), this is the closest we can come to having animals readily available.

Not only can embryo transfer be used to produce many offspring (instead of a few) from a particular male and female that exhibit desirable traits, but the frozen embryos can be readily transported at lower cost than can similar numbers of animals. Also, transporting embryos to a new location and then implanting them into receptive females offers the newly born calves many advantages over calves transported directly from one location to another. First, they are not subjected to the stress of relocation or to environmental differences that may exist in the two locations. Second, since each calf develops in and is reared by a mother at the new location, it gains the benefits of that mother's immune system: Antibodies to prevalent local disease agents will be passed from mother to calf, resulting in healthier offspring.

Recent technical developments allow the sex of embryos to be determined (Figure 16.4), allowing the selection of females for milk production and males for beef production. The technique employs PCR and hybridization. At an early stage of embryo development (eight cells or more), one of the cells is removed. This removal does not alter the development of the embryo. The DNA of the isolated cell is subjected to PCR, using primers designed to amplify a section of the Y chromosome

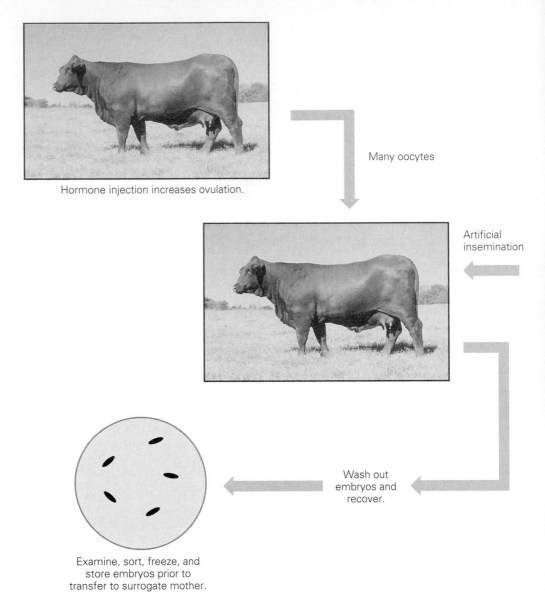

Hormone injection increases ovulation.

Many oocytes

Artificial insemination

Wash out embryos and recover.

Examine, sort, freeze, and store embryos prior to transfer to surrogate mother.

Figure 16.3 Multiple embryo propagation. Multiple eggs are released at the same time in a cow through the action of specific injected hormones. After being artificially inseminated, the resulting embryos are washed out of the mother's uterus, collected, examined, sorted, and frozen. They may be stored indefinitely. When desired, an embryo is thawed and implanted in a surrogate mother.

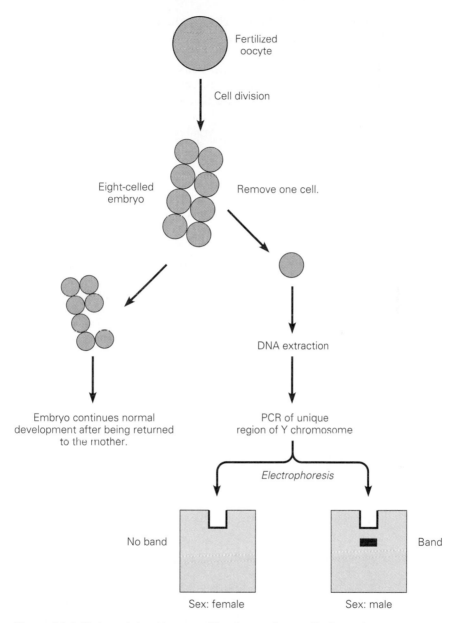

Figure 16.4 Determining the sex of bovine embryos. Bovine embryos are grown for a short time in culture. Typically, at the eight-cell stage, one of the cells is removed by using a micromanipulator. The remainder of the embryo continues growth and normal development when it is returned to the womb. DNA is extracted from the removed cell, and a unique region of the Y chromosome is subjected to PCR amplification. If the region is present, as indicated by the presence of a band on the resulting gel, the embryo is male. If the region is not present, the embryo is female.

(which is specific for males). Hybridization then allows the determination of whether the desired amplification has occurred. If hybridization does occur, the embryo is male; if it does not, the embryo is female. Sexed embryos are of much greater economic value than unsexed ones. Dairy farmers can ensure that their cows are carrying female calves, and beef producers will know positively and far in advance of the birth that their selected and emplanted embryos will be males.

Altered Growth of Animals A number of developments related to increased growth of animals (Figure 16.5) have made the news. The most controversial involve the use of bovine (cow) growth hormones to stimulate increased milk production as well as overall size (Chapter 15). The idea is relatively simple conceptually—as we have learned more about animal physiology, it has become possible to administer extra doses of naturally occurring hormones in order to achieve larger, more productive animals. Such techniques have been applied most often to cattle, but many other animals are also being tested (notably sheep, chickens, hogs, and some varieties of fish).

Commercially, the question is whether such treatment achieves any economic benefit. Cows producing more milk might eat more, and the costs of the extra feed and the hormone treatments might offset the profits from the extra milk. However, tests have consistently shown that animals treated with growth hormones do not require more food to yield more product, and in at least some cases the treated animals show an increased nutritional content. For example, pigs treated with growth hormone grow larger and have higher protein and lower fat content.

The required hormones are, for the most part, produced through the standard recombinant DNA technology we described in Chapters 8 and 9. Production of animal growth hormones from bacterial cells is one of the best examples of the early commercial success of biotechnology companies.

Treatment of Animal Diseases Many diseases threaten large numbers of animals each year. Of these, hoof-and-mouth disease is the most prevalent. It is caused by a virus that can infect many different species of hoofed animals, including cattle, hogs, and sheep. It is highly contagious, and, while it is seldom lethal to an adult animal, infected animals become unhealthy and malnourished and are usually slaughtered to

Figure 16.5 Effect of growth hormones. The smaller mouse on the right is of normal size. The abnormally large mouse on the left is a littermate that has incorporated the gene for the human growth hormone.

prevent its spread. Several vaccines have helped eradicate the disease from some countries. However, in less developed countries, where animal husbandry is conducted in villages and there is often no national program of immunization, the disease remains a significant cause of economic hardship.

Battling a disease in animals is no different in principle from battling a human disease. Improvements will come with refinements in the vaccines and in improved methods of distribution. The latter is a distinctly political and economic, rather than scientific, problem.

The poultry disease of most concern in the world is Newcastle disease, which is caused by a virus and is most prevalent in Asian countries. It is extremely contagious and can infect a variety of birds, with domestic chickens particularly susceptible. In commercial poultry production facilities, it is often possible to prevent outbreaks of the disease by slaughtering whole populations that contain infected birds and then sterilizing the facilities. But in villages these options are usually not viable, and vaccination provides the best possible defense. The challenge is to develop a vaccine that is easily administered by villagers. A possible solution to this problem came in the mid-1980s with development of an oral vaccine that could simply be sprayed on chicken feed. Tests showed that a high percentage of birds became resistant to infection, and there appear to be no side effects. The development of easy, inexpensive methods of vaccination are crucial to the prevention of animal diseases in rural regions of the world.

Medicine and Public Health

We can give no better example of the challenges presented by technological advances to developing nations than those involved in medical care. As is clearly the case in the United States, the cost of medical care is rising at a tremendous rate, with no end in sight. At least part of the reason for this cost escalation is the use of very high tech and innovative tools for the diagnosis and treatment of disease. It is difficult to imagine exporting the U.S. medical care system elsewhere in the world—not only would it be incredibly expensive, but some nations might not want such a system.

So the questions become, for the most part, economic. How much of the medical treatment under development in the United States and other industrialized nations can be exported to the rest of the world? How much will it cost? Nations must choose their investment in medical care wisely. Even in the United States, with its abundant riches, deciding how to allocate money for medical care and research among the various ills afflicting the population is not an easy task. Imagine how difficult it must be to face even more oppressive medical demands with far fewer resources.

Another difficulty faced by developing nations was introduced in Chapter 11: Those companies in the best position to research, develop, and market vaccines and other therapeutic agents are located and serve a clientele in industrially developed countries. The economic incentive to develop medical care applicable to the developing world is largely nonexistent—intervention and support by the World Health Organization and other organizations is required. Also, the ailments that afflict people in developing countries are not always the same as those facing industrialized nations.

On the other hand, much of what we have learned about some diseases and health matters can be easily applied in developing countries. For example, the incidence of cancer due to smoking is declining in most wealthy populations, presumably due to decreased smoking following education efforts. Smoking is increasing, however, in developing countries, as is a preference for the high-fat, low-fiber diets that are tightly linked to the incidence of a number of diseases. Opportunities seem ripe for the transfer of some of our educational experiences to other countries.

Besides introducing medical improvements made in the developed world to less developed nations, we can also directly implement some improvements in medical care and advances in biotechnology in developing nations. In many cases, the premier research institutes that focus on particular diseases are located where the diseases are highly prevalent. Much malarial research is performed in Southeast Asia, for example. With well-established research institutes in existence, it might be simple for a developing country to import the necessary expertise and make the required technological investment. The same argument can be made with regard to the production of at least some pharmaceuticals, particularly those derived from plants. Having the raw materials close to new biotechnological production facilities within a developing country provides both jobs and scientific expertise, helping to improve the economic situation through biotechnology.

Energy

The availability of energy reserves is of such vital importance to technology today that it is no wonder that countries with extensive energy resources are typically wealthy. The corollary to this fact is that countries without energy reserves often do not have the financial resources to purchase energy. This catch-22 situation perpetuates itself: Without money for energy, technological advance is limited; without modern technology, it is usually impossible to accumulate financial resources.

The developing world is caught in this dilemma. Thus, it is not surprising to find many countries actively involved in exploring ways to use alternative fuels, generally those based on abundant natural resources. Thus, biotechnology becomes a factor in securing energy. Later in this chapter, we will explore an energy program based on basic biotechnology: the National Alcohol Program in Brazil.

Bioconversion and Recycling of Materials

As in families, less wealth usually means a greater effort to fully use all available materials, and developing countries have often been leaders in using waste materials and recycling leftovers. Biotechnology, generally through the chemical modification of materials via fermentation, has provided the means to use waste materials effectively. For example, much fibrous plant material is left over following the pressing of sugar from sugar cane. This material was simply discarded in the past, but fermentation now provides the means to convert it to a number of different products, including alcohols and other industrial chemicals. Improvements in paper production technology have allowed similar uses of materials previously considered wastes. It is now possible to use many parts of plants that were once grown and harvested for a single purpose. Current technology allows harvesters to use the entire oil palm plant, rather than simply isolating oil from the seed kernel. Such improvements are inexpensive ways of improving the economics of crop production.

APPLICATION: ETHANOL PRODUCTION IN BRAZIL AND THE UNITED STATES

By some measures, the National Alcohol Program (*Proalcool*), established by the federal government of Brazil in 1975, is an example of what can be achieved through the directed application and financial subsidy of technology. As a developing country, Brazil has a long history of rather serious national debt. Anything that could lower the trade imbalance would help reduce this national debt. A large part of Brazil's import costs go toward oil and other fuels. Research performed during the 1950s and 1960s suggested a way of producing easily renewable fuel and at the same time reducing the dependence on foreign suppliers—promotion of the production and utilization of ethanol, or agricultural alcohol, as a fuel. Originally, the fuel alcohol industry was a small offshoot of the world's largest sugar producer. But in 1974, when crude oil prices drastically increased, Brazil's National Alcohol Program was launched.

Agricultural alcohol (ethanol) can be produced from the microbial fermentation of a wide variety of plant materials (Figure 16.6). In Brazil, the most abundant material was and still is sugar cane, grown and harvested annually to produce sugar. Much of the sugar cane plant is disposed of as waste. However, these leftovers from sugar production can be fermented to produce ethanol.

The ethanol produced in Brazil (1996 production of 14.5 billion liters, or about 46 percent of the global total ethanol production) goes primarily toward fueling private automobiles and commercial vehicles. To understand the past, present, and future of the Proalcool program requires distinguishing between anhydrous and hydrous ethanol. **Anhydrous** (without water) ethanol is used as a gasoline additive. Presently, in Brazil, 22 percent anhydrous ethanol is mixed with gasoline, and the mixture can be used in any type of automobile, truck, or bus engine without modification. On the other hand, **hydrous** alcohol, containing about 5 percent water, can be used in its pure state (without being mixed with gasoline), but only in specially designed engines. While ethanol is not the most efficient fuel, mixed with gasoline it provides a viable alternative, effectively extending the gasoline. In Brazil, the original plans were to produce enough ethanol to reduce dependence on petroleum-based fuels by at least half.

Such a country-wide transformation requires extensive effort. Not only would ethanol-producing facilities need to be built, but the appropriate amount of sugar cane would need to be planted, cultivated, and harvested. In addition, systems would need to be developed for the distribution of the ethanol, and vehicles would have to be modified in order to use ethanol more efficiently as fuel. Perhaps most important, the population would need to be educated about the advantages of the program and would have to accept ethanol-gasoline as a viable alternative fuel. Complete independence from foreign-derived gasoline would require the use of hydrous ethanol–powered vehicles.

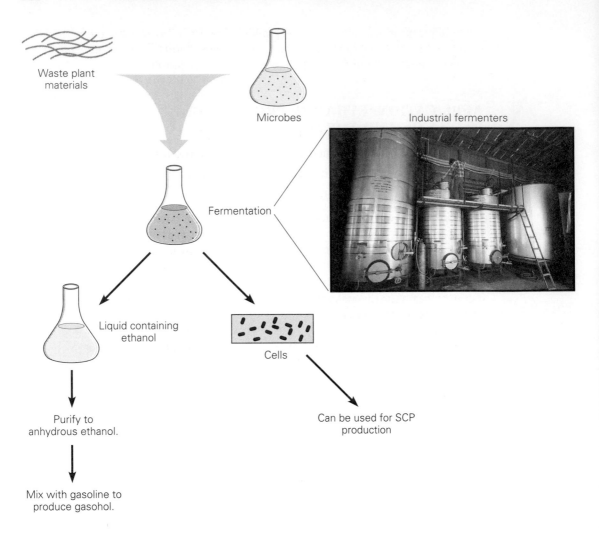

Waste plant
materials

Microbes

Industrial fermenters

Fermentation

Liquid containing
ethanol

Cells

Purify to
anhydrous ethanol.

Can be used for SCP
production

Mix with gasoline to
produce gasohol.

Figure 16.6 Ethanol fermentation. Many complex materials can serve as microbial nutrients. Some microbes produce ethanol by fermentation. After the cells are removed from the liquid, the ethanol can be recovered and purified. It may then be mixed with gasoline, resulting in a mixture called gasohol. Note that microbes removed from the fermentation could be used as single-cell protein (SCP).

The government of Brazil has strongly supported this project. In recent years, the subsidy amounted to $1.5 billion annually. Much of that money was spent to boost the role of hydrous ethanol as a complete substitute for gasoline, even though ignition problems and excessive engine corrosion complicated the specifically designed hydrous ethanol–powered engines.

What has happened? In the past few years, sugar production would have won out in the absence of the government subsidies and regulatory policies. Currently, hydrous ethanol production is decreasing. The alcohol-only vehicles did not sell as well as forecast. On the other hand, the production of anhydrous ethanol in the 1990s has increased to its highest total ever (4.6 billion liters in 1996–97).

Turning our attention to the United States, we see that ethanol production in 1996 was about 7.6 billion liters. Pro-ethanol, or gasohol, supporters suggested that increasing the use of crops such as corn, ultimately fermenting them to produce ethanol, would help solve the following complex problems:

- Dependence on foreign oil
- Decreasing performance due to lower octane (no-lead) gasoline
- Lower farm incomes caused by substantial grain surpluses
- Increased air pollution

Compared with fossil fuels such as gasoline, ethanol was advertised as being renewable and cleaner burning and producing no greenhouse gases on combustion. While few would argue that these issues would not be aided by increased emphasis on corn production/ethanol fermentation and use as a vehicular fuel, two significant problems arose. The first was that less than abundant corn crops have been produced in the past few years (increasing the feedstock price). The second difficulty is the likely reduction in federal government support.

Do the history and the present status of ethanol in Brazil (and, to a lesser degree, the United States) mean that fermentation technology did not work? Certainly not. What we want to convey is that linking social, economic, and political policies with new technology may not solve all our problems—but at least it allows us to explore alternatives.

APPLICATION: MALAYSIA AND PALM OIL

Malaysia is generally considered to be a developed nation, a reflection of the tremendous effort put forth by the government and the people of Malaysia to increase their standard of living by becoming more technologically capable. While many industries are certainly important to the Malaysian economy, two that will be greatly affected by present and future developments in biotechnology are the palm oil and rubber industries. We will look closely at the palm oil industry.

Malaysia is one of the world's largest suppliers of palm oil. Palm oil—or specific oils derived from it—is used in the production of cooking oils, margarines, soaps, and detergents and is obtained from the kernel of the palm oil seed. Once separated, the individual oils can be purified to whatever degree is necessary in order to be useful in production (Figure 16.7).

One of the more valuable oils derived from palm kernels is *lauric acid*, an ingredient in soaps and detergents. Both Malaysia and the Philippines export great quantities of lauric acid. The Malaysian oil is derived from palm kernels, the Philippine version is derived from coconut oil. Lauric acid is a short-chain (12-carbon) fatty acid. Oils isolated from plants contain many different fatty acids, some with short chains, others with longer chains. Typically, the fatty acids are separated before

Figure 16.7 Malaysian palm oil.
(**a**) Selling oil palm seeds (kernels) in an African market. (**b**) Drawings of the palm oil tree, a typical fruit cluster, and a seed (kernel) cross section.

(a)

Fruit cluster

Stony coat

Pulp
(yields palm oil)

Seed (yields palm
kernel oil)

(b)

use in industry, since each displays properties that might be desirable for a particular use. For example, shorter-chain fatty acids are typically liquid at room temperatures, whereas those with longer chains are solid or semisolid. The judicious use of different fatty acids can result in the production of, say, a margarine with a specific softness.

Palm oil is obtained by harvesting seeds from massive palm tree plantations. Most of these plantations were established in Malaysia during the period of British rule before and after World War II and were worked by inexpensive labor, making them very profitable for the British owners. Since gaining independence, Malaysia has maintained the productivity of its palm plantations. The national effort to enter the ranks of the developed nations has raised the standard of living; however, the palm oil industry has struggled to keep expenses low in the face of rising labor costs. It is a tribute to Malaysian inventiveness and perseverance that the country has maintained its competitiveness in the world market for oil in spite of the economics involved in oil production. The Palm Oil Research Institute of Malaysia (PORIM), the government research institute concerned with the continued development of the palm oil industry, has done an outstanding job of exploiting palm oil potential.

However, biotechnology may directly affect Malaysia's ability to remain competitive in the world market. Calgene, a U.S. plant biotechnology company, has developed a genetically engineered rapeseed plant that produces an oil composed of a large amount of lauric acid. This plant was developed through the insertion of a gene that shuts off the synthesis of fatty acids with chains longer than 12 carbons, preventing the synthesis of any fatty acid longer than lauric acid. The resulting rapeseed plants are healthy and produce nearly as much seed per acre as nonengineered plants. Following approval for the growth of this genetically engineered plant, Calgene planted sufficient rapeseed in Georgia to produce enough oil to test the product's marketability.

Producing lauric acid from this modified rapeseed plant is potentially much cheaper than producing it from palm kernel oil, in large part because of the abundant production of this single fatty acid in the modified plants. Calgene is also working with a rapeseed plant that produces high levels of myristate oil and may make raw materials for soaps and personal care products less expensive and much more abundant. Widespread introduction of these engineered plants will most likely have an adverse effect on the palm oil industry in Malaysia. The key question for Malaysia is how to respond to this threat.

Options range from aggressive marketing to wholesale conversion of Malaysian palm oil plantations to rapeseed growth. Aggressive marketing campaigns have been quite successful in shaping public opinion about oil sources. In essence, both palm oil and soybean oil positioned themselves favorably in response to increased consumer demand for partially hydrogenated oils that would be healthier and still serve a variety of cooking purposes. It might be possible to mount another aggressive campaign designed to bolster the sales of palm oil products.

At the other end of the spectrum, the decision could be made to take advantage of the new rapeseed plant and use it to replace palm oil production in Malaysia. This conversion might be easily accommodated—most of the postharvesting isolation and purification of the oil would not be greatly altered. On the other hand, destroying palm plantations just to make room for planting rapeseed seems patently inefficient. Perhaps rapeseed could be phased in as palm plants are naturally replaced. Extensive negotiations have taken place between Calgene and the Malaysian palm oil industry, presumably to determine the feasibility and economic desirability of such a conversion. Malaysia, as a small country with few resources compared to the United States, might not be able to effectively diversify its commitment to oil production. Given its limited resources and acreage, Malaysia is unlikely to be able to quickly respond to market changes—particularly since palm trees must be grown for long periods of time before they become productive.

SUMMARY

In the United States, where both resources and financial backing are plentiful, diversification and changes driven by the market are commonplace and relatively easy to accommodate, though certainly not without personal upheaval. In smaller, poorer countries, such diversification and availability of funding are simply not possible. Developing countries often must make decisions that commit them to courses of action that last for decades, for example, the decision to rely on the palm industry that has benefited Malaysia for many years. However, making such decisions is difficult, and the ever-increasing rapidity with which biotechnology is developing may make any decision obsolete in a short time. As developing nations struggle to obtain the funding and technical resources necessary to pursue biotechnological ventures, they must carefully weigh the available options. One factor that makes planning exceedingly difficult is the uncertainty over how new biotechnologies will be used by developed countries. Will the applications of biotechnology in industrialized nations make it impossible for smaller countries to capitalize on their existing resources and expertise, or even to maintain the status quo? Industrially developed wealthy nations may see the need to exercise restraint to avoid adversely affecting the larger global community.

REVIEW QUESTIONS

1. What are some of the significant questions that developing nations need to ask before they bring in biotechnology?

2. What values does plant tissue culture bring to agriculture that traditional crop planting, cultivating, and harvesting lack?

3. What regional and potentially global problem is generally intensified when a new crop strain with one or more particularly outstanding qualities is introduced?

4. List three foods that are actually produced by microbial fermentation of natural carbohydrates from plant or animal sources.

5. Even though single-cell protein has not met with commercial success, it offers much-needed nutritional supplementation in countries where many people are starving. How could this technology be employed?

6. List at least three different ways that embryo transfer in cattle is said to have been of distinct agricultural benefit in the United States.

7. How have PCR and hybridization been used to allow the determination of the sex of embryos in the beef and dairy industries? What is the value of sex determination prior to implantation of an embryo?

8. Give two examples of animal diseases that could be prevented if not for problems of adequate (speedy) distribution or ease of administration.

9. What does the lack of energy reserves (coal, oil, gas) have to do with the inability to attract

biotechnology firms or efforts to developing countries?

10. Explain how the following are currently interrelated: Calgene, lauric acid, palm kernels, rapeseed, Malaysia, palm oil industry, and easy-to-spread margarine.

Now What?

One early goal of this text was to help you understand how living cells and organisms work. We moved from there to applying that basic knowledge to the developments in genetic engineering and immunology that are revolutionizing medicine as well as many other aspects of our lives.

Science, like any other endeavor of discovery, is a human enterprise, subject to the same human influences as other aspects of our lives. We have seen examples of how social, political, and economic forces directly affect science. Science costs money, and the bills must be paid. We hope we have convinced you that science, as a human endeavor, is not shielded, not separate, and definitely not the ivory tower that some people claim it to be.

Predicting where biotechnology will take us in the next few years, let alone the next decades, is difficult. By the time this book is printed, some of the information will be dated and new developments will make other aspects take on new meaning. We hope that this background will enable you to continue to examine new developments in biotechnology with enhanced awareness and sharpened skills.

Most of the nontechnical considerations in this book involve questions that we, as individuals or as part of society, must address. We feel strongly that biotechnology has furthered our understanding of life, and we have seen how it has helped create opportunities for better products and services. But as with any technology, it raises many questions that are difficult to answer. We will focus on these questions in this final chapter.

THE TECHNICAL ASPECTS OF BIOTECHNOLOGY

There is no doubt that biotechnology is a technical revolution. We now possess the ability to manipulate genetic information, and therefore living organisms, in ways that were not even imagined in science fiction in the recent past. The rapidity of technical developments has rivaled that of our entry into the computer age. Such rapid development involves both risks and rewards and can happen only if the promise of significant breakthroughs and scientific wonders is present.

Summary of Risks

A certain element of risk is involved in any new endeavor. As genetic engineering was first being developed, great concern arose about the potential risks being taken. The concern was so great and the outcry so strong that scientists themselves took the unprecedented action of calling a moratorium on their activities while some of the risks were explored. Although there was never any actual resolution to the moratorium, the simple fact that it happened and slowed progress for even a short while indicates the concern that many felt over our ability to alter genetic information.

In hindsight, none of the disastrous scenarios imagined and loudly predicted have occurred. We are not suggesting that no more risks exist, but simply that the technology itself has yet to become a vehicle for mass destruction. Certainly, the potential for disaster remains and perhaps will always be a factor, but the general safety of the technology seems evident.

But biotechnology is different from other technologies. The direct manipulation of genes involves the control of information, and, in general, information can be used to create, modify, and perpetuate ideas. Therefore, in addition to the potential for physical disaster warned of by early critics, there is also the potential for ideological disaster.

So the question remains: How much risk is acceptable? Anything new involves risk, and many things we have come to accept as commonplace involve risk as well— sometimes a significant level. As individuals and as a society, we have determined levels of risk with which we are comfortable. This does not mean that new risks are any more acceptable.

Another risk involved in the development of new technology is the financial risk assumed by the investors who provide developmental fuel. Huge technologies such as space exploration were funded, in large part, by the federal government. Biotechnology is a bit different. It was not funded as a "big science" project until well after it was developed, but initially most researchers still depended on government grants to fund their work. Now, however, with the commercialization of biotechnology a major goal, funding is usually provided through traditional business ventures. Investment capital is now the raw material of biotechnology, and such investments involve considerable risk for investors. To date, the potential for significant financial gain has provided the capital necessary for continued development.

Potential Breakthroughs?

The claims made under the banner of biotechnology have been astounding. Some of the products already available are listed in Table 17.1. Since much of biotechnology involves medical care, many of the most dramatic promises have been made in this area: Cancer will be cured; the Human Genome Project will help us understand every

Table 17.1 Some Pharmaceutical Products of Genetic Engineering

Product	Comments
tissue plasminogen activator (Activase®)	Dissolves the fibrin of blood clots; therapy for heart attacks; produced by mammalian cell culture.
erythropoietin (EPO)	Treatment of anemia; produced by mammalian cell culture.
human insulin	Therapy for diabetics; better tolerated than insulin extracted from animals; produced by *Escherichia coli*.
interleukin-2 (IL-2)	Possible treatment for cancer; stimulates the immune system; produced by *E. coli*.
alpha-interferon	Possible cancer and viral disease therapy; produced by *E. coli*, *Saccharomyces cerevisiae* (yeast).
gamma-interferon	Treatment of chronic granulomatous disease; produced by *E.coli*.
tumor necrosis factor (TNF)	Causes disintegration of tumor cells; produced by *E. coli*.
human growth hormone (hGH)	Corrects growth deficiencies in children; produced by *E. coli*.
epidermal growth factors (EGF)	Heals wounds, burns, ulcers; produced by *E. coli*.
prourokinase	Anticoagulant; therapy for heart attacks; produced by *E. coli* and yeast.
factor VIII	Treatment of hemophilia; improves clotting; produced by mammalian cell culture.
colony-stimulating factor (CSF)	Counteracts effects of chemotherapy; improves resistance to infectious disease such as AIDS; treatment of leukemia; produced by *E. coli*, *S. cerevisiae*.
superoxide dismutase	Minimizes damage caused by oxygen free radicals when blood is resupplied to oxygen-deprived tissues; produced by *S. cerevisiae* and *Pichia pastoris* (yeast).
monoclonal antibodies	Possible therapy for cancer and transplant rejection; used in diagnostic tests; produced by mammalian cell culture (from fusion of cancer cell and antibody-producing cell).
hepatitis B vaccine	Produced by *S. cerevisiae* that carries hepatitis-virus gene on a plasmid.
bone morphogenic proteins	Induces new bone formation; useful in healing fractures and reconstructive surgery; produced by mammalian cell culture.
taxol	Plant product used for treatment for ovarian cancer; produced in *E. coli*.
orthoclone	Monoclonal antibody used in transplant patients to help suppress the immune system, reducing the chance of tissue rejection; produced by mouse cells.
relaxin	used to ease childbirth; produced by *E. coli*.
Pulmozyme® (DNase)	Enzyme used to break down mucous secretions in cystic fibrosis patients; produced by mammalian cell culture.

disease that afflicts humans; every genetic disease will be numbered among the battles won. At present, most if not all of these claims are more dramatic than the reality on which they are based. Cancer provides the best example. Claims have been made at all levels: Cancer will be completely curable as a result of biotechnology, certain forms of cancer will become treatable, and so on. It is true that developments within biotechnology have led to a far greater understanding of the nature and development of cancer, but this monumental boost in knowledge has had little direct effect on the success of treatment or preventative efforts for most cancers. What success there has been in treating cancer has come about through better surgical treatments and improved chemotherapy and radiation therapy. In the future, giant strides may well be made in treatment and prevention, but the most dramatic promises of biotechnology as applied to medicine have not yet been realized.

Why were such claims made initially? In large part, they may have been just overzealous predictions made by scientists who truly believed in the unlimited potential of this new technology. Those directly involved in the science, especially scientists achieving experimental successes, were likely to imagine that even greater successes were equally possible and could happen at any time. While it is easy to extrapolate from a limited success to a much larger one, this often ignores the far greater complexities presented by larger problems. The simple truth is that things often simply don't work and we don't understand why. No matter how much we understand about cellular workings, there seems to be a great deal more that we don't know.

The very beginnings of biotechnology were a time of great exuberance—everything seemed to be within reach. Today, claims are more conservative because the realities have become clearer, but at times a small step forward is still heralded as a soon-to-be-available cure. The recent identification of a gene associated with obesity is one such example. When this discovery was announced, it was portrayed as a cure for obesity. The reality, however, now seems much less dramatic.

GENETIC INFORMATION: POTENTIAL USES AND ABUSES

As we have seen, much of biotechnology revolves around a basic understanding of genetics. The early portions of this book provided background to help you understand basic genetics as well as some of its complexities. These complexities seem to set up roadblocks to much of what we would like to do in biotechnology. We will briefly review several aspects of genetics that lead to a more comprehensive understanding of the flow of biological information and the implications of our increased knowledge.

Inheritance

Inheritance is strictly defined and usually strictly determined. In humans and many other sexually reproducing organisms, the genes received by offspring come from both parents through the reproductive cells. The reproductive cells, in turn, arise through a mechanism of cell division that allows for the shuffling of genetic information—recombination. The result is that each offspring is distinct and unique, an advantage to the species—especially in a dynamic or changing environment. The genetic traits an individual possesses are determined by the information that individual inherits.

But genetic information does not by itself determine all characteristics. Many also depend to some degree on environmental influences; in fact, some characteristics are determined entirely by environmental influences. Influences may be exerted by the immediate family—in other words, they are dependent on the family environment in which the individual is raised—or by larger societies. Beliefs and practices prevalent in the United States will typically become a part of the makeup of any children raised here, due to socialization. Therefore, especially when we consider human beings, it is often difficult to determine whether a particular characteristic is dependent on genes or environment.

Causality

Some characteristics are definitely determined solely by genes. Many metabolic diseases, for example, are a direct result of changes in particular genes. Corresponding treatments and cures are often possible after sufficient study of the gene and its product.

Identifying the genes responsible for diseases is paramount. In many instances, genes are initially identified by studying **correlations**, or tendencies for particular characteristics to be inherited along with particular genes. But identifying a correlation is a far cry from demonstrating that the particular gene is responsible for the characteristic. Correlation should not be confused with causality. Many dramatic claims of biotechnology have been made because of correlations that have proved difficult to translate into actual cause and effect.

Once a direct function has been shown and cause can be ascribed to a particular gene product, biotechnology can begin to exploit what is known in order to find ways to alter or prevent its abnormal function. All new technologies go through phases, and it is certainly true that at this time scientists are making many discoveries about correlations of genes with particular characteristics and diseases. Whether these discoveries will result in increased understanding of the underlying reasons for the diseases remains to be seen.

Genetic Determinism

Medical applications of biotechnology have fostered numerous reports about genetic causes of diseases as well as genetic causation of other characteristics. Many claims are based on correlations, but often the distinction between correlation and causality is lost in the reporting. The result is that most people believe that genes are directly responsible for most characteristics, giving little thought to the role that nongenetic factors play in the process. Many people would like to believe that there are genetic explanations for particular human characteristics—then they can blame the genes instead of focusing on social institutions or environmental influences. Thus, we seem to be living in another period of **genetic determinism**—the belief that who we are and what we become is determined solely by our genetic information at birth.

What are the effects of determinism? Like various ideas about predestination, deterministic beliefs result in the institution responsible for the beliefs being raised above other institutions. For those who subscribe to predestination, religion is the most important institution of their life, since it is religion that determines the entire life of the individual. For those who believe genes determine who and what we are, science becomes equivalent to religion. It is often said that science is the religion of twentieth-century humankind, and genetic determinism reflects this idea. We have tried to describe enough about science to convince you that science is a human way to learn and is filled with all the risks associated with other human endeavors. Just as atrocities and mistakes have been committed for religious reasons, so too can mistakes be made for scientific reasons. Eugenics (Chapter 6) is probably the best example of the misuse of scientific information, whether deliberate or not, to further particular social ideals (Figure 17.1).

Genetic determinism also allows others to stigmatize individuals due to factors completely beyond their control. Having genes that result in a particular skin color has led to such stigmatization that we have yet to eliminate racism from our society. The possibility of increased stigmatization based on genes is real.

If genes determine many of our characteristics, is it possible to change these characteristics? Gene therapy is being developed for exactly this reason—to alter the genetic makeup of an individual to prevent or eliminate a particular defect. Ultimately, questions will be asked about what constitutes a defect and what does not. If it becomes possible to prevent obesity through gene therapy, it is likely that our efforts (as a society) to exercise and seek balanced nutrition will diminish. Since it is clear that diet and exercise also affect many other human characteristics besides obesity, it is likely that other benefits will be lost.

It is also likely that genetic determinism will result in a lack of desire to change. For example, if alcoholism is found to be determined solely by an individual's genes, then those who like to consume large quantities of alcohol can blame their genes. Individuals will have a built-in excuse—and they will certainly use it. The same rationale was used during the earlier period of eugenics to explain why it would be unwise for wealthy persons to help those in need: The poverty-stricken are so because they are genetically inferior, not because of any circumstance that we might change. It might be enlightening to review the current debate concerning welfare programs in light of public beliefs regarding genetic determinism.

Testing and Prediction

If human characteristics are determined by genes, then it is theoretically possible to test an individual—at any stage of life—to predict that characteristic. It is possible to do this now for some genetic diseases, and the huge amount of genetic research

TEST 8

Notice the sample sentence:

People hear with the   eyes   <u>ears</u>   nose   mouth

The correct word is ears, because it makes the truest sentence.

In each of the sentences below you have four choices for the last word. Only one of them is correct. In each sentence draw a line under the one of these four words which makes the truest sentence. If you cannot be sure, guess. The two samples are already marked as they should be.

SAMPLE { People hear with the   eyes   <u>ears</u>   nose   mouth
         { France is in   <u>Europe</u>   Asia   Africa   Australia

1. America was discovered by   Drake   Hudson   Columbus   Cabot 1
2. Pinochle is played with   rackets   cards   pins   dice.. 2
3. The most prominent industry of Detroit is   automobiles   brewing   flour   packing 3
4. The Wyandotte is a kind of   horse   fowl   cattle   granite 4
5. The U.S. School for Army Officers is at   Annapolis   West Point   New Haven   Ithaca............. 5
6. Food products are made by   Smith & Wesson   Swift & Co.   W.L. Douglas   B.T. Babbitt 6
7. Bud Fisher is famous as an   actor   author   baseball player   comic artist 7
8. The Guernsey is a kind of   horse   goat   sheep   cow 8
9. Marguerite Clark is known as a   suffragist   singer   movie actress   writer.................. 9
10. "Hasn't scratched yet" is used in advertising a   duster   flour   brush   cleanser 10
11. Salsify is a kind of   snake   fish   lizard   vegetable 11
12. Coral is obtained from   mines   elephants   oysters   reefs 12
13. Rosa Bonheur is famous as a   poet   painter   composer   sculptor......................... 13
14. The tuna is a kind of   fish   bird   reptile   insect 14
15. Emeralds are usually   red   blue   green   yellow.. 15
16. Maize is a kind of   corn   hay   oats   rice... 16
17. Nabisco is a   patent medicine   disinfectant   food product   toothpaste.................. 17
18. Velvet Joe appears in advertisements of   tooth powder   dry goods   tobacco   soap 18
19. Cypress is a kind of   machine   food   tree   fabric 19
20. Bombay is a city in   China   Egypt   India   Japan 20
21. The dictaphone is a kind of   typewriter   multigraph   phonograph   adding machine 21
22. The pancreas is in the   abdomen   head   shoulder   neck 22
23. Cheviot is the name of a   fabric   drink   dance   food................................... 23
24. Larceny is a term used in   medicine   theology   law   pedagogy......................... 24
25. The Battle of Gettysburg was fought in   1863   1813   1776   1812 25
26. The bassoon is used in   music   stenography   book-binding   lithography 26
27. Turpentine comes from   petroleum   ore   hides   trees 27
28. The number of a Zulu's legs is   two   four   six   eight................................. 28
29. The scimitar is a kind of   musket   cannon   pistol   sword.............................. 29
30. The Knight engine is used in the   Packard   Lozier   Stearns   Pierce Arrow.............. 30
31. The author of "The Raven" is   Stevenson   Kipling   Hawthorne   Poe.................... 31
32. Spare is a term used in   bowling   football   tennis   hockey 32
33. A six-sided figure is called a   scholium   parallelogram   hexagon   trapezium.......... 33
34. Isaac Pitman was most famous in   physics   shorthand   railroading   electricity.......... 34
35. The ampere is used in measuring   wind power   electricity   water power   rainfall......... 35
36. The Overland car is made in   Buffalo   Detroit   Flint   Toledo 36
37. Mauve is the name of a   drink   color   fabric   food................................... 37
38. The stanchion is used in   fishing   hunting   farming   motoring 38
39. Mica is a   vegetable   mineral   gas   liquid.. 39
40. Scrooge appears in   Vanity Fair   The Christmas Carol   Romola   Henry IV 40

Figure 17.1   A U.S. Army mental test. U.S. Army Mental Test 8, Examination Alpha (for those who could read), administered to almost 2 million men about the time of World War I. Tests such as this were used by eugenicists.

being performed in the Human Genome Project will undoubtedly increase the number of genetic characteristics that we can diagnose.

Genetic testing is in its infancy, both scientifically and socially, and it will not be easy for societies to determine the best possible use of this capability. The tests available now have been easy and straightforward to develop. For example, whether an infant will suffer from phenylketonuria, a disease that is easily treated, can be predicted quickly and accurately. It will not always be possible to predict illnesses with such ease and accuracy or to treat them with such ease and relative comfort. Interpretation of the complications inherent in most genetic characteristics will prove complex, and extremely difficult decisions will need to be made.

In this era of belief in the power of genetics, we can easily imagine being able to test for all sorts of genetic diseases. One difficult question will be to determine what is a defect and what is not. Is farsightedness a defect? Should it be treated as a genetic disease and perhaps prevented through gene therapy, even though a completely effective form of treatment (corrective lenses) already exists? It is theoretically possible to change hair color, for example, through genetic manipulation. Will this capability be explored and marketed? The free-market system ensures that if a demand exists for a product or service, then it will be provided. A society increasingly interested in using plastic surgery to overcome perceived physical defects is unlikely to not make use of another, certainly easier, less painful, and perhaps more permanent method of achieving the same goal.

Of course, as is true for plastic surgery today, future methods of perceived improvement will be preferentially available to those who can most easily afford them. We have seen several instances in which simple economic concerns operate in opposition to what might be viewed as a fair and equitable course of action. Pharmaceutical companies have little or no incentive to provide drugs and vaccines to fight diseases that afflict people in underdeveloped or developing nations. The wealthy within our society often have little incentive to share their wealth with others. As always, opportunities for improvement are more available to those with power—and the availability of these opportunities in turn furthers their ability to maintain their position of power.

Scientific Reductionism

Reductionism is a course of action whereby a problem is continually broken down into smaller and smaller problems until ultimately the system is simple enough and controllable enough to be understood. Science has relied almost completely on reductionism for centuries. It has resulted in our current level of understanding of DNA and genes, but it does not always yield an accurate picture of life—it doesn't tell the true story. By moving molecules, cells, or whole organisms from their natural habitat for purposes of study, we may have altered or lost critical interactions that could be important for understanding how these elements function. Reductionists would contend that finding the smaller pieces of information first enables us to work back up through more complicated systems, increasing our understanding of each level of complexity as we reconstruct it. While this stance may be true, isn't it possible that dissecting systems into bite-size pieces may result in missing information?

Scientific breakthroughs are heralded when a more complete understanding of a phenomenon is discovered. Understanding how two cells interact might form the basis of a new discovery. Not understanding how two cells interact will seldom, if ever, be called a discovery—it will be perceived simply as an inability to find the connection. What if it cannot be found because it is too complex or too dependent upon other processes? Raising the level of complexity—saying that we don't understand something because it is too complex—is not what science is supposed to do for us. We nearly always want to believe more simplistic explanations and may well ignore even obvious complexities to satisfy our urge for simplicity.

Our belief in strictly genetic explanations for many human characteristics is a good illustration—the simple fact that we place such emphasis even today on individual IQ is quite revealing. IQ is a measurement that is inherently designed to distill extraordinarily complex things—cognitive skills, analytical reasoning, and so on—down to a single number. Dealing with IQs and the differences between them is much more palatable, particularly when we believe that IQ is mostly genetically determined. To explain away all sorts of things based on IQ is much easier than to face all the problems that might have contributed to particular human behaviors.

Reductionism has great value and has yielded tremendous scientific successes. But there is also value in espousing the complex and being made aware of what is complex. It is unfortunate that complexities are so often frowned upon and treated with such disdain.

Ownership and Responsibility

The ability to determine and perhaps manipulate an individual's genetic information represents a new way for others to potentially control our lives. We have discussed the possibility of alienation and stigmatization resulting from knowledge of genes. Genetic testing represents another method by which information about us as individuals can be gathered. If we choose to be tested for a particular gene, who else has the right to know the results? Would we be willing to submit to mandatory testing for certain disease genes if such a program becomes economically feasible? We have already incorporated a number of screening programs into our society. Would more be acceptable?

Corporations now spend billions of dollars in efforts to identify and patent particular genes. Does ownership of genetic information, the rights granted by a patent, change the way we view nature or ourselves? Ownership implies responsibility, but responsibility for what and to whom? The Human Genome Project seems destined to result in mass ownership of genetic information and living organisms. Livestock, crops, and pets have long been thought of as commodities—and most people probably do not dispute actual human ownership of these living things. Different breeds of animals and plants, developed through traditional breeding practices, also have clear ownership. Is owning genetically engineered organisms any different? How different is it to own the rights to the information itself? Table 17.2 lists some products of genetic engineering for which such questions are already relevant.

Table 17.2 Some Agriculturally Important Products of Genetic Engineering

Product	Comments
Agricultural Products	
Pseudomonas syringae, ice-minus bacterium	Lacks normal protein product that initiates undesirable ice formation on plants.
Pseudomonas fluorescens bacterium	Has toxin-producing gene from insect pathogen *Bacillus thuringiensis*; toxin kills root-eating insects that ingest bacteria.
Rhizobium meliloti bacterium	Modified for enhanced nitrogen fixation.
Round-Up® (glyphosate)-resistant crops	Plants have bacterial gene; allows use of herbicide on weeds without damaging crops.
Bt cotton and Bt corn	Plants have toxin-producing gene from *B. thuringiensis*; toxin kills insects that eat plants.
FlavrSavr® tomato and carrots	Gene for pectin degradation is removed so fruits have longer shelf life.
Animal Husbandry Products	
porcine growth hromone (PGH)	Improves weight gain in swine; produced by *E. coli*.
bovine growth hormone (BGH)	Improves weight gain and milk production in cattle; produced by *E. coli*.
transgenic animals	Genetically alters animals to produce medically useful products in their milk.
Other Food Production Products	
rennin	Causes formation of milk curds for dairy products; produced by *Aspergillus niger*.
cellulase	Enzymes that degrade cellulose to make animal feedstocks; produced by *E. coli*.

SOCIOLOGY OF SCIENCE

Throughout this book, we have maintained a strong undercurrent of human involvement in science. Economic, political, social, moral, and other forces operate to determine exactly what kind of science is done and how it gets done. Science is a product of our society, so we must, as with all other endeavors, endure improprieties, question motives, and be prepared to justify our decisions. Yet science is still held up as an impartial social force: You cannot argue with science. Science is truth. Science is unimpeachable. Science is beyond reproach. Many in our society today persist in this viewpoint—a view seen through deeply rose-colored glasses.

How has science risen to such a high level of esteem? Is it simply the terminology involved (*fact, proof, evidence*), which makes scientific work sound pure and truthful? Or is it the image that scientists have often cultivated of working in secret laboratories, doing things that most people do not understand? If we cannot understand what scientists say, then it must be true. Scientists are often used as expert witnesses in legal trials, despite the fact that both sides in a case can usually produce numerous scientists who will disagree about the interpretation of the facts. Shouldn't this make us wonder why science is held up as truth?

Perhaps the interpretation of science should not be left to scientists, or perhaps the relationship between scientists and the rest of society should be altered to reflect a union rather than a separation. Scientists are notoriously clumsy when they try to explain to nonscientists what they do. Scientists tend to explain problems in a vocabulary that prevents direct discussion. While scientific vocabulary may be intimidating, we must be quick to point out that other fields (law, art, religion) have their own vocabularies, and somehow we manage to communicate. As science becomes more and more specialized, the distance between scientist and layperson seems to become greater. If you are using this book as part of a college course, the odds are that the course is designed specifically for nonscience majors. Many nonscience college students take such specially designed courses because they feel they do not have the background (or vocabulary) to do well in the normal introductory biology, chemistry, or physics courses. Yet science majors generally take the same introductory history courses as all other students, as well as the same composition, language, religion, and psychology courses. That there is an expanding rift between scientist and nonscientist should come as no surprise—it is cultivated and reinforced by our schooling.

Perhaps responsibility for this dilemma lies primarily with the scientists. If scientists would do a better job of communicating what they do, others would have a better understanding of science and how it gets done. This would provide much better perspective and might reduce some of the blind dependence on science that now exists. Once dependence is removed, nonscientists can assume more responsibility for the science.

Scientists and nonscientists alike must answer the myriad of questions being raised by biotechnology. Scientists must work harder at explaining and accurately conveying what they do, and nonscientists must work harder at understanding the science as well as the issues behind it. Scientific literacy works both ways—one group cannot improve its role in the dialog without the help of the other.

BIOTECHNOLOGY: RISKS AND ETHICS

From an ethical perspective, biotechnology is challenging for three reasons:

1. Risks to worldviews and "slippery slope" risks.
2. Risks associated with social and economic impacts.
3. Risks to animal and human health and to the environment.

With regard to risks to worldviews, it is well to remember that trampling on traditions or worldviews integral to people's self-perception can produce powerful resistance. Fears about the "slippery slope" are often fueled by a concern that such shifts in our thinking may allow tampering with the genetic inheritance of species in an irreversible slide leading to unchecked modification of animals and finally to genetic manipulation of humans. Sometimes, instead of the slippery slope, the image of a burst dam is invoked. The concern is that if we genetically alter a bacterium, plant, or animal, the process will be actively pursued to alter the human genotype.

Genetic manipulation of viruses, bacteria, plants, animals, fish, and birds has important impacts in agriculture, in the manufacture of chemicals and pharmaceuticals, in medicine, and in the use of animals for medical research and therapy. Adverse social and economic impacts are bound to result from the introduction of genetically modified plants and animals in agriculture, from the new concentrations of power in large chemical and pharmaceutical companies, and from relationships between universities and industry (academic and company researchers).

The risks to animal and human health and to the environment are real and are made even more complicated because, as a society in the United States, we are generally optimistic about science and technology, fairly averse to risk taking, financially dependent on scientific progress and technology, and obsessed with our (human) health. The relationship between health care and biotechnology is strong and continues to develop. Several critical questions concerning biotechnology and human health must be addressed.

First, is whatever offers medical help for some individuals worthy of pursuit? Some would suggest that any medical advance is good and ought to be pursued because every life is of infinite value. Others would say if it's going to help even a few people, go ahead and pursue it, even if there may be some financial problems. Still others would say that the social benefits and how they compare with the costs should be assessed to determine whether the pursuit is desirable.

Second, should we employ only those biotechnologies that will hold down or reduce health care costs? How do we think about cost control and the justice involved in the allocation of resources? Some would not want to include technologies in our health care system unless they have been proved in advance to be of financial benefit. Others would want to get the technologies out there so we can then evaluate them. A more conservative view would be that we need to be much tougher gatekeepers, because it's difficult to install an expensive new technology, test it for a time, and then decide it is too expensive to use. People with this viewpoint would argue that turning back the clock is impossible.

Third, do we believe that nature, including human nature, knows best? Medical bioethics seems disinterested in nature and doesn't see nature as a useful guide. We might well ask of nature, Should we intervene at all, regardless of the technology?

Fourth, should we think only of bringing individual members of our population up to some level of normalcy, or should our aim be to reduce illness, pain, and suffering for all? Might we go further and consider using biotechnologies to "improve" on human nature? Should we think of these technologies as a means to normalize or optimize?

Finally, in matters of great uncertainty—and many biotechnologies certainly qualify as uncertain—should we proceed aggressively? Some people will say let's go ahead and seize the possibilities. Others will counsel to proceed cautiously. In matters of uncertainty, they would claim, we should go very slowly, step by step. We frankly don't know whether we will be better or worse off.

Each of these questions is very difficult to address, but as they arise we will have to make decisions. If we are optimistic about scientific progress, we can take a set of values and move biotechnology forward. Background values, perspectives, and biases make a tremendous difference when we get down to deciding specific ethical issues. We might expect to find a range of acceptable and unacceptable activities and thereafter let experience teach us. Perhaps biotechnology can be considered a dynamic experiment. We don't know exactly what it will do for us or to us, so we will have to learn as we go along. If there are spectacular benefits to be gained, we will get to them—sooner or later. Figure 17.2 shows many of the potential applications of biotechnology, some already realized and others still in the future.

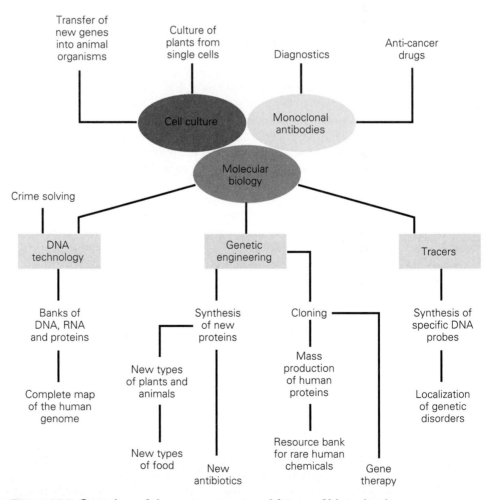

Figure 17.2 Overview of the past, present, and future of biotechnology.

SUMMARY

As the interrelated fields of genetic engineering, immunology, and cell and molecular biology are applied to the rapid development of biotechnology, we understand them more and more. A cautious examination of current emphases indicates that health care, agriculture (including the food industry), and environmental biotechnology will be most intensively explored in the near future. Energy conversion, metal recovery, and chemical manufacturing will also gain prominence. Even though there seems to be a wide distribution of areas of biotechnological interest, most of us will be drawn to the application of genetic engineering to genetic diseases and the potential slippery slope that may lead to genetic improvements in individuals.

One important question involves ownership of genetic information, including who has the right to have access to such information. Individuals, employers, private or public business, and governmental institutions will, no doubt, argue their cases strongly.

Legal, ethical, moral, and religious struggles erupt frequently, and biotechnology will not enjoy smooth sailing as long as excursions into uncharted waters occur. Given the enormity of the potential changes, we might conclude that the struggles, the friction, and the risks for all concerned behoove our society to look long and hard before leaping on the biotech bandwagon. Be cautious. Go slow. Test the waters. All these suggestions are said to be the mark of mature, conservative, experienced people. Perhaps extreme caution should be the guideline. But stop and think: Curing genetic diseases, AIDS, and cancer; feeding the world; repairing our much abused, damaged environment. These grand objectives are within our biotechnological grasp. Shouldn't we reach out?

REVIEW QUESTIONS

1. Several basic technologies described in this book are between one and two decades old. Why, then, is it still risky to project with any precision at all where biotechnology is headed?

2. The elements of risk and potential reward seem to have propelled biotechnology until now. What justification is there for believing that the promise of significant financial reward will continue to fuel the fires of biotechnology? Are there instead particular reasons that signal the extinguishing of the biotechnology flames?

3. What problems would be raised if genetic determinism were once again upheld scientifically? How might the situation be similar to the eugenics movement in the early 1900s?

4. Describe the types of risk presently associated with biotechnology as well as those risks likely to arise in the future.

5. Why is it difficult to determine whether a particular human characteristic is influenced by the individual's environment or is solely due to an individual's genetic makeup?

6. Earlier, when we examined gene therapy, we noted that it has the potential for curing genetic defects or diseases. However, applications regarding the alteration of the genetic makeup of an individual have been applied only to somatic (body) cells, not to eggs or sperm. Why? Would you recommend that we continue to treat only somatic cells, or should we include the sex cells? Why?

7. Orphan diseases are sometimes described as those in which few people are interested in terms of study, prevention, or treatment. Generally, orphan diseases occur either infrequently or mostly among people of poorer

developing countries. Given the rapid advances in medicine over the past two decades due to biotechnological efforts, should we support the study of orphan diseases with the twin objectives of prevention and cure?

8. Have biotechnologists failed to communicate to the public what they do and why they do it? If your answer to that question is even partly yes, what could scientists do to eliminate the problem?

9. Genes, biotechnical techniques, patents, and patent laws are tied closely together, especially in a capitalistic society. Should biotechnology continue to be funded privately, or should a major governmental effort be made to support its continued development?

10. Which of the five human health–related questions raised in the text is most important, and why?

Glossary

acrylamide gel electrophoresis   the use of acrylamide gels to separate molecules by forcing them to migrate through the gel under the influence of an electrical field

activation energy   energy necessary to get a chemical reaction underway

adoptive immunotherapy   cancer treatment that requires removing anticancer cells from the blood, increasing their growth *in vitro*, and returning the much enlarged population to the same patient

affinity chromatography   a type of chromatography in which the matrix contains chemical groups that bind selectively to the molecule being purified

agar   a complex polysaccharide extracted from red algae and used as a solidifying agent in microbial growth media

agarose gel electrophoresis   use of agarose, a gelatinous material extracted from seaweed, in solidifying gel used to separate DNA fragments during electrophoresis

AIDS (acquired immune deficiency syndrome)   a disease caused by a retrovirus, HIV, that causes significant depletion of CD4$^+$ T cells, resulting in diminished immune responsiveness and increased susceptibility to a variety of infections and cancers

allele   alternative forms of a gene; different alleles of a gene each have a unique nucleotide sequence, and their activities are all concerned with the same biochemical and developmental processes, although their individual phenotypes may differ

amino acids   monomers containing a carboxyl group and an amino group that provide the raw material for all protein production in all organisms

amniocentesis   technique of sampling amniotic fluid to determine genetic abnormalities by the presence of certain chemicals or defective fetal cells

anabolic   energy-requiring reactions that build up complex molecules from simpler ones; synthetic reactions

aneuploidies   abnormal conditions in which one or more whole chromosomes of a normal set of chromosomes either are missing or are present in more than the usual number of copies

angioplasty   repair of blood vessels by inserting a small inflatable balloon into a blood vessel and expanding it to allow unobstructed blood flow

anhydrous   without water

anneal   the process of heating and cooling nucleic acids to separate strands and induce base pairing at a lower temperature, especially with complementary strands from different species

antibodies   proteins produced by B lymphocytes in response to the entry of specific foreign antigens into the body. The antibodies will recognize and bind to the antigens that induced the immune response

anticodon   a specific three-base sequence on a tRNA that forms complementary base pairs with a codon in mRNA during translation

antigenic determinant   the site on an antigenic molecule that is recognized and bound by an antibody

antigens   foreign substances that bind specifically to an antibody and elicit an immune response

anti-idiotype antibodies   antibodies that specifically recognize and bind to a single antigenic determinant of the variable region of an antibody

antiparallel strands   description of the complementary strands in the typical DNA molecule, in which the two strands come together oriented in opposite directions

assay   any method for determining the presence or quantity of a component

autoantibodies   antibodies produced by an organism against any of its own components

autoimmune diseases   diseases that result from autoimmune activities

autoimmunity   an abnormal immune response against a body's own components

autoradiography   a method for locating clustered radioactive molecules by covering a sample with photographic film

autosomal dominant diseases   dominant diseases caused by one or more genes found on an autosome

autosomes   all chromosomes other than sex chromosomes

B cells   lymphocytes that mature in the bone marrow and are precursors of antibody-secreting cells (plasma cells)

β-cells   insulin-producing cells within islets of Langerhans in the pancreas

β-galactosidase   a lactose-digesting enzyme encoded by a well-known operon found in a bacterial cell

basic biological research   biological research with the ultimate aim of gathering knowledge for knowledge's sake

behavioral trait   any type of behavior that can be identified and measured; can be instinctive or learned, genetic or environmentally induced

benign   descriptive of tumors that are incapable of invasion and metastasis

bilayered vesicle   a spherical structure whose boundary is composed of a lipid bilayer

bioaugmentation   the addition of microorganisms to a hazardous waste site in order to react with hazardous wastes and render them harmless

biodestruction destroying hazardous chemical wastes using living microorganisms

biofiltration using microorganisms to remove complex waste products from industrial or domestic sewage

biomass utilization innovative use of "waste" material that is left over from agricultural and food preparations

biopesticides toxic chemicals produced by living organisms that specifically kill only particular pest species

bioremediation the process of using prokaryotes to break down or recycle environmental pollutants

biostimulation addition of specific nutrients to increase the growth of naturally-occurring microbes that convert toxic chemicals to non-toxins

biotechnology the application of biological principles, organisms, and products to practical purposes; applied biological science

bonds stable interactions between atoms

bovine growth hormone (BGH) a hormone produced in cattle that may be obtained from genetically engineered *Escherichia coli*, given to cattle to increase milk production or weight gain

broth any fluid microbial growth medium

cancer any malignant tumor; a malignant, invasive cellular neoplasm that has the capability of spreading throughout the body or body parts

carbohydrates the class of organic molecules characterized by a ratio of 1:2:1 (carbon: hydrogen: oxygen); include sugars such as glucose and lactose, polysaccharides such as starches, glycogen, and cellulose

carcinogenic refers to the ability to cause cancer

carriers in genetics, heterozygous individuals who do not express a recessive trait but are capable of passing it on to their offspring

catabolic reactions that break down complex molecules and release energy

catalysis increase in the rate of reaction caused by a material unchanged at the end of the reaction

catalyst something that causes a reaction to occur more readily by lowering the activation energy; an agent that increases the rate of a reaction without being consumed in the process

cDNA complementary DNA; a single strand of DNA copied by reverse transcriptase from mRNA

cell lines animal or plant cells that can be cultivated under laboratory conditions

cell membrane the selective membrane boundary (a lipid bilayer) of every living cell that regulates cytoplasmic contents

cells the smallest units of life

cell surface receptors molecules on cell membranes that can bind other specific molecules

centrifugation the process of separating materials from solution by rotating the vessels containing the solution at high speed, thereby creating artificial gravitational forces

centrioles cylindrical structures found in animal cells that help organize assembly of the spindle apparatus during cell division

centromere   the constricted region of a replicated eukaryotic chromosome where the two sister chromatids remain joined

chain-termination sequencing (dideoxy-sequencing)   a DNA sequencing technique involving the use of a modified base known as dideoxynucleotide as an artificial chain terminator

cholera   an acute gastrointestinal disease caused by large numbers of *Vibrio cholerae* growing in the upper part of the small intestine

chorionic villus sampling (CVS)   sampling a tiny portion of the fetal part of the placenta to determine genetic or congenital defects

chromatin   the aggregate mass of dispersed genetic material comprised of DNA and protein and observed during interphase in eukaryotic cells

chromosome   a single DNA molecule, complexed with protein, that carries part of the genetic information of a cell

chromosome-level mutations   genetic changes that affect large regions of a chromosome or entire chromosomes

chromosome walking   a base-sequencing method in which a chromosome tip is extended a little at a time and the base sequence of the extended section is determined

cleavage   the process of cytokinesis in animal cells, characterized by pinching of the plasma membrane and formation of a cleavage furrow

cleavage furrow   a shallow groove in the plasma membrane formed by constriction of the contractile ring during cleavage in an animal cell

codominance   a phenotypic situation in which both alleles are expressed in a heterozygote

codon   a sequence of three adjacent nitrogenous bases on either DNA or mRNA that is used to code for an amino acid or for termination of polypeptide synthesis

commensals   two organisms that live in a close association in which one is benefited while the other is neither benefited nor harmed

competent   a condition of bacterial cells indicating that they have the ability to take in DNA

complement system   a system of serum proteins that participate in an enzymatic cascade that ultimately results in the formation of a membrane attack complex that destroys invading pathogens

construct   a recombinant plasmid

contractile ring   a band of microfilaments that forms beneath the plasma membrane of dividing animal cells and mediates the process of cytokinesis

control experiment   a check of a scientific experiment based on keeping all factors the same except for the one in question

correlations   demonstration of causal relations

covalent bonds   strong chemical bonds made by the sharing of a pair of electrons between two atoms

crossing over   the reciprocal exchange of chromosome segments between corresponding regions of nonsister chromatids during synapsis of meiosis I

culture medium   nutrients prepared for microbial growth in a laboratory; may be fluid (broth) or solidified with agar

cyclic adenosine monophosphate (cAMP)   a nucleotide employed as a second messenger in a variety of signaling pathways; generally exerts its effects by activating a kinase molecule

cytokines a group of secreted proteins that regulate the intensity and duration of an immune response by stimulating or inhibiting the proliferation of various immune cells

cytokinesis division of the cytoplasm to form two separate daughter cells following mitosis or meiosis

cytoplasm in prokaryotes: everything inside the cell membrane; in eukaryotes: everything between the cell membrane and the nucleus

cytotoxic T cells (T_C) T lymphocytes that are capable of mediating lysis of target cells

degenerates persons with "bad blood," according to eugenics theory

deletions mutations involving the loss of a base from a gene or a deficiency in a chromosome resulting from the loss of a fragment through breakage

denaturation the complete unfolding of a protein or the separation of two complementary strands of a DNA molecule

deoxyribose a five-carbon sugar used in the construction of DNA, lacking an oxygen on its number two carbon atom

dideoxynucleotide a nucleotide that lacks two oxygen atoms in the carbohydrate portion

dihybrid cross a mating between two individuals that differ only in two traits

diploid containing two sets of chromosomes

disease any change from a state of health; a specific state of malfunctioning of a plant or animal body

DNA (deoxyribonucleic acid) a polymer of nucleotides (deoxyribose and a phosphate linked to either adenine, thymine, cytosine, or guanine) that forms a double helix by complementary base-pairing between adenine and thymine (two hydrogen bonds) and between cytosine and guanine (three hydrogen bonds)

DNA fingerprinting a technique for the identification of individuals based on small differences in DNA fragment patterns detected by electrophoresis

DNA polymerases enzymes responsible for replicating DNA

DNA sequencers machines that carry out the entire process of determining the specific sequence of nucleotides in a given DNA molecule

DNA synthesizers "gene machines," computer-controlled machines that take the sequence given to them and go through the ordered reactions, producing the desired DNA molecule

dominant an allele or corresponding phenotypic trait that is expressed in either the homozygous or heterozygous state

electronegativity the tendency for an atom to pull electrons toward itself; measure of the ability of an atom to attract the shared electrons in a covalent bond

enzyme-linked immunosorbent assay (ELISA) technique of detecting very small amounts of specific proteins utilizing antibodies linked to enzymes that catalyze the formation of colored products

enzymes proteins that change the rate of a chemical reaction without being changed into a different molecule in the process

eugenics controlling the gene pool of the human species by selective breeding; "the science of improving human stock by giving the more suitable races or strains of blood a better chance of prevailing speedily over the less suitable"

eukaryotic   descriptive of cellular organisms whose nuclei are bounded by a lipid bilayer or nuclear membrane

exons   in eukaryotic genes, sequences that specify amino acids in a protein

expression vectors   DNA cloning vectors designed to promote the expression of cloned genes in a host cell

ex vivo **gene therapy**   gene therapy involving the removal of affected cells from a patient, correction of the genetic defect, then reintroduction of the corrected cells into the patient

familial hypercholesterolemia   an inherited disease characterized by a very high level of cholesterol in the blood and caused by defective low-density lipoprotein (LDL) receptors

familial traits   traits shared by members of a family for whatever reason, including social, behavioral, and environmental influences as well as heredity

fermentation   enzymatic decomposition of organic molecules without oxygen

first filial (F_1) generation   the first filial or hybrid offspring produced by crossing two parental strains

fluorescent   giving off light of one color when exposed to light of another color (for example, glowing orange when exposed to UV light)

frameshift mutation   insertion or deletion of one or more bases in a gene, thereby altering the reading of all codons beyond the point of insertion or deletion

γ-interferon   a cytokine, released by T lymphocytes, that stimulates the maturation of macrophages

G-protein   a family of proteins that bind to GTP (a compound similar to ATP) and are involved in signal transduction

gametes   haploid egg or sperm cells that unite during sexual reproduction to produce a diploid organism

gel electrophoresis   use of gels to separate molecules by forcing them to migrate under the influence of an electrical field

gene   a linear sequence of DNA that contains all of the information necessary for the production of one separate string of amino acids or one polypeptide; may include the formation of rRNA and tRNA

gene-level mutations   changes in the base sequence of DNA that alter individual genes

gene therapy   adding genetic material to a person so that a defect or disease can be corrected; to date, applied only to somatic cells in humans

genetic code   the base-pair information that specifies the amino acid sequence of a polypeptide; in a practical way, this refers to a printed table of triplets of nitrogen bases from mRNA that enables us to determine what specific amino acid is required for emplacement in a protein

genetic counseling   a medical specialty with responsibility for educating people about genetic diseases, treatments, and potential reproductive choices

genetic determinism   concept or belief that an individual's genes predetermine every aspect of his or her physical, mental, and emotional being

genetic diagnosis   methods used to determine whether a particular genetic defect is present in an individual

genetic disease a disorder that is caused by one or more defective genes and can be passed on to offspring

genetic engineering manipulating genetic information *in vitro*

genetic linkage a term describing genes located close to each other on the same chromosome

genetic screening highly organized attempts to use accurate and reliable genetic diagnostic tests on a specific target group for genetic disorders

genome the total complement of genes carried by a cell or an organism

genotype the genetic makeup of an organism

glycogen a large polysaccharide composed of glucose molecules; the main storage form of glucose in animals

haploid containing one set of chromosomes

haploid insufficiency a condition in which one or more genes are lost from one member of a pair of homologous chromosomes due to breakage so that the sole copy of the affected gene(s) is supplied by the intact homologue; a single copy of a gene may not be able to direct the synthesis of sufficient protein, and any recessive gene(s) on the intact homologue will now be expressed since it is the only copy present

helper T cells (T$_H$) T lymphocytes that, when activated by antigen-presenting cells, release growth and differentiation factors that enhance both cell-mediated and humoral immune responses

hemizygous the condition of X-linked genes in males; descriptive of males that have an X chromosome with an allele for a particular gene but do not have another allele of that gene in the gene complement

heritability the proportion of phenotypic variation in a population attributable to genetic factors; assignment of a numerical value to the relative contribution of genes to the variability of the trait, and therefore to the trait itself

heritable traits characteristics under the control of the genes and thus transmitted from one generation to another

heterotrophic a mode of nutrition that requires consumption of preformed organic food (cells or their products)

heterozygous having two different alleles for a given trait

histones small proteins with a high proportion of positively charged amino acids that bind to the negatively charged DNA and play a key role in its packaging into chromatin

homologous chromosomes the members of a chromosome pair that are identical in the arrangement of genes they contain and in their visible structure

homozygous having two identical alleles for a given trait

hormones signaling macromolecules that are secreted into the bloodstream by endocrine cells and act on target cells possessing receptors for the hormone

human chorionic gonadotropin a hormone produced by the developing placenta

human immunodeficiency virus (HIV) a retrovirus that infects human CD4$^+$ T cells and causes acquired immunodeficiency syndrome (AIDS)

hybridization producing offspring or hybrids from genetically dissimilar parents

hybridoma a clone of hybrid cells produced by fusing an antibody-producing cell with a cancerous cell that grows well in culture; each hybridoma produces only one type of monoclonal antibody

hydrogen bonds   relatively weak noncovalent bonds formed between a hydrogen atom in groups such as—NH or—OH and an oxygen or nitrogen atom in an acceptor group

hydrophilic   said of charged molecules that tend to dissolve readily in water (or interact readily with water) but not in organic solvents such as ethanol; "water-loving"

hydrophobic   said of molecules that tend to dissolve readily in organic solvents such as ethanol but not in water; "water-hating"

hydrophobic force   the clumping together of lipid molecules caused by the presence of water molecules that are tightly associated; this force drives the formation of cellular membranes

hydrous   with water

idiotype   a single antigenic determinant of the variable region of an antibody

immune system   the network of cells, tissues, and organs that defends the body against foreign invaders

immunization   the process of rendering a state of immunity, in one of two ways: either by inoculation with a specific antigen or by administration of specific antibodies from immune individuals

incomplete dominance   the genetic condition resulting when one allele is not completely dominant to another allele so that the heterozygote has a phenotype intermediate to that shown in individuals homozygous for either individual allele involved; example: a cross between homozygous plants with red flowers and homozygous plants with white flowers that gives rise to heterozygous plants with pink flowers

inducer molecule   a molecular agent for bacterial operons that brings about the transcription of an operon; for example, lactose when it binds to the *lac* repressor protein and causes a change in the shape of the *lac* repressor protein, which then is unable to bind to the operator or, if previously bound, is released from the operator

inducer T cells   T lymphocytes that oversee the development of T cells in the thymus

inflammatory response   a tissue response to injury or other trauma characterized by pain, heat, redness, and swelling; consists of altered patterns of blood flow, an influx of phagocytic and other immune cells, removal of the foreign antigen, and healing of the damaged tissue

insert   name given to any gene that through genetic engineering is placed into a plasmid

insertions   mutations involving the addition of one or more bases to a gene

insulin   a hormone, produced by β-cells of the pancreas, that promotes the uptake of glucose from the blood by most body cells and the synthesis and storage of glycogen by the liver

insulin-dependent diabetes   an autoimmune disorder in which the immune system attacks the cells of the pancreas, resulting in insulin deficiency and, thus, diabetes

interacting genes   a situation in which alleles present at one genetic locus interact with alleles at another locus, yielding more possible phenotypes than the individual genes would allow

interferons   a family of glycoproteins, produced by a variety of cell types, that help cells resist viral infection and help regulate the immune response

interleukin-1   a cytokine secreted by macrophages that have ingested a pathogen or foreign molecule and have bound with a helper T cell; stimulates T cells to proliferate and elevates body temperature

interleukin-2   a cytokine, secreted by activated T cells, that stimulates helper T cells to proliferate more rapidly

interphase   the period in the eukaryotic cell cycle when the cell is not dividing, situated between successive M phases

introns   in eukaryotic genes, regions of noncoding DNA (sequences that do not specify amino acids in a protein)

ionic bonds   attractive forces between oppositely charged ions; electrostatic bonds

ions   electrostatically charged atoms or molecules

karyotyping   a method of photographing the complete set of chromosomes for a particular cell type and organizing them into homologous pairs based on size and shape

ketone bodies   the type of chemical compound resulting from the catabolism of fats, for example, acetone

kinase   an enzyme that modifies other proteins by removing a phosphate group from ATP and attaching it to them

kinetochores   disk-shaped structures located at the centromere of a metaphase chromosome that attach the chromosome to spindle fibers

lac **repressor protein**   in *E. coli*, a protein that can detect and bind with lactose, the substrate for β-galactosidase

lactose   milk sugar; a disaccharide composed of the monosaccharides glucose and galactose

lambda (λ) phage   a bacteriophage often used as a vector in genetic engineering

latent   a state in which a virus, which is integrated into a host chromosome, is inactive; that is, its genome is not directing the production of new viruses

law of equal segregation   Mendel's first law, stating that allele pairs separate during gamete formation and then randomly re-form pairs during the fusion of gametes at fertilization

law of independent assortment   Mendel's second law, stating that each allele pair segregates independently during gamete formation; applies when genes for two traits are located on different pairs of homologous chromosomes

leukocytes   "white cells"; refers to any blood cells that are not erythrocytes (red blood cells)

linkers   short DNAs of known sequence that are covalently joined to the ends of DNA molecules during the process of cloning; or, synthetic oligonucleotides containing a restriction site attached to the end of DNA fragments to create a restriction site for cloning

lipids   a chemical group that includes fats and oils, insoluble in water

locus   the location of a gene on a chromosome

lymph   the intercellular tissue fluid that circulates through the lymphatic vessels

lymphokines   a generic term for cytokines produced by activated lymphocytes, especially T cells, that act as intercellular mediators of the immune response

lysogenic cycle   a path other than the lytic cycle that a phage can follow in which the phage chromosome inserts into a specific place in the bacterial chromosome without replicating

lytic cycle   a sequence of bacteriophage replication that results in host cell lysis

M-13 phage   a common bacteriophage used as a vector in genetic engineering

macrophages   large cells, derived from monocytes, that can function as phagocytic cells and antigen-presenting cells

magnetic resonance imaging   a specialized medical technology designed for the observation of internal organs and structures without invasive procedures

malignant   descriptive of any tumor that is capable of spreading by invasion and metastasis

meiosis   a type of cell division in sexually reproducing organisms that produces haploid gametes; consists of a series of two cell divisions preceded by a single round of DNA replication

memory B cells   clonally expanded progeny of B cells formed following a primary immune response and responsible for the speed and heightened levels of the secondary immune response

memory T cells   clonally expanded progeny of T cells formed following a primary immune response and responsible for the speed and heightened levels of the secondary immune response

merozoites   the vegetative form of *Plasmodium* found in infected red blood cells or liver cells

messenger RNA (mRNA)   single-stranded RNA that codes for the synthesis of one or more polypeptide chains

metabolism   all of an organism's chemical processes

metaphase plate   the plane along which chromosomes align during metaphase

metastasize   the spreading of malignant tumorous cells throughout the body so that tumors develop at new sites

micelle   spherical structure formed by a lipid in which the hydrophobic ends are associated with one another and the hydrophilic ends are attracted to water molecules

microfilaments   a type of cytoskeletal fiber made up of the protein actin and involved in cell shape, movement, growth, and formation of the contractile ring during cytokinesis

microinjection   a means of delivering DNA to animal cells by using a microscopic needle to pierce the nuclear membrane

migration inhibition factor   a cytokine that prevents macrophages from migrating away from a site of infection

mitosis   the process of nuclear division that produces two daughter nuclei, each of which contains identical complements of chromosomes; consists of five stages: prophase, prometaphase, metaphase, anaphase, and telophase

mitotic nondisjunction   an accident of mitosis in which both sister chromatids fail to move apart properly, resulting in daughter cells with abnormal numbers of chromosomes

molecular biology   the science dealing with DNA and protein synthesis of living organisms; the study of the storage and flow of information within a cell

molecules   two or more atoms held together by a chemical bond; under normal circumstances these atoms cannot be separated and still maintain the original molecule's characteristic(s)

monoclonal   refers to a homogeneous population of antibodies produced by a hybridoma

monohybrid cross   a mating between individuals that differ in only one trait

monolayer   single molecular layer; for example, some lipids form a monolayer on water's surface, with the hydrophilic ends sticking in the water and the hydrophobic ends poking into the air

multifactorial   influenced by multiple genes and environmental factors

multivalent vaccines   vaccines that combine two or more subunits from an infectious organism or virus

mutagen   any agent that causes mutations

mutations   changes in the sequence of bases in the DNA that are heritable

mutualism   a symbiotic relationship in which both associates benefit

myasthenia gravis   an autoimmune disease in which the body produces antibodies against acetylcholine receptors, resulting in neuromuscular disorders

natural killer cells   large lymphocytes that have the ability to kill foreign or abnormal cells

natural selection   differential success in the reproduction of different phenotypes resulting from the interaction of organisms with their environment

negative eugenics   emphasis on limiting reproduction of persons with "bad blood" so that undesirable traits are lost from the population

neutrophils   circulating, phagocytic cells involved in early inflammatory responses

nodule   as in "root nodule"; a tumorlike growth on the roots of legumes (beans, peas, alfalfa, peanuts) containing symbiotic nitrogen-fixing bacteria

nucleic acids   linear polymers of nucleotides that are responsible for the storage and transmission of biological information (ribonucleic acid and deoxyribonucleic acid, RNA and DNA, respectively)

nucleotide analogs   molecules that are chemically similar to, and thus can take the place of, nucleotides

nucleotides   molecules made up of three simpler chemical constituents: a five-carbon sugar (either ribose or deoxyribose), a phosphate, and a nitrogenous base; the fundamental building blocks of nucleic acids

nucleus   the compartment, enclosed by the nuclear membrane, that stores the biological information used by the cell

oncogene   a gene whose action promotes cell proliferation; altered form of proto-oncogenes

operator genes   the controlling sites adjacent to promoters responsible for controlling the transcription of genes contiguous with the promoter

operon   a cluster of genes whose expression is regulated together by operator-regulator protein interactions, plus the operator itself and the promoter

oxidation   the loss of electrons from a substance

passive immunization   the process of conferring immunity by transferring antibodies or sensitized T cells produced by another individual; usually contrasted with *active immunization*, in which the particular individual responds by making his or her own antibodies or sensitized T cells

pathogen   a disease-causing organism

phagocytes   cells capable of engulfing and digesting particles that are harmful to the body

phagocytosis   the ingestion of solids by cells

phenotype   the physical and physiological traits of an organism controlled by its genes interacting with the environment

phytoremediation   technological use of plants to convert contaminants in soil or water to non-toxic substances

placebo   a pharmacologically inactive substance or a sham procedure administered as a control in testing the efficacy of a drug or a course of action

plasma cells   differentiated, antibody-producing cells derived from activated B cells

plasmid   a small double-stranded DNA molecule that can replicate independently of chromosomal DNA; used as a cloning vector

plate   a petri plate containing a solidified culture media (usually agar) and one or more nutrients that promote bacterial growth

pleiotropic   refers to genes that have multiple effects

polar   readily soluble in water; having one molecular region with a partial positive charge and another region with a partial negative charge; polar bonds have a partial ionic character

polyclonal   refers to the variety of products of many different clones of lymphocytes

polygenic   determined by more than one genetic locus

polygenic traits   traits conferred by more than one gene; may display a continuum of phenotypes

polymerase chain reaction (PCR)   a technique of amplifying DNA *in vitro* by incubating with special primers, DNA polymerase, and nucleotides

polymers   a chain of small molecules chemically linked to form a larger one; examples include polypeptides and polysaccharides

positional cloning   isolation of a gene associated with a genetic disease on the basis of its approximate chromosomal position

positive eugenics   informative and educational programs with an emphasis on reproduction by "good" (middle- and upper-class) people; belief that people who exhibit desirable traits should reproduce with each other so that "good blood" becomes more prevalent

preproinsulin   in the β-cells of the pancreas, the first-formed molecule of what will become insulin; it is inactive but contains chemical signals that direct the molecule to storage

primary immune response   the initial cellular and humoral immune response to antigen exposure

probe   a single-stranded DNA or RNA molecule used to detect the presence of a complementary nucleic acid

proinsulin   in the β-cells of the pancreas, the second-formed molecule of what will become insulin; the inactive, stored form that has had its storage signal removed

prokaryotic   descriptive of the bacteria, single-celled organisms that lack a membrane-bound nucleus

promoter   as in promoter site, a DNA region to which RNA polymerase binds before initiating transcription

prophage bacteriophage DNA inserted into the bacterial host cell's chromosomal DNA

proteases enzymes capable of hydrolyzing proteins

proteins generally lengthy and specifically shaped biological polymers made up of specific sequences of amino acids

proto-oncogenes genes that in normal cells function to control the normal proliferation of cells and that when mutated or changed in any other way become oncogenes

pure-breeding in Mendelian genetics, organisms that always produce offspring with traits identical to theirs

pyrogens fever-causing substances released by activated white blood cells

radioactive the property of some elements of emitting alpha, beta, or gamma radiation by nuclear disintegration of atoms

reading frames in a DNA molecule, potential protein-coding sequences identified by an initiation codon in frame with a chain-terminating codon

recessive an allele or corresponding phenotypic trait that is expressed only in the homozygote

recombinant DNA DNA molecule created by joining two or more DNA fragments from two or more sources

recombination the exchange of genetic information between two different DNA molecules, such as that occurring between homologous chromosomes during synapsis in prophase I of meiosis

reduced (as reduction) the gain of electrons by a substance

reductionism procedures that reduce complex data or phenomena to simple terms

redundancy condition descriptive of the genetic code, wherein most of the twenty amino acids are encoded by more than one mRNA codon, some by as many as six

regulator genes genes responsible for the production of a repressor protein; indirectly control the operator

replication the formation of a new copy of the DNA for the daughter cell

repressor proteins proteins that bind to the promoter regions of particular genes and block the access of RNA polymerase

restriction digestion the process of cutting DNA with restriction enzymes

restriction enzymes enzymes (restriction endonucleases) that cut DNA molecules at specific restriction sites that are four to eight bases long

restriction fragment length polymorphism (RFLP) a difference in restriction fragment length between very similar DNA molecules; RFLP analysis is used to detect relatively small differences between DNA molecules; used in DNA typing and genetic disease detection

restriction mapping creation of a diagram of sites on a DNA molecule that are cut by different restriction enzymes

retrovirus single-stranded RNA viruses that replicate by using reverse transcriptase to make a DNA intermediate

reverse transcriptase an enzyme that synthesizes double-stranded DNA on an RNA template

rheumatic fever an inflammation of heart valves caused by autoantibodies produced in response to *Streptococcus pyogenes* infection

rheumatoid arthritis   the progressive destruction of joint tissue caused by an immune reaction against molecules present in joints

ribonucleic acid (RNA)   a nucleic acid polymer similar to DNA except that the base uracil (U) is used instead of T, ribose is used as the sugar unit instead of deoxyribose, and its molecules are generally single-stranded

ribosomes   protein synthesis occurs on ribosomes and consists of RNA and proteins organized into two subunits

RNA polymerase   the enzyme responsible for the synthesis of RNA molecules using DNA as a template

secondary immune response   a more rapid and heightened immune response that occurs following a second exposure to an antigen

septic shock   a sudden drop in blood pressure caused by an infection by a toxin present in the cell walls of many infectious bacteria

seroconversion   a change in a person's response to an antigen in an antigen-antibody test; usually used to indicate that a person has developed antibodies to a particular antigen

serum   the fluid part of the blood free of all cells, platelets, and clotting factors

severe combined immunodeficiency disease (SCID)   a genetic disease characterized by a deficient immune system. and caused by a defective gene for the enzyme adenosine deaminase (ADA)

sex chromosomes   the pair of chromosomes responsible for determining the sex of an individual (In many organisms, one sex possesses a pair of visibly different chromosomes: one an X chromosome, the other a Y chromosome. Commonly, the XX sex is female and the XY sex is male.)

signal transduction   the process by which the binding of a signal molecule to the cell surface triggers changes within the cell

single-cell protein (SCP)   a food substitute made up of microbial cells

sister chromatids   replicated forms of a chromosome joined together by the centromere and eventually separated during mitosis or meiosis II

somatic disorders   disorders caused by mutations in somatic cells, that is, cells that are not germ (sperm or egg) cells

spindle apparatus   an assemblage of microtubules that orchestrates chromosome movement during eukaryotic cell division

spindle fibers   fibers, comprised of microtubules, that make up the spindle apparatus

sporozoite   the vegetative form of *Plasmodium*, found in mosquitoes, that is infective for humans

stem cells   cells from which differentiated cells derive

sticky ends   cohesive ends; single-stranded strings of nucleotides that extend from the end of a fragment of DNA; or, the cohesive DNA termini produced with overlapping 5' or 3' ends, after restriction endonuclease digestion

structural genes   genes that code for an mRNA that will later be translated into a polypeptide chain

subcloning   breaking a large cloned gene into smaller parts and making a new clone from each of the DNA pieces

suppressor T cells   T lymphocytes that are thought to end an immune response after an antigen is no longer present

symbiotic   refers to ecological relationships in which two different species of organisms live together

synapsis   the close pairing of replicated homologous chromosomes during prophase I of meiosis

T cells   lymphocytes that are dependent on the thymus for their differentiation and play a role in cell-mediated immune responses

technology   applied science; scientific methods for achieving practical purposes

terminator   a DNA sequence located after the gene that tells the RNA polymerase to stop making RNA accessory proteins required by RNA polymerase for binding at the promoter to initiate RNA synthesis (transcription)

Ti plasmids   plasmids found in *Agrobacterium tumifaciens* that cause a disease known as crown gall; used to join with, transport, and cause the integration of foreign genes into plant cells in tissue culture

tissue culture   propagation of plant or animal cells or tissues under laboratory conditions

tissues   groups of similar cells organized to perform specific functions

topoisomerases   enzymes that alter DNA supercoiling by introducing transient breaks in one or both DNA strands

transcription   the process by which RNA polymerase uses one strand of DNA to assemble RNA

transcription factors   proteins that initiate or regulate transcription

transfer RNA (tRNA)   small molecules that have two important features: (1) at one end is a three-base sequence of RNA (the anticodon), (2) at the other end is a chemical group to which a specific amino acid can be attached (the particular amino acid that gets attached to an individual tRNA depends on the sequence of the anticodon)

transformation   a change in the hereditary properties of a cell brought about by the uptake of foreign DNA; changing of a normal cell into a cancerous one due to a deregulation of growth, in most instances marked by an increase in the rate of growth or cell proliferation

transgenic   describes an organism that has been altered to contain a gene from a different species

translation   the process of protein synthesis; what is accomplished is a translation from one language, the nucleic acid (RNA and DNA) language of four different bases arranged in codons, into the protein language of twenty different amino acids

translocations   chromosome-level mutations resulting from the attachment of a chromosomal fragment to a nonhomologous chromosome

trisomy   an aberrant, aneuploid state in a normally diploid cell or organism in which there are three copies of a particular chromosome instead of two

tumor (neoplasm)   a swelling or enlargement, especially one due to pathologic overgrowth of tissue; a growing tissue mass created by the loss of normal growth control

tumor infiltrating lymphocytes   T cells that can invade and destroy tumors

tumor suppressor genes   genes whose absence or loss of function can cause a cell to become malignant or cancerous

vaccination   the intentional introduction of a harmless or less harmful version of a pathogen to induce a specific immune response that protects the individual against a later exposure to the antigen

vector a plasmid or virus used in genetic engineering to insert genes into a cell

Western blot a technique for identifying electrophoretically separated proteins (that have been transferred to nitrocellulose filters) by labeling protein bands with radioactive- or enzyme-attached specific antibodies

X-gal 5-bromo-4-chloro-3-indoylgalactoside; a synthetic substrate molecule for β-galactosidase

X-linked referring to genes located on the X chromosome

zygote a diploid cell produced by the fusion of a sperm cell with an egg cell

Index

Credits

ILLUSTRATION CREDITS

Figures 1.01a, b, 1.02a, 1.02b, 2.08, 3.02 1, 3.03 from *Biology, 5/E* by Neil A. Campbell. Copyright © 1999 by The Benjamin/Cummings Publishing Company.

Figures 1.03, 1.10, 2.01a, 2.06, 3.01 1, 3.01 4 from *Biology: Concepts and Connections, 3/E* by Neil A. Campbell, Lawrence G. Mitchell, and Jane B. Reece. Copyright © 2000 by The Benjamin/Cummings Publishing Company.

Figures 2.01a, 2.04, 8.09, 8.10, 8.11 from *Biochemistry, 2/E* by Christopher K. Mathews and K. E. Van Holde. Copyright © 1996 by The Benjamin/Cummings Publishing Company.

Figures 1.06, 2.03, 2.05a, 2.05b, 2.06, 2.09, 2.10, 3.08, 4.01, 4.02, 4.03, 4.04, 5.04, 5.05, 5.10 1, 8.05, 8.07b, 9.01a, 9.01 b1, 9.03, 9.05, 9.06, 9.14, 10.02, 10.08, 11.01, 11.05, 11.07, 11.08, 11.12, 13.02, 13.06, 14.03a, 14.03b, 15.03, 16.01 from *Biology, 4/E* by Neil A. Campbell. Copyright © 1996 by The Benjamin/Cummings Publishing Company.

Figures 2.12, 3.05a, 3.06, 3.07, 3.09, 4.07, 9.11, 11.06, from *Biology: Concepts and Connections, 2/E* by Neil A. Campbell, Lawrence G. Mitchell, Jane B. Reece. Copyright © 1997 by The Benjamin/Cummings Publishing Company.

Figures 2.11, 5.01, 8.01a, 8.04, 8.06, 8.07a, 9.09, 10.07, 11.10a,b, 11.11, 12.07, 12.08, 13.03, 15.04 from *Microbiology: An Introduction, 5/E* by Gerard J. Tortora, Berdell R. Funke, and Christine L. Case. Copyright © 1995 by The Benjamin/Cummings Publishing Company, Inc.

Figures 2.02 and 5.08 by Irving Geis. Copyright © by Irving Geis. Reprinted by permission of The Estate of Irving Geis.

Figures 11.03 and 11.09 from *Biology, 4/E* by Raven and Johnson, 1996. Copyright © 1996. Reprinted by permission of Wm. C. Brown, a division of McGraw-Hill Publishers.

Figure 11.04 adapted from Lennart Nilsson and Jan Lindberg, *The Body Victorious* (Delacorte Press, NY, 1987), p. 27. Illustration by Urban Frank. Reprinted by permission of Urban Frank.

Figure 14.02 by J. Adams from *Nature*, 315 (1985):542. Copyright © 1985 by Macmillan Magazines Ltd. Reprinted by permission of *Nature*.

Figures 4.09, 10.01, 10.03, 10.05, 10.06 from *Genetics, 5/E* by Peter J. Russell. Copyright © 1998 by Peter J. Russell. Published by Benjamin/Cummings, an imprint of Addison-Wesley Longman, Inc.

Figures 13.01 and 13.05 from *Biology: The Science of Life, 4/E* by Robert A. Wallace, Gerald P. Sanders, Robert J. Ferl. Copyright © 1996 by HarperCollins Publishers Inc.

Figure 2.01b from *Genetics: Analysis and Principles* by Robert J. Brooker. Copyright © 1999 by Benjamin/Cummings, an imprint of Addison Wesley Longman, Inc.

Figures 9.07, 12.03, 12.05 from *Principles of Cell and Molecular Biology, 2/E* by Lewis J. Kleinsmith and Valerie M. Kish. Copyright © 1995 by HarperCollins College Publishers.

Figures 9.15, 12.01, 12.04, 13.07 from *The World of the Cell, 3/E* by Wayne M. Becker, Jane B. Reece, Martin F. Poenie. Copyright © 1996 by The Benjamin/Cummings Publishing Company, Inc.

Fig. 11.16 from "An Advanced Case of Tetanus" from The Bettman Archives.

Fig. 13.T01 from "Classification of Human Viruses" adapted from R. I. B. Francki, et al. from "Classification and Nomenclature of Viruses." Fifth Report of the International Committee on Taxonomy of Viruses. Springer-Verlag, 1991.

Figure 17.02 found on Genentech Access Excellence Site, adapted from BIO "Biotechnology in Perspective," Washington DC, Biotechnology Industry Organization, 1990. Reprinted by permission.

PHOTOGRAPHIC CREDITS

Figure 3.5b1 J. Richard McIntosh, University of Colorado at Boulder.

Figure 3.5b2 Matthew Schibler, *Protoplasma* 137(1987):29–44, 11.9a. Published by Springer-Verlag New York, Inc. © Matthew Schibler.

Figure 5.10.2 Copyright Boehringer Ingelheim International GmbH. Photo Lennart Nilsson/Albert Bonniers Forlag AB.

Figure 6.01.2 Oriel Incorporated, formerly Joiner Associates.

Figure 6.02c Photodisc experiment.

Figure 9.12 Courtesy of Cellmark Diagnostics, Inc., Germantown, Maryland.

Figure 9.13 Ralph L. Brinster, University of Pennsylvania.

Figure 11.2 Copyright Boehringer Ingelheim International GmbH. Photo Lennart Nilsson/Albert Bonniers Forlag AB.

Figure 11.10c Arthur J. Olson, Molecular Graphics Laboratory, Scripps Research Institute.

Figure 11.17, inset Masamichi Aikawa, Tokai University School of Medicine, Japan. Published in *Science* 137:43, January 1990.

Figure 12.7 e.p.t. is a registered trademark of Warner-Lambert Company.

Figure 15.1 Ananda M. Chakrabarty, College of Medicine, University of Illinois at Chicago.

Figure 15.5 © Ken Graham/Ken Graham Agency.

Figure 15.6 Ralph L. Brinster, University of Pennsylvania.

Figure 16.5 Ralph L. Brinster, University of Pennsylvania.

Figure 16.7a B.L. Turner II, Clark University.